METEOROLOGICAL MONOGRAPHS

VOLUME 21   DECEMBER 1986   NUMBER 43

PRECIPITATION ENHANCEMENT— A SCIENTIFIC CHALLENGE

Roscoe R. Braham, Jr., William A. Cooper, William R. Cotton, Robert D. Elliot,
John A. Flueck, J. Michael Fritsch, Abraham Gagin, Lewis O. Grant, Andrew J. Heymsfield,
Geoffrey E. Hill, George A. Isaac, John D. Marwitz, Harold D. Orville, Arthur L. Rangno,
Bernard A. Silverman, Paul L. Smith
Contributing Authors

Roscoe R. Braham, Jr.
Editor

Published by the American Meteorological Society
45 Beacon St., Boston, MA 02108

ISBN 0-933876-65-3
ISSN 0065-9401

American Meteorological Society
45 Beacon Street, Boston, MA 02108

Printed in the United States of America
by Lancaster Press, Lancaster, PA

Preface

This volume had its origin in a workshop held at Park City, Utah, 23–25 May 1984. The objective of the workshop was to provide an opportunity for cloud scientists to assess the status of knowledge and understanding about the physics of precipitation formation in clouds and the response of clouds to glaciogenic seeding. All papers were by invitation. Half of the time was given to discussion and debate sparked by 4 major review papers and 14 short reports. Restriction of the workshop to scientific aspects of glaciogenic seeding for precipitation enhancement was intentional.

Following the workshop, all authors were invited to prepare their manuscripts for possible publication. All papers were subjected to anonymous peer review, as is normal for scientific journal publication. This Monograph is the result. In the review process it was decided that the volume required an introductory chapter to provide the nonspecialist with an overview of the subject and to help interrelate detailed materials of individual chapters. Thus, Chapter 1 was created.

The Park City workshop was organized by the Committee on Weather Modification of the American Meteorological Society. Committee members A. S. Dennis, W. L. Woodley, and R. R. Braham, Jr., arranged the program. As editor of this monograph, I wish to acknowledge with sincere appreciation the help of A. S. Dennis throughout the editorial phases of this work.

The entire community is indebted to the 29 reviewers, some of whom handled more than one manuscript, for their careful and incisive reviews. Special thanks are given to Ms. Theresa Gawlas and Ms. Maureen Dungey for editorial assistance and Ms. Julia Jacobs for secretarial assistance.

The workshop and publication of this Monograph were supported in part by the National Science Foundation, Program in Experimental Meteorology, and the Department of the Interior, Bureau of Reclamation, Division of Atmospheric Resources Research. This Monograph reflects the views of individual authors and does not necessarily reflect the views of either of these agencies of the United States government.

R. R. Braham, Jr.
Editor

Chicago, Illinois
January, 1986

PRECIPITATION ENHANCEMENT—
A SCIENTIFIC CHALLENGE

TABLE OF CONTENTS

CHAPTER 1

Precipitation Enhancement—A Scientific Challenge

ROSCOE R. BRAHAM, JR.

University of Chicago, Chicago, Illinois

ABSTRACT

Schaefer's 1946 cloud seeding experiment initiated a quest for weather modification techniques. Progress has been slow; but there are several reasons for believing that useful precipitation augmentation may be possible.

1.1. Introduction

On 13 November 1946, Vincent Schaefer scattered a little Dry Ice into the top of a supercooled stratified cloud over some low mountains east of Schenectady, New York. Within minutes the seeded portion of the cloud was transformed into a mass of snow crystals. But this was not the only result of Schaefer's experiment. The nature of meteorological inquiry itself was changed; a new, experimental dimension was added to a science that had previously encompassed only theory and field observation.

Few experiments have so fired the imagination of both the scientific and the nonscientific world. Scientific interest in the properties of clouds and the physics of precipitation formation quickly expanded. Cloud physics emerged as a multidisciplinary science involving classical meteorology, physical chemistry, statistics, and many areas of engineering. With strong federal and state funding, cloud physics rapidly became an important part of meteorology, making significant contributions on a broad front.

A commercial cloud seeding industry also emerged as a by-product of the early experimental results and the increased understanding of cloud physics. This development was motivated by several factors, which can be identified as follows:

a. Conventional meteorological wisdom, at the time, held that virtually all rain develops from ice crystals in supercooled clouds;

b. The Schaefer experiment gave clear-cut evidence that crystals could be created in nonprecipitating supercooled clouds by introducing artificial ice nucleants;

c. Artificial ice nucleants were relatively inexpensive.

In addition to the purely technical reasons favoring the plausibility of cloud seeding, there were economic and social considerations as well:

d. Relatively small increases in rain can have great economic value, especially in areas receiving marginal amounts of rain;

e. Since many severe weather effects are associated with the development of rain, it was hoped that successful intervention in the processes of rain formation might also provide some measure of relief from other weather adversities;

f. At that time there was little scientific evidence to suggest that cloud seeding would ever result in an unwanted, or "negative," effect.

It is not surprising that a technology for weather modification developed more rapidly than the science of weather modification.

The majority of all experiments and commercial operations in weather modification have used ice nucleants (such as Dry Ice and silver iodide) as the cloud seeding agent. In cloud physics and weather modification circles this method is known as "glaciogenic seeding."

Research scientists and commercial operators have explored cloud seeding as a possible means for modifying a wide variety of weather phenomena (including precipitation, hail, fog, lightning, hurricanes). However, as of 1986, glaciogenic seeding to enhance rain and snow is the only cloud seeding goal for which there is both a reasonably well developed scientific foundation and a potential for substantial economic benefits. This monograph reviews the current status of that research as seen by 15 scientists who have been involved in it for a number of years. This introductory chapter presents a broad background to the subject and an overview of the science for the benefit of nonspecialists.

1.2. Technology outran science

Perhaps the most difficult task faced by weather modification researchers and practitioners is to determine a causal link between the acts of man (cloud seeding) and the acts of nature (precipitation). Establishment of this causal link is hampered by two difficulties: 1) the great complexity of cloud systems and of their reactions to seeding, and 2) the difficulty in obtaining adequate experimental controls by which to judge and understand experimental results. This "causal link" factor has emerged as a major difference as to how weather modification is viewed by basic research scientists and by those more applications oriented.

Scientists from university and government laboratories have conducted a limited number of basic research experiments to establish this causal link. These experiments have had two general characteristics: 1) seeded clouds (days) were randomly selected from a larger group, all of which were judged to be "suitable" for seeding, and 2) extensive measurements were made of cloud and precipitation parameters (using airplanes, radars, and augmented surface raingage networks) in an effort to describe in detail the clouds and to verify their response to seeding. Some of these experiments used elaborate schemes to avoid the introduction of experimenter biases. Typically, these experiments amassed fewer than 200 test cases even though they were carried out over several seasons. As a whole, these experiments gave inconclusive results; some suggested rainfall increases due to seeding, others suggested decreases. Ex post facto analyses of the detailed measurements were used to sort out the conditions under which the different results were obtained. However, these analyses generally have not been followed up with new, confirming field experiments.

The number of commercial cloud seeding operations greatly exceeds the number of research experiments. Typically, these operations were of short duration, had very limited measurement (verification) components, lacked randomization, and often were carried out during periods of atypical weather. Generally speaking, the results from individual commercial seeding operations could not be rigorously established.

A few commercial operations, however, were repeated using the same methods and location over several consecutive seasons, thus contributing to a growing body of experiences. These experiences have led some scientists to believe that, under appropriate conditions, cloud seeding will produce useful effects.

Differences in views between basic research scientists and scientists in the operational sector as to the status of the science of cloud seeding persist to this day. It has led to seemingly contradictory statements in the literature and confusion in the minds of the public. It has given rise to several high-level reviews of the field of weather modification. These reviews all reached essentially the same conclusions, viz., that cloud seeding is promising, unproven, and worth pursuing (Advisory Committee on Weather Control, 1957; National Academy of Sciences, 1964, 1966, 1973; Weather Modification Advisory Board, 1978).

The experiments most often cited as providing strong evidence for rainfall increases are those carried out in Israel. Two separate, randomized experiments (the first lasting 6½ seasons and accumulating 364 test days, and the second 6 seasons with 388 days) indicated that winter rainfall in the target area of northern Israel had been increased 13%–15%. These results appear to have good statistical significance. Cloud physics measurements made by the project scientists appear to give the results physical plausibility (Gagin and Neuman, 1974, 1981; Tukey et al., 1978). These experiments have not been replicated elsewhere, both because scientists have not yet identified another location providing a high frequency of clouds similar to those reported by Israeli scientists and because the nature of the experiments precludes a detailed cause-and-effect analysis.

The studies of basic cloud physics, begun in the 1940s–1950s, have continued to the present day, though decreased in intensity because of decreased funding support. Included in this effort is research to establish the microphysical characteristics of clouds in a variety of meteorological/geographical areas, to elucidate physical processes that govern precipitation efficiency in natural clouds, and to construct better numerical models of cloud/precipitation processes.

It is this author's view that research has gradually reduced the range of probable benefits from cloud seeding. At the same time, it has sharpened understanding of how clouds and cloud systems behave and how they react to seeding, with the result that the prospect for *some forms* of useful cloud seeding is still very real.

1.3. Glaciogenic seeding—dynamic and static modes

The scientific foundation of glaciogenic seeding for precipitation enhancement rests upon a few simple physical principles and two postulates.

A large fraction of natural precipitation in middle latitudes begins as ice crystals in supercooled clouds. Supercooling, the condition of water drops remaining liquid at temperatures below 0°C, is very common in convective clouds and middle-level stratified clouds down to temperatures of −20°C, or so. An ice crystal in the presence of supercooled drops will grow by vapor deposition, while nearby drops evaporate, because the saturated vapor pressure over ice is less than that over supercooled liquid at the same temperature. In this way a crystal can gradually acquire the mass of many cloud drops. This allows it to fall through the cloud, perhaps growing by collision with some of the drops, to reach the ground as a precipitation particle. The growth of ice from supercooled cloud releases latent heat of fusion, which augments cloud buoyancy.

Artificial ice nucleants, effective at temperatures of −5°C and colder, are obtainable at modest cost. Acceptable means for releasing them into clouds are available. Adding such nucleants to supercooled clouds induces ice crystal formation at locations and in concentrations completely consistent with theory and laboratory findings.

The greatest amounts of precipitation come from convective clouds. But observations and simple calculations show that convective clouds are inefficient in converting cloud water to rain. For example, the rain from a typical thunderstorm may be only 10%–40% of the total cloud water condensed.

The first postulate of cloud seeding states that the precipitation efficiency of some clouds is limited by a shortage of natural ice nuclei effective at the extant cloud temperatures. Adding artificial nucleants to these clouds should enhance the precipitation process and increase precipitation on the ground.

The second postulate states that increased buoyancy, resulting from seeding-induced conversion of supercooled drops into ice particles, will invigorate cloud updrafts. This will enable clouds to grow larger, process more water vapor, and yield more precipitation.

The liquid water in clouds comes from water vapor condensing in upward-moving air parcels. In large-scale synoptic systems, upward motions typically are about 1–4 cm per second and are driven mainly by large-scale dynamical forcing. Updrafts in convective clouds are largely driven by thermal buoyancy and range up to tens of meters per second. In orographic clouds, wind-driven upslope motions may reach a few meters per second. Since the cloud condensate source-rate is a function of updraft speed, precipitation processes differ markedly in these classes of clouds. Similarly, their response to seeding appears to be different. For this reason, they are treated separately in this monograph.

Production of significant amounts of precipitation involves synergistic interactions of physical systems of three different scales: (i) large-scale atmospheric motion systems, (ii) clouds and mesoscale systems, and (iii) cloud particle systems. The only direct effect of ice-phase seeding is to alter cloud microphysics by creating ice crystals and releasing latent heat of fusion. Thus, cloud seeding will be most effective if applied in places and at times where these rather small effects will augment, or set into motion, natural physical processes to yield a more desirable final state.

The two seeding postulates, previously outlined, have been used as the basis for two different concepts of how seeding could increase precipitation. The so-called *static-mode* seeding concept is to add small concentrations of ice nuclei to clouds in which the precipitation efficiency is limited by a shortage of natural ones. In this concept, any effect of the concomitant heat release is assumed to be small and of little consequence. The *dynamic-mode* seeding concept aims at maximizing the effect of latent heat release by glaciating cloud updrafts earlier in time, at warmer temperatures, and more completely than can be accomplished by the less efficient natural nuclei. The additional heat release should invigorate the updraft, resulting in bigger clouds and more precipitation.

1.4. Factors that have limited research progress

More than anything else, our experiences since 1946 have taught us that, regardless of how easy it might be to produce important changes in rain by cloud seeding, it is very difficult to prove that such changes come as a result of seeding and not from natural causes.

On the occasion of the 100th anniversary of the founding of the National Academy of Science, President J. F. Kennedy remarked, "Wisdom is the child of experience." From the experiences of the past four decades, one can identify several reasons why the science of cloud seeding has not progressed as rapidly as the application of seeding technology.

• The physical mechanisms of precipitation development in natural clouds are much more complex than was realized in 1946. In addition to the ice crystal mechanism, we now recognize that drop collision and coalescence often plays an important role in rain formation. In most clouds these two mechanisms operate simultaneously, though at different rates. The response of a cloud to glaciogenic seeding seems to depend upon the reaction rates of the two natural precipitation mechanisms.

• Ex post facto analyses of seeding experiments suggest that seeding may have increased rainfall during certain meteorological conditions and decreased it in others. Experiments that include both favorable and unfavorable situations may end up with a net effect that is quite small.

• The variability among natural clouds was not fully appreciated in the early days of cloud seeding, and it is not yet fully understood. The larger the natural variability, the more difficult it is to establish relatively small changes due to seeding.

• Seeding-induced changes in rainfall, too small to be predicted by models or measured by conventional rain-gage networks (except by averaging over large areas and long times), can be of economic importance. For use in cloud seeding research, techniques for modeling and measuring rainfall are inadequate.

• Experiments capable of detecting seeding-induced changes in rainfall at acceptable significance levels must be very elaborate and run for long periods of time. Thus, they are difficult to organize and very expensive.

In addition to these scientific and technical issues, there are several social, legal, and political ones that make cloud experimentation more difficult and, as a consequence, slower. The fact that weather moves with the wind and knows no political boundaries means that a cloud experiment generally affects many people who may have varying degrees of enthusiasm for it. Some may oppose cloud experimentation on religious grounds. Others may worry about "losing some of *their* water," or about "paying for someone else's benefits." One group may need rain at a time when others need dry weather. These are but examples—readers wishing more detail could consult the Weather Modification Advisory Board (1978), Cooper (1973), Davis (1978), and Sewell (1966).

1.5. Looking forward

Cloud seeding presents a scientific challenge with few parallels outside of medicine. Some operationally oriented scientists believe that useful techniques for precipitation enhancement exist and should be used. Some research scientists believe that a useful degree of precipitation enhancement may be possible though a reliable technology for doing this has not yet been adequately demonstrated.

Several factors support the view that some changes in weather, through seeding, are possible. It is well established that supercooled cloud conditions are very important for natural precipitation. The fact that ice crystals can be artificially induced in supercooled clouds—both stratified and convective—is well established (Langmuir et al., 1947; Sax et al., 1979; Cooper and Lawson, 1984). Static-mode seeding in the Israeli experiments appears to have increased rainfall by about 15% with good statistical significance and physical plausibility. Project HIPLEX (Cooper and Lawson, 1984; Mielke et al., 1984) showed that static-mode seeding of summer convective clouds in Montana resulted in changes as predicted by theory, up through the stage of production of precipitation embryos. It was reported by Simpson et al. (1967) that dynamic-mode seeding of isolated cumulus congestus clouds resulted in increased cloud-top heights, as predicted by theory. Similar results have been reported by other investigators. It is well established that man's activities in building and occupying cities, and in tilling the soil, are producing weather effects equal to or greater than those indicated for cloud seeding.

In addition to the impediments to research enumerated above, several scientific issues remain to be dealt with. The role of entrainment (mixing of a cloud with its cloud-free environment) in limiting the lifetimes and reducing the precipitation efficiencies of clouds is not adequately understood. Our understanding of precipitation growth processes is in many cases too qualitative and conceptual. We need more reliable quantitative precipitation forecasting techniques. The question of local rain increases versus wide-area redistribution has not been adequately addressed. Even if it turns out that seeding mainly redistributes precipitation, there may be times and places where this would be acceptable. We must find more sensitive ways for following seeding effects from the initial ice crystals, through various and diverse growth processes, until they reach the ground in precipitation. We must find ways

to shorten the time required to reach valid conclusions in field trials.

It is absolutely essential that randomized, exploratory cloud seeding be continued. We cannot afford to give up the power inherent in the experimental method. We also must remember that, until perfect numerical models are developed, randomized field experiments will remain the only way to establish whether or not any given cloud seeding technique will reliably and repeatedly produce a change in rainfall.

There is little question but that new field trials are required. Complicated as these will have to be, they will have to deploy the fullest set of remote sensors, so that the physical processes within the clouds can be monitored to the fullest possible extent, *as the experiment progresses.* There seems to be no other way to abridge the long durations required of experiments that rely principally on statistical testing of hypotheses.

The following chapters of this volume contain authoritative discussions of the scientific issues associated with glaciogenic seeding for precipitation enhancement. They also provide a review of the background of this area of science, a comprehensive report on its current status (strengths and weaknesses), and suggestions about research that must be accomplished before glaciogenic seeding for precipitation enhancement can be accepted as a proven technology based in sound science.

Acknowledgments. In several places these remarks were adapted from an edited version of a public lecture given during the IAMAP/IAPSO Joint Assembly, Honolulu, Hawaii, August 1985 (Braham, 1986a, 1986b). Reviews of the draft manuscript by Professor W. F. Hitschfeld, Dr. A. S. Dennis, and Professor W. Blumen were very helpful and sincerely appreciated. Support of the National Science Foundation under Grant NSF ATM-8310429 is gratefully acknowledged. Views expressed herein are those of the author and not necessarily those of the National Science Foundation.

REFERENCES

Advisory Committee on Weather Control, 1957: *Final Report, Vol. I and II.* Washington, DC, 44 pp. and 432 pp.

Braham, R. R., Jr., 1986a: The cloud physics of weather modification, Part 1: Scientific basis. *WMO Bull.,* **35**(3), 215–222.

——, 1986b: The cloud physics of weather modification, Part 2: Glaciogenic seeding for precipitation enhancement. *WMO Bull.,* **35**(4), 307–314.

Cooper, C. F., 1973: Ecological opportunities and problems of weather and climate modification. *Modifying the Weather: A Social Assessment,* University of Victoria, British Columbia, 99–134.

Cooper, W. A., and R. P. Lawson, 1984: Physical interpretation of results from the HIPLEX-1 Experiment. *J. Climate Appl. Meteor.,* **23**, 523–540.

Davis, R. J., 1978: Weather modification, stream flow augmentation, and the law. *Mineral Law Institute,* **24**, 833–863.

Gagin, A., and J. Neumann, 1974: Rain stimulation and cloud physics in Israel. *Weather and Climate Modification,* W. H. Hess, Ed., Wiley and Sons, 454–494.

——, and ——, 1981: The second Israeli randomized cloud seeding experiment: Evaluation of the results. *J. Appl. Meteor.,* **20,** 1301–1311.

Langmuir, I., V. J. Schaefer, B. Vonnegut, R. E. Falconer, K. Maynard and R. Smith-Johannsen, 1947: *First Quarterly Progress Report, Meteorological Research.* General Electric Research Lab., Schenectady, 37 pp.

Mielke, P. W., Jr., K. J. Berry, A. S. Dennis, P. L. Smith, J. R. Miller, Jr. and B. A. Silverman, 1984: HIPLEX-1: Statistical evaluation. *J. Climate Appl. Meteor.,* **23,** 513–522.

National Academy of Science, 1964: *Weather and Climate Modification.* NAS/NRC Publ. 1236, Washington, DC.

——, 1966: *Weather and Climate Modification, Problems and Prospects, Vol. I and II.* NASA/NRC Publ. 1350, Washington, DC, 39 pp and 210 pp.

——, 1973: *Weather and Climate Modification, Problems and Progress.* NAS, Washington, DC, 258 pp.

Sax, R. I., J. Thomas, M. Bonebrake and J. Hallett, 1979: Ice evolution within seeded and nonseeded Florida cumuli. *J. Appl. Meteor.,* **18,** 203–214.

Sewell, W. R. D., Ed., 1966: Human dimensions of weather modification. Res. Paper 105, Dept. of Geography, University of Chicago, 423 pp.

Simpson, J., G. W. Brier and R. H. Simpson, 1967: Stormfury cumulus seeding experiments, 1965: Statistical analysis and main results. *J. Atmos. Sci.,* **24,** 508–521.

Tukey, J. W., D. R. Brillinger and L. V. Jones, 1978: The role of statistics in weather resources management. *The Management of Weather Resources, Vol. II.,* Dept. of Commerce, Washington, DC, 97 pp.

Weather Modification Advisory Board, 1978: *The Management of Weather Resources, Vol. I,* Dept. of Commerce, Washington, DC, 229 pp.

CHAPTER 2

Static Mode Seeding of Summer Cumuli—A Review

BERNARD A. SILVERMAN

Bureau of Reclamation, Denver, Colorado

ABSTRACT

A review of the state of knowledge of the physics of the static mode seeding hypothesis for convective clouds is presented. The central thesis of the review is that the results of past experimental work are diverse but valid and that credibility of the science depends on understanding the physical reasons for the diverse results. Areas of uncertainty and conflicts in evidence associated with the statement of physical hypothesis, the concept of seedability, the seeding operation, and the chain of physical events following seeding are highlighted to identify what issues need to be resolved to further progress in precipitation enhancement research and application.

It is concluded that the only aspect of static seeding that meets scientific standards of cause-and-effect relationships and repeatability is that glaciogenic seeding agents can produce distinct "seeding signatures" in clouds. However, the reviewer argues that a body of inferential physical evidence has been amassed that provides a better understanding of which clouds are seedable (susceptible to precipitation enhancement by artificial seeding) and which are not, even though the tools for recognizing and properly treating them are imperfect. In particular, the inferred evidence appears to support the claims of physical plausibility for the positive statistical results of the Israeli experiments.

It is suggested that future work continue to be designed for physical understanding and evaluation through comprehensive field studies and numerical modeling. Duplicating the Israeli experiments in another location should receive high priority but, in general, future experiments should move upscale from cumulus congestus to convective complexes. In doing so, a new, more complex physical hypothesis that accounts for cloud–environment and microphysical–dynamical interactions and their response to seeding will have to be developed.

2.1. Introduction

Ice phase seeding for microphysical effects, the so-called static mode seeding hypothesis, has been the physical basis for most precipitation enhancement programs on convective clouds. Evaluations of the results of these programs, mainly statistical in nature, report precipitation increases in a few cases, precipitation decreases in a few cases, and no significant change in precipitation in most cases (Dennis, 1980, 1984). A few of these programs had a physical component that attempted to gain an understanding of natural precipitation processes and their response to glaciogenic seeding. This paper focuses on the observations and results of these physical studies in an assessment of the state of knowledge of the static mode seeding hypothesis for convective clouds.

The central thesis of this review is that the results of past experimental work are diverse but valid and that credibility of the science depends on understanding the physical reasons for the diverse results. Consequently, areas of uncertainty and apparent conflicts in evidence are highlighted in an effort to identify what issues need to be resolved to further progress in precipitation enhancement research and applications. The review begins with a discussion of the evolution of the physical hypothesis. Using the physical hypothesis as a framework for further discussion, research findings related to the selection of suitable clouds, seeding agents and their delivery to the clouds, and the chain of physical events following seeding are presented. Some of the key issues raised in this review paper are discussed in greater detail in other papers in this volume.

2.2. The physical hypothesis

The static seeding concept is rooted in the classical work of Bergeron (1933) on the role of ice in the initiation of precipitation in supercooled clouds. Invoking the description of physical processes in mixed-phase clouds that was put forth earlier by Wegener (1911), Bergeron postulated that the inherent colloidal stability of supercooled liquid clouds could be upset by the presence of a few ice particles, which would rapidly grow by deposition at the expense of the water droplets to sizes that were large enough to fall. He believed that the ice particles most likely formed by the freezing of a few cloud droplets. Findeisen (1938) expanded on Bergeron's precipitation initiation concept and suggested that the ice crystals formed by the "sublimation" of vapor on special nuclei rather than by the freezing of droplets. This concept of precipitation initiation in mixed-phase clouds is commonly referred to in the scientific literature as the Wegener–Bergeron mechanism, the Bergeron–Findeisen theory, or simply the Bergeron process.

Findeisen recognized the potential for weather modification in these findings and wrote, "It can be boldly stated that, at comparatively moderate expense, it will, in time, be possible to bring about rain by scientific means, to obviate the danger of icing, and to prevent the formation of hailstorms." However, it was not until the next decade, when Schaefer (1946) conducted his historic dry ice experiment and Vonnegut (1947) discovered the ice-nucleating ability of silver iodide, that a practical method became available for artificially introducing the required concentrations of ice nuclei into clouds and the scientific exploration of the weather modification possibilities envisioned earlier could begin. Field experiments on artificially stimulating rain from convective clouds based on the Bergeron process began almost immediately (e.g., Kraus and Squires, 1947; Leopold and Halstead, 1948; Squires and Smith, 1949; Smith, 1949).

The static seeding hypothesis for convective clouds is based on observations that precipitation development in these clouds is frequently inefficient and on the expectation that the natural precipitation process can be made more efficient by the introduction of additional precipitation embryos through glaciogenic seeding. The presence of supercooled water and concomitant low concentrations of ice particles at temperatures warmer than about $-20°C$ are generally taken as evidence of the inefficiency of the precipitation process. The static seeding strategy is to provide an optimum concentration of ice particles for the available liquid water and to initiate the precipitation process earlier and, perhaps, lower in altitude in the developing cloud than would occur naturally. Attaining on the order of 1 to 10 ice crystals per liter that eventually would become 0.1 to 1.0 graupel embryos per liter is the usual goal of the seeding operation. If too many ice crystals for the available water are produced, then none may grow large enough to fall out and the cloud is said to be "overseeded" (Bergeron, 1949).

Static seeding is intended to improve the precipitation efficiency of a cloud by affecting its microphysical properties only. Changes in the dynamical properties of the clouds may occur, but they are neither intended nor part of the cause-and-effect relationship in the seeding hypothesis. Partly because of this, most physical studies and experiments designed to investigate or test the static seeding hypothesis have been conducted on single, isolated convective clouds and cloud clusters in which dynamic effects from seeding are expected to be negligible.

Since the goal of static seeding is to improve the efficiency of a cloud's natural precipitation process, a brief review of our understanding of natural precipitation processes in convective clouds is appropriate before proceeding. Results of cloud physics research have shown that all important convective rain from mixed-phase clouds involves graupel as a dominant precipitation type (Mason,

1957; Braham, 1981). It has also been shown that graupel in these clouds develops through two main processes (Braham, 1968): ice crystal growth by vapor diffusion followed by riming into graupel (hereafter referred to as the IRG mechanism) and coalescence-grown drizzle drops followed by freezing and subsequent riming into graupel (hereafter referred to as the CRG mechanism). Both processes may be operative in some clouds, but one is usually dominant. For nearly two decades after Bergeron (1933) presented his precipitation initiation concept to the scientific community, many scientists believed that the IRG mechanism was responsible for all rains of consequence, especially those producing large raindrops. In the early 1960s the University of Chicago Project Whitetop group (Koenig, 1963; Braham, 1964) documented the existence of the CRG mechanism in Missouri summer cumuli through in situ measurements from aircraft and, thereby, established it as an important rain-producing process in mixed-phase clouds. These results confirmed the coalescence development of precipitation embryos and, therefore, the probable existence of the CRG mechanism in cold-top summer cumuli that radar meteorologists had inferred from data on the height of first echo development (Battan, 1953, 1963; Clark, 1960).

Cloud-base temperature appears to be a good indicator of whether ice crystals (IRG) or large drops (CRG) will likely be the dominant precipitation embryos in a cloud. Warmer cloud-base temperatures generally favor the CRG mechanism. Other factors, such as liquid water content, cloud droplet concentration, cloud depth, and updraft speed, are also important in this determination, but they are generally related to, influenced by, or correlated with cloud-base temperature. Consideration of factors other than cloud-base temperature is most important in this determination at cloud-base temperatures near the value that appears to be the transition between the IRG and CRG precipitation growth processes. MacCready et al. (1957a) developed a precipitation initiation model and calculated that the cloud-base temperature separating these two processes was 14°C. Based on theoretical calculations and an examination of observations of cloud-base temperature in different geographical areas, Johnson (1982) showed that the separation cloud-base temperature is about 10°C. Knight (1981) examined hail embryo types in different geographical areas and found that the frequency of occurrence of graupel and frozen drop embryos correlated rather well with average cloud-base temperatures calculated from rawinsonde soundings on hail days. Knight showed that the percentage of frozen drop embryos increases and the percentage of graupel embryos decreases as the average cloud-base temperature increases, with the 50 percent value for both occurring at an average cloud-base temperature of about 9°–10°C. The results of model calculations by Nelson (1979) of hail embryo type in different geographical areas are consistent with the cloud-base temperature relationship deduced by Knight.

Clouds with an active coalescence process appear to be associated with the development of secondary ice particles, a process usually referred to as SICP (secondary ice crystal production). However, an active coalescence process has not been established as a prerequisite for SICP. This type of SICP has been found to occur in the laboratory by two main processes, freezing–splintering and rime–splintering. The freezing–splintering process may generate secondary ice particles by ice splinter ejection in the course of spike formation, which frequently accompanies the symmetric freezing of large drops (Johnson and Hallett, 1968; Hobbs and Alkezweeny, 1968; Pruppacher and Schlamp, 1975). The rime–splintering process, usually referred to as the Hallett–Mossop mechanism, may generate secondary ice particles during the growth of graupel by riming at a specific temperature range (Hallett and Mossop, 1974) and with specific cloud droplet sizes and concentrations (Mossop, 1976; Mossop, 1978a). Mossop (1978b) has shown that cloud-base temperature and cloud drop concentration are useful in separating cloud conditions in which SICP by the rime–splintering process takes place from those in which it does not. Considering cloud-base temperature only, the "ice multiplication boundary" appears to be at about 5°C, with SICP occurring when cloud-base temperatures are warmer.

SICP may also occur in other ways. Gagin and Nozyce (1984) found that the production of secondary ice particles resulting from the freezing of 1–2 mm drops in the laboratory seemed to be attributable to nucleation at relatively high, transient supersaturations and not to ice splinter ejection. Another possible way for SICP to occur is by ice particle fragmentation during collisions between graupel and dendrites (Hobbs and Farber, 1972), between graupel and rimed ice particles (Vardiman, 1978), and between filamentary, low density rimed ice particles (Vali, 1980). Cloud studies have also revealed SICP processes that do not seem to fit any of the above processes (Hobbs et al., 1980; Hobbs and Atkinson, 1976; Cooper and Saunders, 1980; Paluch and Breed, 1984). The reader is referred to the comprehensive review of SICP by Mossop (1985) for a detailed discussion of this subject. Since SICP contributes to the natural ice crystal concentration of a cloud, it has a direct bearing on whether such clouds are statically seedable. This matter will, therefore, be discussed further in the following section on seedability. For the purposes of this review, only the freezing–splintering and rime–splintering processes will be considered, but it should be remembered that SICP can occur in many other ways over a wide range of conditions and they will have direct bearing on the seedability of clouds as well.

The HIPLEX-1 experiment (Bureau of Reclamation, 1979; Smith et al., 1984) was the first one to state explicitly, in advance, the specific precipitation growth process that was to be optimized by glaciogenic seeding. It was hypothesized that the efficiency of the IRG mechanism would be improved by the prescribed seeding and thereby lead to both additional precipitation and an increase in the proportion of clouds that precipitate. Statements of physical mechanism were, however, implicit in all static seeding experiments conducted previously. In experiments conducted up to about the mid-1960s, before the importance of the CRG mechanism was first established, the implied strategy had to be the improvement of the efficiency of the IRG mechanism. In experiments conducted after that date the implied strategy is less clear, but their references to the Bergeron process indicate that the physical hypothesis continued to be based on the IRG mechanism. In fact, no cause-and-effect argument has been presented that details how the efficiency of the CRG mechanism can be improved by glaciogenic seeding.

2.3. Seedability

According to the static seeding hypothesis, a cloud is postulated to be seedable (WMO, 1982) if it contains supercooled water that is or will be underutilized by the cloud's natural precipitation process, that will not be eroded by competitive depletion processes, and that will last long enough in sufficient quantities to permit the growth of additional precipitation particles induced by seeding to sizes that can reach the ground. If the amount and persistence of supercooled water in a cloud is high, then the depletion rate of water associated with natural precipitation development, cloud ice evolution, and entrainment is likely to be low, and the opportunity for seeding tends to be high. The coexistence of ice in the cloud is only a deterrent to seeding if it exists in sizes and concentrations that cause the supercooled water to be depleted faster than seeding can exploit it.

Seedability criteria are a practical expression of the physical hypothesis. They are used in an attempt to distinguish and select among those clouds that result in increases, no effect, and decreases in precipitation when seeded. They are used to help control variability in an experiment. They are the basis for estimating the technical and economic feasibility of seeding activities for an area.

The criteria for static seedability established by the World Meteorological Organization (WMO, 1982) also include the requirement that the coalescence process in the cloud be inefficient. This requirement is supported by Gagin (1981), who contends that clouds with an efficient condensation–coalescence process are not colloidally stable, are capable of generating precipitation particles quite efficiently, and are, therefore, less amenable to augmenting rainfall by static seeding. Moreover, since the presence of an active coalescence process is associated with the likely development of SICP, it is possible that SICP can generate ice crystals in concentrations and at rates comparable to those intended by seeding. The addition of more ice crystals by seeding in these situations is generally regarded as contributing to "overseeding."

In considering the effects of SICP on static seedability, two types of situations need to be considered—case 1: clouds in which the IRG mechanism is dominant with an inefficient coalescence process, and case 2: clouds in which the CRG mechanism is dominant with an efficient coalescence process. In case 1, it has been commonly assumed (but not proven) that SICP occurs mainly by the rime–splintering process. Gagin (1981) argues that the occurrence of SICP in such cases is not a deterrent to static seeding. He claims that the initial ice concentration in the cloud determines the graupel concentration that will develop (Gagin, 1975) and, since the rime–splintering process occurs only after graupel is present, some 10–20 min after ice crystals first appear, there is time to influence the precipitation efficiency of the cloud through static seeding. In case 2, SICP can occur by both the rime–splintering and freezing–splintering mechanisms, and sufficient time may not be available to influence the precipitation efficiency of the cloud. When the coalescence-grown drops freeze to form graupel embryos, SICP by the freezing–splintering process occurs right away. In addition, since the graupel embryos that develop in this way are initially rather large, time is fairly short before the rime–splintering process becomes active. Therefore, ice phase seeding of such clouds may contribute to "overseeding."

It is interesting to note that those experiments which resulted in statistically significant decreases or redistribution of precipitation involved clouds in which the CRG mechanism was probably dominant. Koenig (1963) and Braham (1964) documented its existence in Whitetop (Flueck, 1971); Battan (1963) inferred from the height of first radar echoes in Arizona that the coalescence process was the dominant precipitation initiation mechanism in convective clouds that grew well above the freezing level, from which it can reasonably be assumed that the CRG mechanism was operative during the Arizona Project (Battan, 1966; Battan and Kassander, 1967). The occurrence of the CRG mechanism in Necaxa (Perez-Siliceo, 1970) clouds is assumed to be likely based on reasoning related to cloud-base temperature that was discussed above. In fact, the negative results of Whitetop have been attributed to "overseeding" (Braham, 1979). The results of the model seeding experiments by Nelson (1979) also support this contention. He found that seeding of clouds that naturally develop hail with frozen drop embryos resulted in both hail and rain decreases. Seeding those with graupel embryos resulted in hail and rain increases. The model seeding experiments of Farley and Orville (1982), on the other hand, resulted in both hail and rain decreases from a cloud in which graupel embryos were dominant.

Recognizing or forecasting the above-stated seedability conditions, which are rate-dependent processes of clouds, has proven to be extremely difficult, and it is likely to remain so. Thus far, the approach has been to identify a set of cloud characteristics which is believed, according to a physically plausible conceptual or numerical model, to represent a cloud that will respond positively to seeding. A cloud is considered seedable or has seeding potential if it has the requisite characteristics at the time of selection.

Early applications of this approach to the selection of seedable clouds were based on observations that many convective clouds extending well above the freezing level did not precipitate (Battan and Kassander, 1960). This was taken as an indication of the deficiency of ice nuclei, which are required for precipitation formation by the Bergeron process, and, therefore, such clouds were considered to be seedable. Based on this reasoning, early experimenters selected clouds for seeding based on visual appearance alone; if a cloud grew to at least the −5°C level and had no visible signs of glaciation, it was considered to be a suitable candidate for glaciogenic seeding (Smith, 1970). As time progressed other specification criteria that were learned from earlier experiences were included. The principal additions, still based on the experimenter's judgment, involved indications that the cloud was growing, that it would last at least 30 min, that it was reasonably isolated, that it would not be affected by surrounding clouds, and that it did not rain prior to selection (Bethwaite et al., 1966).

The University of Chicago Cloud Physics Project's experiments under the Artificial Cloud Nucleation program (Braham et al., 1957) appear to be the first time cloud physics data from an inspection pass with an aircraft was used in determining cloud eligibility. A cloud was accepted if the inspection pass provided evidence that the cloud contained liquid water and if the lack of an echo on the calibrated nose radar indicated it did not contain precipitation. With the advent of sophisticated airborne cloud physics instrumentation, in situ microphysical data collected during a pretreatment aircraft pass were incorporated into the cloud selection criteria (Bureau of Reclamation, 1979, 1983; Hobbs and Politovich, 1980; Marwitz and Stewart, 1981; English and Marwitz, 1981; Isaac et al., 1982). Final selection of a cloud for treatment (either seeding or placebo) was made if such characteristics as liquid water content, ice crystal concentration, vertical velocity, and cloud depth were within specific bounds. Rangno and Hobbs (1983) have presented evidence that the passage of an aircraft through supercooled clouds can produce high concentrations of ice crystals, which they called APIPs (Aircraft Produced Ice Crystals). If APIPs do, in fact, occur as a result of a pretreatment aircraft pass or even the treatment pass, its effects would have to be determined in the evaluation of such seeding experiments.

Research on the relative effects of seeding clouds with different water-to-ice ratios has been undertaken in an attempt to establish the limits of seedability (Bureau of Reclamation, 1983). Water and ice concentration measurements from aircraft were also used in assessing the seeding potential of various cloud types at the PEP (Precipitation Enhancement Project) site in Spain (WMO, 1982). General guidance on susceptibility to intervention was derived from studies based on a MAY-B index, which represented the proportion of flight time the water and ice concentrations jointly satisfied specific threshhold values. To identify cloud regions which had characteristics that were more consistent with the concept of seedability, PEP scientists developed a parameter called ROP (regions of potential), which required that the averaged liquid water content exceed particular values and that these regions persist for more than 10 min.

It is interesting to note that the decision to undertake areawide experiments to establish the effects of seeding was prompted by cloud studies (Braham, 1960) that showed only a fraction of otherwise promising clouds lasted long enough to be seedable and did not develop precipitation naturally. With no means available to preselect only this fraction of clouds for an experiment, it was concluded (Braham, 1981) that a very large sample of clouds would be needed to test the seeding concept. By conducting an areawide experiment involving numerous clouds, it was hoped that the cumulative effects of seeding the favorable fraction of clouds could be detected in the precipitation at the ground. The occurrence of suitable conditions, numerous convective clouds that would reach heights of at least $-5°C$, was predicted on the basis of thermodynamic and synoptic parameters such as precipitable water, stability, and wind direction (Braham, 1966). Similar criteria were used in the Arizona Project (Battan and Kassander, 1960) areawide experiment.

2.4. Seeding

Proper implementation of the static seeding hypothesis is dependent on initiating the precipitation process with an ice crystal concentration that is optimum for the subject cloud. Therefore, delivery of the seeding material must be designed to produce the required ice crystal concentration at the time and place in the cloud as specified by the physical hypotheses. In doing so, care must be taken not to "overseed" the cloud.

Attempts to achieve the seeding objective have involved the use of dry ice and a variety of silver iodide agents, a wide range of seeding rates, cloud-by-cloud and areal broadcast seeding, and ontop, in-cloud, cloud-base and ground-based delivery systems. It has proven to be an extremely difficult objective to achieve, involving the joint solution of such complex processes as seeding-agent effectiveness and nucleation rates, and transport and dispersion of seeding material and/or ice crystals. It is a problem that has not yet been satisfactorily solved. In this reviewer's opinion, the diverse results of past research and tests of the static seeding hypothesis are due in large part to the inability to execute the seeding operation as intended and are not necessarily a reflection on the validity of the hypothesis. It is beyond the scope of this general review of static mode seeding to thoroughly discuss this complex subject; however, a few of the major seeding issues will be mentioned briefly to illustrate some of the uncertainties that need to be resolved.

2.4.1. Dry ice

Airborne seeding with crushed dry ice was used in many experiments conducted in the decade following Schaefer's discovery of dry ice's nucleating ability. It was replaced by silver iodide when logistic and cost considerations caused experimenters to conduct seeding operations near cloud base or from the ground. Dry ice came into use again in the 1970s when ontop cloud seeding to produce a vertical curtain of ice crystals was thought to be the most effective seeding strategy. Some scientists preferred using pellets of dry ice for this purpose (Holroyd et al., 1978) rather than silver iodide pyrotechnics because dry ice produces ice crystals nearly instantaneously at temperatures of $-2°C$ and colder and because its effectiveness was thought to be virtually independent of temperature.

Despite its extensive use as a seeding agent, several questions concerning the activity and effectiveness of dry ice remain unanswered. The first is its mode of nucleation. It is generally agreed that the number of ice crystals produced by dry ice cannot be explained solely by the freezing of preexisting cloud droplets or by heterogeneous nucleation of droplets followed by homogeneous freezing. However, there is no consensus as to whether its predominant nucleation mode is the homogeneous deposition of ice, the homogeneous nucleation of droplets followed by homogeneous freezing, or a combination of both mechanisms (Vonnegut, 1981; Mason, 1981).

The second question concerns its effectiveness. Schaefer (1946, 1949) estimated that dry ice produces at least 10^{16} crystals per gram. Braham and Seivers (1957) made a theoretical study of the production and growth of ice crystals in dry ice seeded clouds and concluded that an effectiveness of 10^8 to 10^9 crystals per gram was consistent with their observational data. Other investigators have reported values, based on laboratory studies, in the range of 10^{10} to 10^{11} crystals per gram (Weickmann, 1957; Eadie and Mee, 1963; Fukuta et al., 1971). Holroyd et al. (1978) conducted dry ice seeding experiments in supercooled convective clouds and obtained empirical nucleation effectiveness values of $(2 \text{ to } 5) \times 10^{11}$ crystals per gram. Hobbs et al. (1978) deduced values of 2×10^{10} to 10^{11} crystals per gram from field measurements. Horn et al.

(1982) determined from theoretical and laboratory re-search that the effectiveness of dry ice is at least 10^{13} crystals per gram and could approach values in excess of 10^{14} crystals per gram. Most recently, Morrison et al. (1984) reported that laboratory and model studies indicated that dry ice effectiveness was moderately temperature-dependent, ranging from about 10^{11} crystals per gram at $-2°C$ to almost 10^{13} crystals per gram at $-20°C$. Thus, estimates of dry ice effectiveness have almost come full circle since its inception as a seeding agent, spanning a range from 10^8 to 10^{16} crystals per gram in the process.

The third question, related to the first two, concerns the appropriate seeding rate or dosage. A wide range of seeding rates have been used with no indication that a specific seeding rate is most effective in initiating precipitation. Squires and Smith (1949) and Bowen (1952), among others, claimed successful precipitation inducement using about 100 kg of dry ice in a single turret. However, claims of success were also reported for experiments using a seeding rate of about 1 kg km^{-1} in Project Cirrus (Langmuir, 1950; also see Havens, 1981, for a review of Project Cirrus activities). Braham et al. (1957) seeded supercooled cumulus clouds in the central United States with two different seeding rates of about 5 and 13 kg km^{-1}. Better, though not statistically significant, results were obtained with the higher seeding rate. There were indications that not enough dry ice was dropped in some of the clouds and that many of the clouds were seeded at too warm a temperature ($-5°C$). Marwitz and Stewart (1981) seeded supercooled convective clouds over the Sierra Nevada with two different seeding rates also, 0.1 and 1.0 kg km^{-1}. They reported that the high dry-ice seeding rate produced too many ice particles to develop a naturally precipitating cloud. English and Marwitz (1981) reported the occurrence of similar effects when an Alberta cumulus cloud was seeded at a rate of 0.2 kg km^{-1}. The results of HIPLEX-1 (Cooper and Lawson, 1984) also indicate that a seeding rate of 0.1 kg km^{-1} in Montana cumulus congestus produced more ice crystals than desired or required to enhance the IRG precipitation process. Care has been taken not to refer to the effect of producing excessive ice particles by seeding as "overseeding" since it does not strictly satisfy the classical definition of "overseeding" as given earlier. However, the effect is an important consequence of seeding and will be discussed in the context of the physical hypothesis in a subsequent section of this review.

2.4.2. Silver iodide

Silver iodide has also been used extensively as a seeding agent in precipitation enhancement experiments on supercooled convective clouds, but the state of knowledge of the various processes associated with its activity as a seeding agent does not appear to be in any better shape.

Most of what is known on the nucleation characteristics of silver iodide agents has been learned from theoretical (e.g., see Fletcher, 1962) and isothermal cloud chamber studies (e.g., see Blair et al., 1973; Garvey, 1975; DeMott et al., 1983), but the applicability of these results to real cloud conditions has not been established. One of the major difficulties in making this transfer is the fact that ice nucleation on silver iodide particles can occur by four different mechanisms, that is, deposition, condensation-freezing, contact–freezing, and immersion–freezing, and existing ice nuclei counters are not able to simulate or distinguish among them. While ice nucleus counters are useful in confirming the existence of a silver iodide plume in the atmosphere, they cannot indicate with reasonable accuracy how many ice particles will be nucleated in a cloud.

Soon after his discovery of silver iodide's ice nucleating ability, Vonnegut (1949) found that the number of ice crystals formed in a cloud chamber by a given quantity of silver iodide increased with decreasing temperature and that the nucleation rate decreased with increasing temperature and decreasing aerosol size. In the years that followed, theoretical and laboratory studies were conducted to explain the nucleating behavior of silver iodide with activity focused on establishing the relative importance of such factors as aerosol size (Fletcher, 1958a; Gerber, 1972, 1976), surface active sites (Edwards and Evans, 1968; Fletcher, 1969), mode of nucleation (Fletcher, 1959a; Weickman et al., 1970; Sax and Goldsmith, 1972; Demott et al., 1983; Blumenstein et al., 1983), chemical complexing (Burkardt et al., 1970; Sax et al., 1979a; Finnegan et al., 1984), activation time (Fletcher, 1958b; Isaac and Douglas, 1972; DeMott et al., 1983; Blumenstein et al., 1983), deactivation by photolysis (Reynolds et al., 1951; Vonnegut and Neubauer, 1951; Fletcher, 1959b), and warm temperature dissolution (St. Amand et al., 1971; Mathews et al., 1972). These studies have shown that the ice nucleation characteristics of silver iodide are quite complex, with the production of ice crystals depending in a complicated manner on the physicochemical properties of the silver iodide aerosols and the environmental conditions in which the silver iodide operates. Considerable progress has been made in understanding the influence of these complex processes, with conflicting trends and major shifts in thought along the way, but precisely how many ice crystals any given silver iodide aerosol will produce in natural cloud situations is not yet predictable with confidence.

The best estimates of the potential effectiveness of silver iodide formulations and generating systems have been derived from cloud chamber tests. The results of these tests (Blair et al., 1973; Garvey, 1975) have shown that, depending on mode of generation and chemical complexing, the ice nucleating activity of silver iodide aerosols begins at about $-5°$ to $-8°C$, increases by three to four orders of magnitude with decreasing temperature to about −16°C, and increases by less than one order of magnitude more with further decreases in temperature. Aerosols of silver iodide complexes produced by acetone generators tend to nucleate more ice particles than those produced by pyrotechnics, and increasing the burn rate of acetone generators tends to increase their activity at warm temperatures. Silver iodide–ammonium iodide complexes are considerably more active at warm temperatures than silver iodide–sodium iodide and silver iodide–potassium iodide complexes, and silver iodide–silver chloride complexes appear to be more active yet (DeMott et al., 1983).

The real effectiveness of silver iodide as a static mode seeding agent, however, depends on the number of ice crystals it produces at the cloud temperatures specified by the seeding hypothesis. The ice crystal yield of the seeding material, as distinguished from its cloud chamber calibration of effectivity (potential effectiveness), is mainly determined by its nucleation rate and residence time in the portion of the cloud of interest. The nucleation rate is, in turn, determined by the particular silver iodide agent's mode of nucleation and the cloud environment factors that affect it. Seeding agents such as silver iodide–silver chloride and silver iodide–ammonium iodide, which appear to act by the contact–freezing nucleation mode, have long time constants of nucleation, their nucleation rate being dependent on Brownian coagulation, which is a function of temperature and the size and concentration of both the silver iodide aerosols and the cloud droplets. Using chemical kinetic theory and experimental methodology, DeMott et al. (1983) showed that at temperatures warmer than −16°C it takes from about 15 min to 1 h for 90 percent of the silver iodide–silver chloride aerosols to nucleate ice crystals by the contact–freezing mode, the exact times being dependent on the specific chemical composition and cloud chamber conditions used. De Mott et al. concluded that at temperatures colder than −16°C the deposition nucleation mode was dominant with even longer nucleation time constants. These findings are generally consistent with field observations (Dye et al., 1976; Strapp et al., 1979; Marwitz and Stewart, 1981; English and Marwitz, 1981), which report activation of ice crystals following silver iodide seeding for 15 min and longer.

Seeding agents such as silver iodide–sodium iodide and silver iodide–potassium iodide that appear to act by the condensation–freezing mode of nucleation have somewhat shorter nucleation time constants. Blumenstein et al. (1983) reported that a silver iodide–sodium iodide nucleant produced 90 percent of its ice crystals in about 14 to 20 min, depending on temperature, in a cloud chamber at water saturation. However, when subjected to transient supersaturations with respect to water, both the nucleation rate and effectivity of the nucleant increased. In supersaturated conditions this nucleant produced 90 percent of its ice crystals in about 4 min, and its effectivity in-

creased by almost one order of magnitude. Rilling et al. (1984) found that a silver iodide–potassium iodide nucleant behaves similarly. Finnegan et al. (1984) showed that the incorporation of a hygroscopic salt, like sodium chloride, into the silver iodide–silver chloride nucleus composition changed its mode of nucleation from contact to condensation–freezing, thereby increasing both its effectivity and nucleation rate, especially under transient supersaturation conditions.

The combined effect of a silver iodide seeding agent's nucleation rate and residence time in a cloud is to make the nucleant's ice crystal yield less than its effectivity at all temperatures, the amount of reduction depending on the mode of nucleation and the cloud conditions where and when nucleation actually occurs. In effect it lowers the temperature threshold of activity for most silver iodide seeding agents to about −9° to −10°C. These factors may help to explain the diverse results of some of the past cloud seeding experiments (Kerr, 1984; Blumenstein et al., 1984).

2.4.3. Transport and dispersion

Both dry ice and silver iodide seeding material are initially dispensed in highly concentrated dosages, either as vertical lines produced by airborne drops of dry ice pellets or silver iodide pyrotechnics, as lines in the horizontal plane produced by airborne silver iodide–acetone generators and end-burning silver iodide flares, or as point sources produced by ground-based silver iodide–acetone generators. Transport and dispersion by natural air motions of the seeding material and/or the ice crystals they produce are then relied on to achieve the proper concentration of ice crystals in the targeted cloud volume at the appropriate time in the evolving cloud as required by the static mode physical hypothesis (see WMO, 1980, for a general review of the state of knowledge of the dispersion of cloud seeding agents). It is implicitly assumed that there is no penalty from the transient high concentrations of seeding material/ice crystals that occur during the seeding process. It will be shown in section 2.5 that this assumption is frequently invalid.

Cloud-top or in-cloud seeding provides the greatest assurance that the seeding material/ice crystals will be introduced at the appropriate levels in the cloud in a timely manner, but the time and vertical distance available for dispersion throughout the volume to be affected are quite limited. Due to the nearly instantaneous nucleation rate of dry ice (Morrison et al., 1984), its seeding signature is very dramatic, but relatively rapid dispersion of growing ice crystals is required. It is not uncommon to find peak ice crystal concentrations on the order of 1000 L^{-1} or more following dry ice seeding (Cooper and Lawson, 1984; Hobbs and Politovich, 1980; Kochtubajda and Rogers, 1984), which reduce in time as dispersion processes exert their influence. HIPLEX-1 studies (Lawson, 1978; Bureau of Reclamation, 1979; Rodi, 1981) showed that the ice crystals from dry ice seeding spread through small cumulus congestus in about 5 min. Silver iodide seeding, on the other hand, produces ice crystals over a period of time starting with its introduction into the cloud, so a time-varying combination of silver iodide aerosols and ice crystals are involved in the dispersion process. However, the peak concentration of ice crystals that droppable silver iodide flares produce initially are also quite high, reaching values in excess of 1000 L^{-1} in some cases (Marwitz and Stewart, 1981; Sax et al., 1979b). Provided that the required seeding concentration can be maintained, the timed release feature of silver iodide may be beneficial (Marwitz and Stewart, 1981; English and Marwitz, 1981) since it can have an effect on clouds where the liquid water continues to be replenished. Otherwise, additional doses of seeding material are required under these conditions, as is the case with dry ice. Experimental evidence (Jiusto and Holroyd, 1970; Weickmann, 1974) indicates that dry ice seeded regions spread two to three times faster than similar volumes seeded with silver iodide. This was presumably due to the enhanced convection and turbulence in the cloud resulting from the greater heat released more rapidly by the induced phase change. Dry ice seeding had a faster nucleation rate and likely produced a higher concentration of ice crystals at the seeding temperature levels than the silver iodide.

If the seeding material is dispensed farther from the intended zone of effect, by either cloud-base or ground-based silver iodide delivery systems, more time is available for volume dispersion and dilution. In addition, much of this dispersion can take place prior to activation of the silver iodide. However, targeting and timing of the seeding effect become more difficult and uncertain. Holroyd and Super (1977) and Hobbs et al. (1980) seeded small cumulus congestus clouds and convective complexes with silver iodide from cloud base and reported difficulty in detecting seeding effects at the targeted levels in the clouds. Dye et al. (1976), on the other hand, reported success in intercepting the silver iodide plume from cloud-base seeding of small cumuli in most cases. Maximum concentrations of ice crystals in the plume of about 400 to 800 L^{-1} produced by cloud-base silver iodide seeding have been observed (Dye et al., 1976; Hobbs and Politovich, 1980). Linkletter and Warburton (1977) and Warburton et al. (1982) used silver content of the precipitation as an index of the targeting effectiveness of the cloud-base seeding techniques used in NHRE (National Hail Research Experiment) and found uneven distributions both temporally and spatially such that only about 10 percent of the samples had observed concentrations comparable to those expected by theoretical calculations and those required by the physical hypothesis. Foote et al. (1979) es-

timated that the seeding coverage in NHRE was only about 50 percent on the average, which was attributed to finite aircraft reaction time and unpredictable storm evolution. They concluded "that seeding convective clouds using aircraft flying near cloud base is more difficult than is widely acknowledged."

With ground-based seeding the problems of seeding coverage, controlling seeding concentrations, and targeting and timing of seeding effects become even more difficult to overcome. Early studies (MacCready et al., 1957b) report finding silver iodide particles from ground generators in sufficient concentrations to produce seeding effects as far as 48 km from their source in particular situations. McPartland and Super (1978) and Heimbach and Stone (1984) found that silver iodide plumes from ground generators ascended to cloud-base altitudes in concentrations up to several orders of magnitude above background during unstable conditions favorable for convection. However, Heimbach and Stone (1984) showed that targeting was difficult and that the rising plumes did not disperse as much as expected from theory. Admirat and Buscaglione (1982) measured ice nuclei concentrations in updrafts just below cloud base of a storm being seeded with a dense network of ground-based silver iodide–acetone generators and found values from one to two orders of magnitude higher than those outside the experimental zone. They also proposed the concept of a SSF (Storm Seeding Factor), defined as the ratio of the artificial ice nuclei concentration below the cloud base of seeded storms to the natural concentrations below unseeded storms, to express the concentration of effective silver iodide nuclei from ground-based seeding systems that enter the storm base after diffusion, transport, and deactivation. The SSF was computed for a large number of ground-based seeding experiments, yielding values from 1 to 500, with most experiments being close to 1.

Another significant problem with ground-based seeding and, in fact, airborne patrol seeding at cloud-base altitudes is that it is nonselective, either among clouds in the target area or between the target and downwind areas. This implies that either (i) all clouds are seedable, (ii) clouds that are not seedable will not effect precipitation adversely, or (iii) if negative effects do occur, they are outweighed by the positive effects. The results of the Israeli experiments (Gagin and Neumann, 1981) seem to indicate that condition (ii) prevailed. However, the results of Project Whitetop (Flueck, 1971) and the Arizona Project (Battan, 1966; Battan and Kassander, 1967) seem to represent situations where none of these conditions are met; rather, the nonselective seeding was counterproductive. The inability to predict and isolate the occurrence of seedable clouds (see section 2.3) further compounds the aforementioned engineering problems in ground-based seeding that must be overcome.

2.5. Chain of physical events

HIPLEX-1 (Bureau of Reclamation, 1979) was a randomized, double-blind experiment to test the static mode seeding concept for convective clouds. It specified in advance and attempted to verify by observations during the course of the experiment each step leading to additional precipitation at cloud base. It was similar in approach to experiments conducted by Braham et al. (1957) and Bethwaite et al. (1966), but more of the steps in the chain of physical events could be included because of recent developments in observing, measuring, and real-time data processing capabilities. Therefore, the HIPLEX-1 design (Bureau of Reclamation, 1979; Smith et al., 1984) and physical results (Cooper and Lawson, 1984) will be used as the framework for the discussions to follow. They will also serve to illustrate the concepts discussed in the foregoing sections.

The experimental units of HIPLEX-1 were semi-isolated cumulus congestus clouds. Such clouds usually de-

TABLE 2.1. Cloud selection criteria for HIPLEX-1.

Class A-1 Cloud Criteria
1. Average cloud liquid water concentration greater than 0.5 g m^{-3} over approximately a 1-km-long cloud region determined by 10 s of flight at approximately 100 m s^{-1}
2. Average ice crystal concentrations less than 1.0 L^{-1} in the 1-km-long (10 s of flight) cloud region of maximum average liquid water concentration
3. Maximum ice crystal concentration less than 5.0 L^{-1} for any 1-km-long (10 s of flight) cloud region (defined by FSSP liquid water concentration greater than 0.01 g m^{-3}) during the test pass
4. Vertical air velocity greater than −1.0 m s^{-1} in the region defined by item 1, but if the vertical velocity is greater than 10.0 m s^{-1} and the buoyancy is greater than 1°C, reject the candidate
5. Length of the test penetration more than 2 km and less than 8 km as defined by an FSSP liquid water concentration greater than 0.01 g m^{-3}
6. No radar echo detectable on the aircraft weather radar
7. Cloud-top temperature lower than −6°C but higher than −12°C
8. Cloud-base temperature higher than 0°C
9. Minimum separation between the current test cloud and previous test clouds greater than 15 km to insure the meteorological independence of the clouds
Class A-2 Cloud Criteria
1. Items 1 through 9 of Class A-1 Cloud Criteria
2. An average wind direction between the surface and 800 kPa from 250° to 040° true
3. A 30-kPa-thick stable layer present with its base between 0° and −10°C and its top temperature at least 1.5°C higher than the temperature extrapolated from the base of the layer to the top using pseudoadiabatic ascent
4. A 10°C dewpoint depression present somewhere within the 30-kPa layer of B.3 above
Class B Cloud Criteria
1. Items 1, 2, 3, 5, 6, 8 and 9 of Class A-1 Cloud Criteria
2. Cloud-top temperature lower than −6°C but higher than −20°C
3. Vertical air velocity greater than −1.0 m s^{-1} in the region defined by Item A.1, but no other vertical velocity or buoyancy restrictions

TABLE 2.2. HIPLEX-1 primary response variables.

1.	CIC2	Cloud ice concentration, 2 min after treatment
2.	CIC5	Cloud ice concentration, 5 min after treatment
3.	CCR5	Concentration of crystals rimed, 5 min after treatment
4.	PIC8	Precipitating ice number concentration, 8 min after treatment
5.	MVD8	Mean volume diameter of precipitating ice particles, 8 min after treatment
6.	AWC8	Average liquid water concentration, 8 min after treatment
7.	TFPI	Time to first precipitating ice (particles with diameters >0.6 mm in concentrations >0.1 L^{-1})
8.	TFE	Time to first SWR-75 radar echo (15 dBZ)
9a.	TIPA	Time to initial precipitation at +10°C level, aircraft measurement
b.	TIPR	Time to initial precipitation at +10°C level, SWR-75 radar (15 dBZ)
10a.	RERC	Radar-estimated rainfall at +10°C level, using a constant Z–R relationship
b.	AER	Aircraft-estimated rainfall at +10°C level

velop precipitation naturally through the IRG mechanism (Cooper, 1978; Hobbs et al., 1980). Detailed criteria and procedures for selecting and classifying test cases, for performing the seeding, and for collecting the observations to be used in calculating response variables were prescribed. Based on preliminary exploratory studies, the cloud selection criteria (see Table 2.1) were expected to result in a sample of clouds that would be amenable to seeding according to the static hypothesis and last at least 30 min after treatment for the seeding to be effective. It was expected that the seeding effects would be most easily detected in clouds with tops in the −6° to −12°C range (type A clouds), but the experimental design permitted selection of rain clouds with tops in the −12° to −20°C range (type B clouds) as test cases when no suitable type A clouds were present. Type A-2 clouds were included to account for the occurrence of situations where SICP was found (Hobbs et al., 1980; Hobbs and Cooper, 1981). The qualifying variables were measured during a pretreatment pass by a cloud physics aircraft flying through a visually promising cloud at the −8°C level and immediately evaluated by an onboard computer to determine whether or not the selection criteria for any of the specified cloud types were met. The seeding was conducted by dropping a line of dry ice pellets from a jet aircraft at a rate of 0.1 kg km^{-1} near the −10°C level within 2 min after a suitable cloud was selected. Following the treatment, dry ice or placebo, both the seeding and cloud physics aircraft made repeated passes at specified times and specified levels in and below the cloud to document the subsequent chain of physical events as represented by the response variables (see Table 2.2). During the course of the two-year experiment, 55 clouds were tested for acceptance as experimental units, but only 20 met all the selection criteria. This sample size was considerably less than the 30–45 clouds per year that were expected from the preliminary field investigations. The years in which the seeding experiments were carried out were considerably drier than those used to develop the cloud selection criteria (Smith et al., 1984). Of the 20 test cases, 7 were type A-1 clouds, 4 seeded and 3 not seeded, and 13 were type B clouds, 8 seeded and 5 not seeded. The statistical results (Mielke et al., 1984) showed that the postulated increases in cloud ice concentrations associated with the seeding and the subsequent onset of riming were unequivocally established despite the limited sample size. For all response variables beyond 5 min after treatment, except the average liquid water content at 8 min, changes in the sample average values of the response variables were consistent with those suggested by the physical hypothesis, but it was clear that many of the clouds were not behaving as expected. Because of the experimental approach, the complementary physical evaluation was able to determine where, how, and why the seeded clouds behaved differently than expected and, thereby, provide a better understanding of precipitation processes in natural and seeded cumulus clouds in spite of the limited data set available.

The physical evaluation (Cooper and Lawson, 1984) revealed that in 4 of the 12 clouds that were seeded precipitation developed in the hypothesized manner, but physically significant departures occurred in the remainder. These studies indicated the following about the behavior of natural and seeded cumulus congestus clouds in Montana and the problems in applying the static mode seeding hypothesis to such clouds:

1) The liquid water content in many of the HIPLEX-1 clouds decayed rapidly due to entrainment, accounting for their low natural precipitation efficiency. The liquid water depleted faster by entrainment than seeding could exploit it to develop precipitation. In this respect, the cloud selection criteria had failed to provide the intended sample of clouds whose liquid water would persist at least 30 min after treatment, as required by the physical hypothesis.

2) Precipitation development in most of the seeded clouds did not proceed via the graupel process as hypothesized. Primarily because of the high ice concentrations produced by the seeding, a combination of aggregation and low-density accretion onto the loose aggregates was the dominant precipitation process. Only small raindrops were produced in this way, which were insignificant in comparison to those produced by the natural accretional growth process. In this respect, the seeding operation failed to produce the target concentrations of 10 L^{-1} without penalty.

3) Precipitation development proceeded as hypothesized in those clouds with sustained updrafts such that the main precipitation growth occurred at temperatures colder than $-10°C$ (above the seeding level). These results are supported by calculations (Cooper and Lawson, 1984;

Cooper, 1984) which show that accretional growth is more rapid and more efficient at about $-12°$ and $-20°C$. In this respect, the seeding hypothesis that emphasized the warm temperature region of the cloud was in error and the choice of seeding level was, perhaps, too low since it failed in most cases to take advantage of the region of rapid development of graupel from ice crystals.

The finding that entrainment exerts a controlling influence on the lifetime of HIPLEX-1 clouds and that it, in turn, limits precipitation development in natural and seeded clouds is consistent with the findings from other experiments on similar clouds. For the HIPLEX-1 clouds the exponential decay time constant of liquid water, defined as the time for the maximum liquid water to reduce to $1/e$ of its initial value, was 14 min. Marwitz (1984) reported that winter postfrontal cumulus congestus clouds over the central Sierra Nevada also had a decay constant of 14 min. Similarly, Hallett (1984) indicated a decay constant of 12 to 14 min for towering cumulus clouds over Florida. Kochtubajda and Rogers (1984) calculated the decay constant of the liquid water in treatment plumes (dry ice, silver iodide, and placebo) in isolated towering cumulus over Alberta. They found that the natural (placebo) clouds had a decay constant of about 12 min and that the concentration of ice crystals produced by the dry ice seeding depleted the liquid water faster than either the ice crystal concentration produced by the silver iodide seeding or entrainment (the natural clouds) for the first few minutes after seeding and, thereafter, entrainment dominated. The decay constants in the dry ice and silver iodide plumes were about 6 and 13.5 min, respectively, with the depletion rate being faster for the dry ice seeding presumably because of the higher ice crystal concentration it produced.

Braham (1960) examined the duration of summer cumuli in the central United States and Arizona; he defined duration as "the total life of the cloud after its top reached the $-5°C$ level." He found that only 25 percent of the clouds lasted longer than 15 min and of these about 60 percent developed precipitation naturally. His Fig. 1 indicates that about 33 percent of the clouds that just reached the $-5°C$ level developed rain naturally, indicating that the condensation–coalescence process was responsible for precipitation initiation in many of the clouds studied. Schemenauer and Isaac (1984) also discussed the importance of cloud-top lifetime as a factor controlling precipitation development in natural and seeded clouds. They defined cloud-top lifetime in terms of the persistence of liquid water at a level of about $-7°C$ and quantified it as "the time between the first aircraft measurements and the measured (or projected) disappearance of the cloud at the penetration level." They found that clouds near Yellowknife, Canada; Thunder Bay, Canada; and Miles City, Montana; had lifetimes of 16 and >20 min, 6 and

8 min, and 4 and 15 min, respectively, in the various years of study. Only cloud width was found to be related to cloud-top lifetime and in a directly proportional sense. For the combined sample of finite calculated lifetimes, only about 33 percent lasted longer than 15 min. The calculations of Cooper and Lawson (1984) indicate that 15 min is about the minimum time to grow graupel particles with reasonably sized melted diameters in favorable conditions. Thus, it can be seen from the above discussion that only a small fraction of cumulus clouds will have liquid water that persists long enough for seeding to have a chance to be effective. The dilemma for conducting cloud seeding experiments is to be able to predict which of the clouds in the population will have liquid water that lasts long enough.

The formation of aggregates as a result of seeding also appears to be a common occurrence. It is so common that it may be the rule rather than the exception, despite the fact that graupel development is usually postulated and expected. Aggregates have been found following seeding with dry ice (Hobbs and Politovich, 1980; English and Marwitz, 1981; Cooper and Lawson, 1984; Rodi, 1984), following cloud-top and in-cloud silver iodide seeding (Holroyd, et al., 1977; Strapp et al., 1979; English and Marwitz, 1981; Warburton and Telford, 1981), and following cloud-base silver iodide seeding (MacCready and Baughman, 1968; Holroyd and Jiusto, 1971; Dennis et al., 1974; Dye et al., 1976). Koenig (1963), Braham (1963), Todd (1965), Weinstein and Takeuchi (1970), and Sax et al. (1979b), on the other hand, did not observe aggregates following silver iodide seeding. Not enough is known about the aggregation process to specify under what conditions it will or will not occur (Hobbs, 1974, and Pruppacher and Klett, 1978, each sum up the state of knowledge of aggregation in about three pages).

The evidence from the cited experiments suggests that aggregation is more likely to occur in clouds where the dominant precipitation process is the IRG mechanism and, perhaps, not occur at all when the CRG mechanism is dominant. The empirical evidence also suggests that aggregation is primarily the result of high ice concentrations following seeding but may also be related to the occurrence of low liquid water contents and weak updrafts. Hobbs et al. (1974) showed that the probability of finding aggregates in a cloud increases with increasing temperature and ice particle concentration. For temperatures above $-10°C$ and ice particle concentrations in excess of about 800 L^{-1}, they found that there is more than a 50 percent chance that aggregates will form. English and Marwitz (1981) found that aggregates were formed following dry ice seeding at $-9°C$ and following silver iodide seeding of a similar cloud at $-8°C$ on the same day. More aggregates were formed following the dry ice seeding, presumably because of the higher concentration of ice crystals it produced.

Lin et al. (1983) and Cotton et al. (1984) have attempted to simulate the aggregation process in cloud models. Lin et al. used a parameterized formulation in which the aggregation rate was proportional to the ice crystal concentration and the rate coefficient varied in accordance with the temperature-dependent collection efficiencies of Hallgren and Hosler (1960) and Hosler and Hallgren (1960), hereafter referred to as the H-H collection efficiencies. The results of cloud simulations with this model showed that aggregation of ice crystals is not an important mechanism in producing "snow" in both natural convective clouds (Lin et al., 1983; Kopp et al., 1983) and dry ice seeded convective clouds (Kopp et al., 1983). Cotton et al. used an aggregation formulation based on coagulation theory (Green and Lane, 1964) in which the aggregation rate was a function of the square of the ice crystal concentration and a collection kernel estimated from the stochastic kernel of Passarelli and Srivastava (1978) for a distribution of particle densities of equal-size crystals. The constant value collection efficiency of 1.4 based on Passarelli and Srivastava's aircraft estimate and H-H's temperature-dependent, laboratory estimates of collection efficiency were used in various combinations in model sensitivity experiments. They found that the model predicted cloud structure that was most consistent with observation when the constant collection efficiency of 1.4 was used over the temperature range from $-12°$ to $-15°C$ and the H-H collection efficiencies were used at warmer temperatures.

Even though the formation of aggregates is not considered in the static seeding hypothesis, it may be a route to getting enhanced precipitation over the natural case in some situations. Vali et al. (1982) examined the microphysical processes of precipitating clouds in the Duero Basin, Spain, to determine which process was responsible for the precipitation and found that aggregation was the dominant process in 20 percent of the cumulus congestus clouds studied. Heymsfield (1982) calculated the growth of particles of various types to determine the rates of development of graupel and hail and found that aggregates that are introduced into a region of updraft are the most likely candidates to become hailstones in High Plains storms. On the other hand, the calculations and experimental results of Cooper and Lawson (1984) suggest that aggregates tend to grow slowly because they are inefficient embryos for accretion. A better understanding of the formation and growth of aggregates is needed to determine if and when aggregation can lead to precipitation in the time available in convective clouds. If it is feasible, then the appearance of aggregates following seeding would, in general, be different from the graupel that characterizes natural precipitation development and, as such, would constitute a physically meaningful seeding signature. Aggregation may be a mechanism that offsets the effect of too many ice crystals competing for the available water, that is, "overseeding." However, if aggregational growth is slow, it would mean that the seeding process is driving the cloud into a less efficient precipitation process than its natural one, with counterproductive consequences. If the latter is the usual case, the formation of aggregates that result from the high ice concentrations produced by seeding can be regarded as another definition of "overseeding."

The conclusion from the results of HIPLEX-1 that seeding would be more effective at a temperature level of $-12°C$ or colder (Cooper and Lawson, 1984) supports the notion that there is a warm-side cloud-top temperature limit for seedability of convective clouds. Additional support is provided by the experimental results of Bowen (1965), Smith (1967, 1970), Bethwaite et al. (1966), Flueck (1971), and Gagin and Neumann (1981). There is no statistical support for a change in rainfall from seeded cumuliform clouds with tops warmer than about $-10°$ to $-12°C$. Grant and Elliott (1974), Gagin and Neumann (1981), and Gagin (1981) hypothesize the existence of the warm cloud-top temperature limit to be the result of a combination of low net effectiveness of silver iodide and slow ice diffusional growth at temperatures warmer than about $-10°C$. Gagin (1975) showed that clouds in Israel and elsewhere do not produce graupel naturally until the cloud-top temperature is $-12°C$ or colder. The above results are also consistent with the explanation by Cooper and Lawson (1984) that accretional growth is inefficient in such clouds at the warmer temperatures.

Grant and Elliott (1974), Gagin and Neumann (1981), and Gagin (1981) point out that there is also a cold-side cloud-top temperature limit to seedability of convective clouds. They estimate it to be at about $-21°$ to $-25°C$, which creates a cloud seeding temperature window ranging from about $-10°$ to $-25°C$. They argue that the concentrations of natural ice nuclei at cloud-top temperatures colder than about $-25°C$ are nearly always sufficient to produce enough crystals for the available liquid water and that the addition of ice crystals by seeding will tend to cause an "overseeded" condition. The results of the Israeli experiments (Gagin and Neumann, 1981) tend to support this physical reasoning; the positive effect of seeding appeared to be confined to clouds with top temperatures ranging from $-12°$ to $-26°C$ with the maximum effect in the range from $-16°$ to $-21°C$. Similarly, the Project Whitetop results (Flueck, 1971) indicate positive treatment effects for clouds with echo tops between $-10°$ and $-40°C$, but, unlike the Israeli experiments, negative treatment effects were indicated for clouds with colder echo-top temperatures.

These results suggest that there is a "cloud seeding temperature window" or, perhaps, a "cloud depth window" that is conducive for seeding. If confirmed, it may be an important criterion for seedability if the depth to which clouds will grow can be predicted reasonably well. Being

a cloud property that is mainly determined by the thermodynamic structure of the environment, attempts have been made to estimate cloud-base, cloud-top, and echo-top heights from radiosonde data. The results of these postanalysis studies have been encouraging. Simpson et al. (1967), Matthews and Henz (1975), Dennis et al. (1975), and Matthews (1981) estimated the height of clouds with one-dimensional, steady state cloud models and showed that they were in reasonable agreement with observations when the appropriate cloud or updraft radius for the day was assumed. Matthews and Henz (1975) and Matthews (1981) also showed that model predictions of cloud-base height were in good agreement with aircraft-observed heights. Dennis et al. (1975) compared observed maximum echo heights with model-estimated cloud-top heights based on aircraft-measured updraft radii and obtained a correlation coefficient of 0.89 that was significant at the 1 percent level. This indicates that the appropriate input radius for the model could be obtained by aircraft measurements during a pretreatment pass. Hartzell and Jameson (1981) took a different approach and estimated the probability of cloud-top heights with a discriminant function based on stability and moisture parameters derived from a radiosonde sounding. They also found that their estimates of cloud-top height agreed reasonably well with observations, the postclassification accuracy of their estimates being in the range of 71 to 84 percent. If these techniques applied in a predictive sense are equally encouraging, they should be incorporated into the cloud eligibility criteria of future experiments.

The Numerical Models Group of the South Dakota School of Mines and Technology has actively pursued the simulation of static mode seeding of convective clouds for many years. They have made a series of modifications to their model in response to advances in theory and observation in an attempt to improve the realism of their seeding simulations. The most recent developments added conservation equations for glaciogenic seeding material (Hsie et al., 1980; Kopp et al., 1983) and for snow (Lin et al., 1983) to the two-dimensional, time-dependent cloud model with bulk water microphysics formulated by Orville and Kopp (1977). The latest version of this model attempts to simulate the introduction of the seeding material, to trace it as it advects and diffuses along the cloud's flow field, and to examine its effect on natural precipitation development (the unseeded model case) as it activates and interacts with the cloud's microphysical and dynamical processes. Results of model cloud seeding experiments have indicated both the potential to augment precipitation by seeding with silver iodide (Hsie et al., 1980) and with dry ice (Kopp et al., 1983) and the cloud and seeding conditions required to achieve it.

The seeding sensitivity tests of Hsie et al. (1980) showed that the response of precipitation to silver iodide seeding at cloud base was dependent on the initial seeding region, the time of seeding, and the natural precipitation efficiency of the cloud. They found that the circulation of the cloud influences the transport and dispersion of the seeding material and that seeding seemed to induce maximum precipitating ice in the peripheral regions of the cloud rather than in the upper portion of the updraft core as intended. Seeding of clouds with inefficient precipitation processes produced small increases in total precipitation, while seeding those with moderately efficient processes and strong convection did not increase precipitation significantly but did redistribute the precipitation. Kopp et al. (1983) simulated one of the HIPLEX-1 on-top dry ice seeding experiments and found that precipitation started 6 min earlier in the seeded cloud and that approximately 20 percent more rain fell in a light shower from the relatively weak cloud. However, the model cloud was wetter and developed faster than the actual cloud that was observed. This result raises the interesting prospect that the results of HIPLEX-1 might have been different, perhaps even as hypothesized, if it succeeded in selecting a set of wetter clouds as intended.

2.6. Overview

Braham (1981) examined the information base available for cloud seeding experiments and subjectively categorized the evidence from field studies according to how well they met specified standards of objectivity and repeatability. When these criteria are applied to the evidence related to static seeding, there does not appear to be anything of a positive nature that should be added to Braham's lists. The only aspect of static seeding that is known with reasonable certainty is that glaciogenic seeding agents can produce distinct "seeding signatures" in clouds in the form of above-background ice crystal concentrations that can be followed in time and space. However, it is not yet possible to routinely achieve the hypothesized time and space distribution of seeding-produced ice crystals in near-optimum concentrations for the available water without penalty.

The foregoing review mainly focuses on conflicts in physical evidence, difficulties in execution of static seeding as intended, and unexpected and diverse results. That, of course, was done intentionally to identify problem areas needing resolution, but such an exercise is also instructive, albeit through inference. This reviewer believes that the body of inferential evidence that has been amassed provides a better understanding of which clouds are seedable and which are not, even though the tools for recognizing and properly treating them are still imperfect. In particular, the body of inferred physical evidence appears to support the claims of physical plausibility for the positive statistical results of the replicated Israeli experiments I and II (Gagin and Neumann, 1981) and helps to explain in retrospect why these experiments were so successful and others were not (Tukey et al., 1978; Kerr, 1982).

Benefiting from clever research, insight and design, and a large measure of luck, it appears that the Israeli scientists were blessed with the following unique combination of favorable meteorological and operational conditions:

(i) The target clouds are post-cold-front, continental convective bands and clusters that develop precipitation exclusively through the IRG mechanism (Gagin, 1975). The cloud-base temperatures occur in a fairly narrow range from 5° to 8°C. Gagin states that there is no evidence that the coalescence process is active or that SICP occurs in the Israeli clouds.

(ii) A substantial fraction of the total daily cloud population on rain days, especially light rain days, have cloud-top temperatures that are within the cloud-seeding temperature window (Gagin, 1981).

(iii) As the cloud systems move into Israel from the west, they are invigorated by convergence at the coastline and a moderate rise in topography, thereby providing for continued release of condensate and a fairly persistent supply of liquid water in the clouds.

(iv) Airborne cloud-base patrol seeding allowed time for dispersion of the nuclei before activation and produced ice crystals in moderate concentrations at the appropriate time and place in the invigorated clouds (Gagin, 1979). A particularly convincing result is the eastward shift in the maximum effect of seeding in Israeli II (as opposed to Israeli I) by a distance comparable to the shift in the line of seeding (Gagin and Neumann, 1981).

(v) The use of a silver iodide–sodium iodide seeding material may have produced an unexpectedly high ice-crystal yield at the targeted temperature levels (Blumenstein et al., 1984).

(vi) Despite the lack of cloud selectivity by the mode of seeding, it does not appear that there were any negative effects of seeding clouds outside the cloud seeding temperature window (Gagin and Neumann, 1981).

The physical picture that has been presented for the Israeli experiments should be subjected to confirmation by making some critical chain-of-physical-events measurements in a sample of seeded and nonseeded clouds. One wonders, for example, how clouds with top temperatures as warm as -12°C can precipitate unless SICP is occurring. The Israeli experiments must, of course, also be duplicated in another area before the inferential support for the physical results can be considered as "hard" data according to Braham's criteria. Nevertheless, it does provide information on a subset of clouds that may be statically seedable and on the cloud characteristics to be sought in another area for attempting a duplication of the Israeli results. Finding another location with the appropriate combination of meteorological and cloud conditions may be difficult. The WMO Precipitation Enhancement Project attempted to do this without success (WMO, 1982).

2.7. A look to the future

This review ends where it began, that is, with the statement that the results of past experimental work are diverse but valid and that credibility of the science depends on understanding the physical reasons for the diverse results. Areas of uncertainty and apparent conflicts in evidence associated with the static seeding physical hypotheses, the concept of seedability, the seeding operation, and the chain of physical events following seeding have been highlighted. Future work should include studies directed at resolving these issues and, as Braham (1981) suggests, in the context of experiments designed to improve physical understanding of basic cloud processes and their responses to glaciogenic seeding.

The question is how to proceed. Should the main thrust continue to be on relatively small, cumulus congestus clouds that fit our current concepts of the static seeding physical hypothesis or on larger, dynamically more complicated convective clouds? In some respects, the dilemma (discussed at the end of section 2.3) that Braham (1981) confronted when he decided to conduct Project Whitetop is relevant again. Certainly, duplicating the Israeli results in another location should receive a high priority. The smaller clouds also provide a simpler experimental situation for shedding light on relevant physical processes such as nucleant behavior in cloud conditions, riming thresholds, accretional rates, aggregation, etc. But whether attention should be focused on demonstrating net rainfall increases from the general class of clouds to which the static seeding hypothesis strictly applies is questionable. Identifying criteria which would distinguish among clouds in this class that persist from those that do not is scientifically worthwhile but may not be very critical to further progress in the understanding and application of cloud seeding for rainfall augmentation. The dynamical conditions that give rise to clouds that strictly satisfy the static seeding hypothesis also constrain their precipitation efficiency. As such, it imposes an upper limit on their precipitation efficiency that is very much less than 100 percent. If a margin for improvement by static seeding of a subset of these clouds exists, it is likely to be rather small. The Israeli situation is apparently an exception that may be unique.

In moving to larger clouds it should be recognized that the simple static seeding concept, as presently conceived, would not be applicable. A new, more complex physical hypothesis that considers cloud–environment and microphysical–dynamical interactions associated with multicell convective cloud systems and their response to seeding will have to be developed. The concept of seeding for purely microphysical effects will no longer be appropriate.

Indeed, it is reasonable to ask whether this concept was ever appropriate and whether it was ever meaningful to make a distinction between seeding for purely microphysical effects and seeding for purely dynamical effects.

The move upscale should proceed gradually, beginning with convective cloud systems about which a substantially improved understanding of precipitation development can reasonably be expected with current experimental capabilities. Convective complexes larger than cumulus congestus but smaller than mesoscale convective systems would appear to be the best candidates for continued experimentation. Scientific knowledge and seeding know-how derived from past research efforts can serve as a foundation for future experiments. However, formulating a viable seeding hypothesis, defining seedability, achieving a seeding capability, and evaluating the results of seeding through a testable chain of physical events for convective complexes will be difficult tasks. Since convective complexes are manifestations of mesoscale dynamical processes, it will be necessary to consider how the modification of an individual complex affects the total precipitation of a larger area as an integral part of the seeding hypothesis. Larger, more comprehensive, field studies and experiments in concert with more realistic numerical models will be required to accomplish these objectives. Efforts like CCOPE (Knight, 1982) and the WMO International Cloud Modeling Workshop (Silverman et al., 1984; WMO, 1984) are important steps in this direction.

Acknowledgments. This review is founded on the contributions of numerous scientists to the science of weather modification and their dedication and efforts are gratefully acknowledged. Thanks are also due to a number of colleagues at the Bureau of Reclamation, University of Wyoming, and Illinois State Water Survey, especially Arnett Dennis, William Cooper, and Bernice Ackerman, for their stimulating discussions of some of the ideas in this review. I also thank Renate Colloton for assistance in preparing the manuscript.

REFERENCES

Admirat, P., and A. Buscaglione, 1982: Computation of a "storm seeding factor" for hail suppression by ground seeding. *J. Wea. Mod.,* **14,** 18–22.

Battan, L. J., 1953: Observations on the formation and spread of precipitation in convective clouds. *J. Meteor.,* **10,** 311–324.

——, 1963: Relationship between cloud base and initial radar echo. *J. Appl. Meteor.,* **2,** 333–336.

——, 1966: Silver-iodide seeding and rainfall from convective clouds, *J. Meteor.,* **23,** 669–683.

——, and A. R. Kassander, Jr., 1960: Design of a program of randomized seeding of orographic cumuli. *J. Meteor.,* **17,** 583–586.

——, and ——, 1967: Summary of results of a randomized cloud seeding project in Arizona. *Proc. Fifth Berkeley Symp. on Mathematical Statistics and Probability, Vol. V,* University of California Press, 29–33.

Bergeron, T., 1933: On the physics of cloud and precipitation. *Verbal*

Proc. Int. Union Geod. Geophys., Fifth General Assembly, Lisbon, 1933, Paris, 1935, 156–178.

——, 1949: The problem of artificial control of rainfall on the Globe I. General effects of ice-nuclei in clouds. *Tellus,* **1,** 32–43.

Bethwaite, F. D., E. J. Smith, J. A. Warburton and K. J. Heffernam, 1966: Effects of seeding isolated cumulus clouds with silver iodide. *J. Appl. Meteor.,* **5,** 513–520.

Blair, D. N., B. L. Davis and A. S. Dennis, 1973: Cloud chamber tests of generators using acetone solutions of AgI–NaI, AgI–KI and AgI–NH₄I. *J. Appl. Meteor.,* **12,** 1012–1017.

Blumenstein, R. R., W. G. Finnegan and L. O. Grant, 1983: Ice nucleation by silver iodide–sodium iodide—a reevaluation. *J. Wea. Mod.,* **15,** 11–15.

——, —— and ——, 1984: Rates and mechanisms of ice nucleation by silver iodide–sodium iodide aerosols: Implications for the Climax and Israeli programs. *Preprints Ninth Conf. on Weather Modification,* Park City, Amer. Meteor. Soc., 35–36.

Bowen, E. G., 1952: Australian experiments on artificial stimulation of rainfall. *Weather,* **7,** 204–209.

——, 1965: Lessons learned from long-term cloud-seeding experiments. *Proc. Int. Conf. on Cloud Physics,* Tokyo, 429–433.

Braham, R. R., Jr., 1960: Physical properties of clouds. *J. Irrig. Drain. Div., Proc. ASCE,* **86,** 111–119.

——, 1963: Some measurements of snow pellet bulk densities. *J. Appl. Meteor.,* **2,** 498–500.

——, 1964: What is the role of ice in summer rain showers? *J. Atmos. Sci.,* **21,** 640–645.

——, 1966: *Final Report of Project Whitetop, A Convective Cloud Randomized Seeding Project Part I: Design of the Experiment; Part II: Summary of Operations,* Dept. of the Geophysical Sciences, University of Chicago, 77 pp. [NTIS PB-176-622.]

——, 1968: Meteorological bases for precipitation development. *Bull. Amer. Meteor. Soc.,* **49,** 343–353.

——, 1979: Field experimentation in weather modification. *J. Amer. Stat. Assoc.,* **74,** 57–104.

——, 1981: Designing cloud seeding experiments for physical understanding. *Bull. Amer. Meteor. Soc.,* **62,** 55–62.

——, and J. R. Sievers, 1957: Overseeding of cumulus clouds. *Artificial Stimulation of Rain,* H. Weickman and W. Smith, Eds., Pergamon, 250–256.

——, L. J. Battan and H. R. Byers, 1957: Artificial nucleation of cumulus clouds. *Cloud and Weather Modification, Meteor. Monogr.,* No. 11, Amer. Meteor. Soc., 47–85.

Bureau of Reclamation, 1979: *The Design of HIPLEX-1.* Rep., Division of Atmospheric Resources Research, Bureau of Reclamation, U.S. Dept. of the Interior, Denver, 271 pp.

——, 1983: *The Design of SCPP-1.* Rep., Division of Atmospheric Resources Research, Bureau of Reclamation, U.S. Dept. of the Interior, Denver, 66 pp.

Burkardt, L. A., W. G. Finnegan, F. K. Odencrantz and P. St.-Amand, 1970: Pyrotechnic production of nucleants for cloud modification: Part IV. Compositional effects on ice nuclei activity. *J. Wea. Mod.,* **2,** 65–76.

Clark, R. A., 1960: A study of convective precipitation as revealed by radar observation, Texas 1958–1959. *J. Meteor.,* **17,** 415–425.

Cooper, W. A., 1978: Precipitation mechanisms in summertime storms in the Montana HIPLEX area. *Conf. on Cloud Physics and Atmospheric Electricity,* Issaquah, Amer. Meteor. Soc., 347–350.

——, 1984: Accretional growth processes in seeded and unseeded HIPLEX-1 clouds. *Preprints Ninth Conf. on Weather Modification,* Park City, Amer. Meteor. Soc., 100–101.

——, and C. P. R. Saunders, 1980: Winter storms over the San Juan Mountains. Part II: Microphysical processes. *J. Appl. Meteor.,* **19,** 927–941.

——, and R. P. Lawson, 1984: Physical interpretation of results from the HIPLEX-1 experiment. *J. Climate Appl. Meteor.,* **23,** 523–540.

Cotton, W. R., G. J. Tripoli and R. M. Rauber, 1984: A numerical simulation of the effects of small scale topographical variations on the generation of aggregate snowflakes. *Proc. Ninth Int. Conf. on Cloud Physics,* Tallinn, U.S.S.R., 613–616.

DeMott, P. J., W. G. Finnegan and L. O. Grant, 1983: An application of chemical kinetic theory and methodology to characterize the ice nucleating properties of aerosols used for weather modification. *J. Climate Appl. Meteor.,* **22,** 1190–1203.

Dennis, A. S., 1980: *Weather Modification by Cloud Seeding.* Academic Press, 267 pp.

——, 1984: The science and technology of summer cumulus modification. *Preprints Ninth Conf. on Weather Modification,* Park City, Amer. Meteor. Soc., 93–96.

——, P. L. Smith, B. L. Davis, H. D. Orville, R. A. Scheusener, G. N. Johnson, J. H. Hirsch, D. E. Cain and A. Koscielski, 1974: Cloud seeding to enhance summer rainfall in the northern plains. Rep. 74-10, South Dakota School of Mines and Technology, 32 pp.

——, A. Koscielski, D. E. Cain, J. H. Hirsch and P. L. Smith, Jr., 1975: Analysis of radar observations of a randomized cloud seeding experiment. *J. Appl. Meteor.,* **14,** 897–908.

Dye, J. E., G. Langer, V. Tootenhoofd, T. Cannon and C. Knight, 1976: Use of a sailplane to measure microphysical effects of silver iodide seeding in cumulus clouds. *J. Appl. Meteor.,* **15,** 264–274.

Eadie, W. J., and T. R. Mee, 1963: The effect of dry ice pellet velocity on the generation of ice crystals. *J. Appl. Meteor.,* **2,** 260–265.

Edwards, G. R., and L. F. Evans, 1968: Ice nucleation by silver iodide: III. The nature of the nucleating site. *J. Atmos. Sci.,* **25,** 249–256.

English, M., and J. D. Marwitz, 1981: A comparison of AgI and CO_2 seeding effects in Alberta cumulus clouds. *J. Appl. Meteor.,* **20,** 483–495.

Farley, R. D., and H. D. Orville, 1982: Cloud modeling in two spatial dimensions. *Hailstorms of the Central High Plains I. The National Hail Research Experiment,* C. A. Knight and P. Squires, Eds, Colorado Associated University Press, 207–223.

Findeisen, W., 1938: Colloidal meteorological processes in the formation of atmospheric precipitation. *Meteor. Z.,* **55,** 121–133.

Finnegan, W. G., F. DaXiong and L. O. Grant, 1984: Composite AgI–AgCl ice nuclei efficient, fast-functioning aerosols for weather modification experimentation. *Preprints Ninth Conf. on Weather Modification,* Park City, Amer. Meteor. Soc., p. 3.

Fletcher, N. H., 1958a: Size effect in heterogeneous nucleation. *J. Chem. Phys.,* **29,** 572–576.

——, 1958b: Time lag in ice crystal nucleation in the atmosphere. Part II: Theoretical. *Bull. Obs. Pay de Dome,* 11–18.

——, 1959a: On ice-crystal production by aerosol particles. *J. Meteor.,* **16,** 173–180.

——, 1959b: A descriptive theory of the photo de-activation of silver iodide as an ice-crystal nucleus. *J. Meteor.,* **16,** 249–255.

——, 1962: *The Physics of Rainclouds.* Cambridge University Press, 386 pp.

——, 1969: Active sites and ice crystal nucleation. *J. Atmos. Sci.,* **26,** 1266–1271.

Flueck, J. A., 1971: *Final Report of Project Whitetop: Part V—Statistical analyses of the ground level precipitation data.* Dept. of the Geophysical Sciences, University of Chicago. [NTIS N72-13559.]

Foote, G. B., C. G. Wade, J. C. Fankhauser, P. W. Summers, E. L. Crow and M. E. Solak, 1979: Results of a randomized hail suppression experiment in northeast Colorado. Part VII: Seeding logistics and post hoc stratification by seeding coverage. *J. Appl. Meteor.,* **18,** 1601–1617.

Fukuta, N., W. A. Schmeling and L. F. Evans, 1971: Experimental determination of ice nucleation by falling dry ice pellets. *J. Appl. Meteor.,* **10,** 1174–1179.

Gagin, A., 1975: The ice phase in winter continental cumulus clouds. *J. Atmos. Sci.,* **32,** 1604–1614.

——, 1979: Cloud seeding technology. *WMO Training Workshop on*

Weather Modification for Meteorologists, Lecture Notes, PEP Rep. No. 13, WMO, Geneva, 136–152.

——, 1981: The Israeli rainfall enhancement experiment—a physical overview. *J. Wea. Mod.,* **13**, 108–120.

——, and J. Neumann, 1981: The second Israeli randomized cloud seeding experiment: Evaluation of the results. *J. Appl. Meteor.,* **20**, 1301–1311.

——, and H. Nozyce, 1984: The nucleation of ice crystals during the freezing of large supercooled drops. *J. Rech. Atmos.,* **18**, 119–129.

Garvey, D. M., 1975: Testing of cloud seeding materials at the cloud simulation and aerosol laboratory, 1971–1973. *J. Appl. Meteor.,* **14**, 883–890.

Gerber, H., 1972: Size and nucleating ability of AgI particles. *J. Atmos. Sci.,* **29**, 391–392.

——, 1976: Relationship of size and activity for AgI smoke particles. *J. Atmos. Sci.,* **33**, 667–677.

Grant, L. O., and R. E. Elliott, 1974: The cloud seeding temperature window. *J. Appl. Meteor.,* **13**, 355–363.

Green, H. L., and W. R. Lane, 1964: *Particulate Clouds: Dusts, Smokes and Mists.* E. and F. N. Spon Ltd., 471 pp.

Hallett, J., 1984: Modelling the convective cloud environment: Florida. *Annex to the Notes on the Planning Session for the Int. Cloud Modelling Workshop/Conference held at Aspen, Colorado, U.S.A.,* Informal Rep., Weather Modification Programme, WMO, Geneva.

——, and S. C. Mossop, 1974: Production of secondary ice particles during the riming process. *Nature,* **249**, 26–28.

Hallgren, R. E., and C. L. Hosler, 1960: Preliminary results on the aggregation of ice crystals. *Geo. Monogr.,* **5**, 257–263.

Hartzell, C. L., and T. C. Jameson, 1981: Identifying sounding predictor variables for HIPLEX-1 cumulus congestus clouds. *Preprints Eighth Conf. on Inadvertent and Planned Weather Modification,* Reno, Amer. Meteor. Soc., 98–99.

Havens, B. S., 1981: Early history of cloud seeding. *J. Wea. Mod.,* **13**, 14–88.

Heimbach, J. A., Jr., and N. C. Stone, 1984: Ascent of surface-released silver iodide into summer convection Alberta 1975. *J. Wea. Mod.,* **16**, 19–26.

Heymsfield, A. J., 1982: A comparative study of the rates of development of potential graupel and hail embryos in High Plains storms. *J. Atmos. Sci.,* **39**, 2867–2897.

Hobbs, P. V., 1974: *Ice Physics.* Clarendon, 837 pp.

——, and A. J. Alkezweeny, 1968: The fragmentation of freezing water droplets in free fall. *J. Atmos. Sci.,* **25**, 881–888.

——, and R. J. Farber, 1972: Fragmentation of ice particles in clouds. *J. Rech. Atmos.,* **6**, 245–258.

——, and D. G. Atkinson, 1976: The concentration of ice particles in orographic clouds and cyclonic storms over the Cascade Mountains. *J. Atmos. Sci.,* **33**, 1362–1374.

——, and M. K. Politovich, 1980: The structures of summer convective clouds in eastern Montana. II: Effects of artificial seeding. *J. Appl. Meteor.,* **19**, 664–675.

——, and W. L. Cooper, 1981: Field evidence supporting the operation of the Hallett–Mossop mechanism of secondary ice crystal production. *Preprints Eighth Conf. on Inadvertent and Planned Weather Modification,* Reno, Amer. Meteor. Soc., 106–107.

——, S. Chang and J. D. Locatelli, 1974. The dimensions and aggregation of ice crystals in natural clouds. *J. Geophys. Res.,* **79**, 2199–2206.

——, L. F. Radke and M. K. Politovich, 1978: Comments on "The practicability of dry ice for on-top seeding of convective clouds." *J. Appl. Meteor.,* **17**, 1872–1874.

——, M. K. Politavich and L. F. Radke, 1980: The structures of summer convective clouds in eastern Montana. I: Natural clouds. *J. Appl. Meteor.,* **19**, 645–663.

Holroyd, E. W., and J. E. Jiusto, 1971: Snowfall from a heavily seeded cloud. *J. Appl. Meteor.,* **10**, 266–269.

——, and A. B. Super, 1977: Seeding experiments using AgI in convective clouds. *Preprints Sixth Conf. on Inadvertent and Planned Weather Modification,* Champaign-Urbana, Amer. Meteor. Soc., 354–357.

——, —— and B. A. Silverman, 1978: The practicability of dry ice for on-top seeding of convective clouds. *J. Appl. Meteor.,* **17**, 49–63.

Horn, R. D., W. G. Finnegan and P. J. Demott, 1982: Experimental studies of nucleation by dry ice. *J. Appl. Meteor.,* **21**, 1567–1570.

Hosler, C. L., and R. E. Hallgren, 1960: The aggregation of small ice crystals. *Disc. Farad. Soc.,* **30**, 200–208.

Hsie, E-Y, R. D. Farley and H. D. Orville, 1980: Numerical simulation of ice-phase convective cloud seeding. *J. Appl. Meteor.,* **19**, 950–977.

Isaac, G. A., and R. H. Douglas, 1972: Another "time lag" in the activation of atmospheric ice nuclei. *J. Appl. Meteor.,* **11**, 490–493.

——, J. W. Strapp, R. S. Schemenauer and J. I. MacPherson, 1982: Summer cumulus cloud seeding experiments near Yellowknife and Thunder Bay, Canada. *J. Appl. Meteor.,* **21**, 1266–1285.

Jiusto, J. E., and E. W. Holroyd III, 1970: Great Lakes snowstorms, Part I. Cloud physics aspects. ASRC, State University of New York at Albany, 142 pp. [NTIS COM 7100012/cc.]

Johnson, D. A., and J. Hallett, 1968: Freezing and shattering of supercooled water drops. *Quart. J. Roy. Meteor. Soc.,* **94**, 468–482.

Johnson, D. B., 1982: Geographical variations in cloud-base temperature *Preprints Conf. on Cloud Physics,* Chicago, Amer. Meteor. Soc., 187–189.

Kerr, R. A., 1982: Cloud seeding: One success in 35 years. *Science,* **217**, 519–521.

——, 1984: The fine points of cloud seeding. *Science,* **223**, p. 153.

Knight, C. A., 1982: The Cooperative Convective Precipitation Experiment (*CCOPE*), 18 May–7 August 1981. *Bull. Amer. Meteor. Soc.,* **63**, 386–398.

Knight, N. C., 1981: The climatology of hailstone embryos. *J. Appl. Meteor.,* **20**, 750–755.

Kochtubajda, B., and D. C. Rogers, 1984: Seeding effects on liquid water and ice crystal concentrations in Alberta cumulus clouds. *Preprints Ninth Conf. on Weather Modification,* Park City, Amer. Meteor. Soc., 102–103.

Koenig, L. R., 1963: The glaciating behavior of small cumulonimbus clouds. *J. Atmos. Sci.,* **20**, 29–47.

Kopp, F. J., H. D. Orville, R. D. Farley and J. H. Hirsch, 1983: Numerical simulation of dry ice cloud seeding experiments. *J. Climate Appl. Meteor.,* **22**, 1542–1556.

Kraus, E. B., and P. Squires, 1947: Experiments on the stimulation of clouds to produce rain. *Nature,* **159**, 489–492.

Langmuir, I., 1950: Results of seeding of cumulus clouds in New Mexico. Project Cirrus Occasional Rep. No. 24, General Electric Research Laboratory, Schenectady.

Lawson, R. P., 1978: Turbulent rates of dispersion of ice particles in convective clouds. *Conf. on Cloud Physics and Atmospheric Electricity,* Issaquah, Amer. Meteor. Soc., 355–362.

Leopold, L. B., and M. H. Halstead, 1948: First trials of the Schaefer–Langmuir dry-ice cloud seeding technique in Hawaii. *Bull. Amer. Meteor. Soc.,* **29**, 525–534.

Lin, Y-L, R. D. Farley and H. D. Orville, 1983: Bulk parameterization of the snowfield in a cloud model. *J. Climate Appl. Meteor.,* **22**, 1065–1092.

Linkletter, G. O., and J. A. Warburton, 1977: An assessment of NHRE hail suppression seeding technology based on silver analysis. *J. Meteor.,* **16**, 1332–1348.

MacCready, P. B., and R. G. Baughman, 1968: The glaciation of an AgI seeded cloud. *J. Appl. Meteor.,* **7**, 132–135.

——, T. B. Smith, C. J. Todd and K. M. Beesmer, 1957a: Nuclei, cumulus and seedability studies. Part II. Cumulus studies. *Final Report of the Advisory Committee on Weather Control,* Vol. II, 151–168.

——, ——, —— and ——, 1957b: Nuclei, cumulus and seedability

studies. Part V. Silver iodide study. *Final Report of the Advisory Committee on Weather Control,* Vol. II, 185–190.

McPartland, J., and A. B. Super, 1978: Diffusion of ground-generated silver iodide to cumulus cloud formation levels. *J. Wea. Mod.,* **10,** 71–75.

Marwitz, J. D., 1984: Convective clouds in California. *Annex to the Notes on the Planning Session for the International Cloud Modelling Workshop/Conference held at Aspen,* Informal Rep., Weather Modification Programme, WMO, Geneva.

——, and R. E. Stewart, 1981: Some seeding signatures in Sierra storms. *J. Appl. Meteor.,* **20,** 1129–1144.

Mason, B. J., 1957: *The Physics of Clouds.* Oxford University Press, 481 pp.

——, 1981: The mechanisms of cloud seeding with dry ice (Reply to Vonnegut). *J. Wea. Mod.,* **13,** p. 11.

Mathews, L. A., D. W. Reed, P. St.-Amand and R. J. Stirton, 1972: Rate of solution of ice nuclei in water drops and its effect on nucleation. *J. Appl. Meteor.,* **11,** 813–817.

Matthews, D. A., 1981: Natural variability of thermodynamic features affecting convective cloud growth and dynamic seeding: A comparative summary of three High Plains sites from 1975 to 1977. *J. Appl. Meteor.,* **20,** 971–996.

——, and J. F. Henz, 1975: Verification of numerical model simulations of cumulus–environmental interaction in the High Plains. *Pure Appl. Geophys.,* **113,** 803–823.

Mielke, P. W., K. J. Berry, A. S. Dennis, P. L. Smith, J. R. Miller and B. A. Silverman, 1984: HIPLEX-1: Statistical evaluation. *J. Climate Appl. Meteor.,* **23,** 513–522.

Morrison, B. J., W. G. Finnegan, R. D. Horn and L. O. Grant, 1984: A laboratory characterization of dry ice as a glaciogenic seeding agent. *Preprints Ninth Conf. on Weather Modification,* Park City, Amer. Meteor. Soc., 8–9.

Mossop, S. C., 1976: Production of secondary ice particles during the growth of graupel by riming. *Quart. J. Roy. Meteor. Soc.,* **102,** 45–47.

——, 1978a: The influence of dropsize distribution on the production of secondary ice particles during graupel growth. *Quart. J. Roy. Meteor. Soc.,* **104,** 323–330.

——, 1978b: Some factors governing ice particle multiplication in cumulus clouds. *J. Atmos. Sci.,* **35,** 2033–2037.

——, 1985: The origin and concentration of ice crystals in clouds. *Bull. Amer. Meteor. Soc.,* **66,** 264–273.

Nelson, L. D., 1979: Observations and numerical simulations of precipitation mechanisms in natural and seeded convective clouds. Ph.D. dissertation, University of Chicago, 188 pp.

Orville, H. D., and F. J. Kopp, 1977: Numerical simulation of the life history of a hailstorm. *J. Atmos. Sci.,* **34,** 1596–1618.

Paluch, I. R., and D. W. Breed, 1984: A continental storm with a steady, adiabatic updraft and high concentrations of small ice particles: 6 July 1976 case study. *J. Atmos. Sci.,* **41,** 1008–1024.

Passarelli, R. E., and R. C. Srivastava, 1978: A new aspect of snowflake aggregation theory. *J. Atmos. Sci.,* **36,** 484–493.

Perez-Siliceo, E., 1970: 19 years of cloud seeding operations in the Necaxa, Puebla and Lerma, Mexico, watersheds from the periods 1949–1951 plus 1953–1968. *Preprints Second Natl. Conf. on Weather Modification,* Santa Barbara, Amer. Meteor. Soc., 87–90.

Pruppacher, H. R., and R. H. Schlamp, 1975: A wind tunnel investigation on ice multiplication by freezing of waterdrops falling at terminal velocity in air. *J. Geophys. Res.,* **80,** 380–386.

——, and J. D. Klett, 1978: *Microphysics of Clouds and Precipitation.* D. Reidel, 714 pp.

Rangno, A. L., and P. V. Hobbs, 1983: Production of ice particles in clouds due to aircraft penetrations. *J. Climate Appl. Meteor.,* **22,** 214–232.

Reynolds, S. E., W. Hume, B. Vonnegut and V. J. Schaefer, 1951: Effect

of sunlight on the action of silver iodide particles as sublimation nuclei. *Bull. Amer. Meteor. Soc.,* **32,** p. 47.

Rilling, R. R., R. R. Blumenstein, W. G. Finnegan and L. O. Grant, 1984: Characterization of silver iodide–potassium iodide nuclei: Rates and mechanisms, and comparison to the silver iodide–sodium iodide system. *Preprints Ninth Conf. on Weather Modification,* Park City, Amer. Meteor. Soc., 16–17.

Rodi, A. R., 1981: Study of the fine-scale structure of cumulus clouds. Ph.D. dissertation, University of Wyoming, 328 pp.

——, 1984: Dry ice seeding signatures in winter convective clouds over the Sierra Nevada. *Preprints Ninth Conf. on Planned and Inadvertent Weather Modification,* Park City, Amer. Meteor. Soc., 67–70.

Sax, R. I., and P. Goldsmith, 1972: Nucleation of water drops by Brownian contact with AgI and other aerosols. *Quart. J. Roy Meteor. Soc.,* **98,** 69–72.

——, D. M. Garvey and F. P. Parungo, 1979a: Characteristics of AgI pyrotechnic nucleant used in NOAA's Florida Area Cumulus Experiment. *J. Appl. Meteor.,* **18,** 195–202.

——, J. Thomas, N. Bonebrake and J. Hallett, 1979b: Ice evaluation within seeded and nonseeded Florida cumuli. *J. Appl. Meteor.,* **18,** 203–214.

Schaefer, V. J., 1946: The production of ice crystals in a cloud of supercooled water droplets. *Science,* **104,** 457–459.

——, 1949: The formation of ice crystals in the laboratory and the atmosphere. *Chem. Rev.,* **44** (2), 291–320.

Schemenauer, R. S., and G. A. Isaac, 1984: The importance of cloud top lifetime in the description of natural cloud characteristics. *J. Climate Appl. Meteor.,* **23,** 267–279.

Silverman, B. A., L. R. Koenig, G. B. Foote and T. L. Clark, 1984: The international cloud modelling workshop. Results of the planning session. *Proc. Ninth Int. Cloud Physics Conference,* Tallinn, U.S.S.R., IAMAP, 4 pp.

Simpson, J., G. W. Brier and R. H. Simpson, 1967: Stormfury cumulus seeding experiment 1965: Statistical analysis and main results. *J. Atmos. Sci.,* **24,** 508–521.

Smith, E. J., 1949: Five experiments in seeding cumuliform cloud layers with dry ice. *Austral. J. Sci. Res. (A), Phys. Sci.,* **2,** 78–91.

——, 1967: Cloud seeding experiments in Australia. *Proc. Fifth Berkeley Symp. on Mathematical Statistics and Probability,* Vol. V, University of California Press, 161–176.

——, 1970: Effects of cloud-top temperature on the results of cloud seeding with silver iodide in Australia. *J. Appl. Meteor.,* **9,** 800–804.

Smith, P. L., A. S. Dennis, B. A. Silverman, A. B. Super, E. W. Holroyd, W. A. Cooper, P. W. Mielke, K. J. Berry, H. D. Orville and J. R. Miller, 1984: HIPLEX 1: Experimental design and response variables. *J. Climate Appl. Meteor.,* **23,** 497–512.

Squires, P. A., and E. J. Smith, 1949: The artificial stimulation of precipitation by means of dry ice. *Austral. J. Sci. Res. (A),* **2,** 232–245.

St.-Amand, P., L. A. Matthews, D. W. Reed, L. A. Burkhardt and W. G. Finnegan, 1971: Effects of solubility of AgI nucleation effectiveness. *J. Wea. Mod.,* **3,** 106–110.

Strapp, J. W., H. G. Leighton and G. A. Isaac, 1979: A comparison of model calculations of ice crystal growth with observations following silver iodide seeding. *Atmos.-Ocean,* **3,** 234–252.

Todd, C. J., 1965: Ice crystal development in a seeded cumulus cloud. *J. Atmos. Sci.,* **22,** 70–78.

Tukey, J. W., D. R. Brillinger and L. V. Jones, 1978: The role of statistics in weather resources management. The Management of Weather Resources, Vol. II. *Final Report of Weather Modification Advisory Board,* Dept. of Commerce, Washington, DC. [U.S. Government Printing Office No. 003-018-00091-1.]

Vali, G., 1980: Ice multiplication by rime breakup. *Preprints Eighth Int. Conf. on Cloud Physics,* Clermont-Ferrand, 227–228.

——, T. C. Yoksas and P. G. Grube, 1982: Ice evolution versus precipitation in the Duero Basin. *Preprints Conf. on Cloud Physics,* Chicago, Amer. Meteor. Soc., 218–221.

Vardiman, L., 1978: The generation of secondary ice particles in clouds by crystal–crystal collision. *J. Atmos. Sci.,* **35,** 2168–2180.

Vonnegut, B., 1947: The nucleation of ice formation by silver iodide. *J. Appl. Phys.,* **18,** p. 593.

——, 1949: Nucleation of supercooled water clouds by silver-iodide smokes. *Chem. Rev.,* **44,** p. 277.

——, 1981: Misconception about cloud seeding with dry ice. *J. Wea. Mod.,* **13,** 9–11.

——, and R. L. Neubauer, 1951: Recent experiments on the effect of ultraviolet light on silver iodide nuclei. *Bull. Amer. Meteor. Soc.,* **32,** p. 356.

Warburton, J. A., and J. W. Telford, 1981: The capture of cloud seeding aerosols by supercooled cloud particulates. *Preprints Eighth Conf. on Inadvertent and Planned Weather Modification,* Reno, Amer. Meteor. Soc., 22–23.

——, G. O. Linkletter and R. Stone, 1982: The use of trace chemistry to estimate seeding effects in the National Hail Research Experiment. *J. Appl. Meteor.,* **21,** 1089–1110.

Wegner, A., 1911: *Thermodynamik der Atmosphare.* Barth, Leipzig.

Weickmann, H. K., 1957: Current understanding of the physical processes associated with cloud nucleation. *Beitr. Phys. Atmos.,* **30,** 97–118.

——, 1974: The mitigation of Great Lakes storms. *Weather and Climate Modification,* W. N. Hess, Ed., Wiley and Sons, 318–354.

——, U. Katz and R. L. Steele, 1970: AgI-Sublimation or contact nucleus? *Preprints Second Natl. Conf. on Weather Modification,* Santa Barbara, Amer. Meteor. Soc., 332–336.

Weinstein, A. I., and D. M. Takeuchi, 1970: Observations of ice crystals in a cumulus cloud seeded by vertical-fall pyrotechnics. *J. Appl. Meteor.,* **9,** 265–268.

WMO, 1980: Dispersion of cloud seeding reagents. PEP Rep. No. 14, World Meteorological Organization, Geneva, 28 pp.

——, 1982: *Weather Modification Programme, Preliminary Assessment Report of the Site Selection Phase 3 of the Precipitation Enhancement Project,* PEP Rep. No. 28, World Meteorological Organization, Geneva, 168 pp.

——, 1984: Notes on the planning session for the Int. Cloud Modelling Workshop/Conf. held at Aspen. Informal Rep., Weather Modification Programme, WMO, Geneva, 15 pp.

CHAPTER 3

Summer Cumulus Cloud Lifetime—Importance to Static Mode Seeding

GEORGE A. ISAAC

Atmospheric Environment Service, Downsview, Ontario, Canada

ABSTRACT

Any observing program studying summer cumulus clouds should attempt to measure cloud lifetime. This parameter is important for determining whether a cloud will last long enough for precipitation to form by either natural or artificially stimulated mechanisms. When reporting cloud lifetime, the definition used and the method of calculation should be clearly specified. In North America, after a summer cumulus cloud has been identified and selected, lifetimes, at temperatures below $-5°C$, of approximately 10 to 12 min are being reported. This lifetime must be considered marginal for static mode seeding to produce precipitation by artificial ice nucleants.

3.1. Introduction

Cloud lifetime is a useful parameter for describing the physical character of a summer cumulus cloud. For example, the lifetime can indicate whether a cloud will last long enough for a cold or warm rain process to develop by either natural or artificial means. However, recent textbooks on clouds physics such as those by Mason (1971), Pruppacher and Klett (1978), Rogers (1979), Ludlam (1980) and Dennis (1980) do not discuss cloud lifetime in detail, and they report few lifetime observations describing any cloud type. This should not be taken as indicating the unimportance of cloud lifetime. Houghton (1968) discusses the relevance of this parameter using the earlier work of Braham (1964) and Saunders (1965). He concluded that "simple single cell clouds have effective lifetimes about equal to the time required to grow raindrops." That time was estimated to be approximately 25–30 min. Houghton (1968) stressed that this conclusion implied that the precipitation process must start near cloud base at an early stage of cloud development. He also stated that more direct measurements of cloud lifetime in a variety of convective situations were desirable.

There are many possible ways of defining cloud lifetime and a few are given in Table 3.1. Although Definition 1, the time from initial condensation to total dissipation, is the most logical definition, it is difficult to measure in the real atmosphere. Definitions 6 and 7 have been used in weather radar and cloud photography studies (e.g., Braham, 1958; Knight et al., 1983; Warner, 1976; Johnson and Dungey, 1978; Ludlam, 1980), but they have a limited usefulness in weather modification studies, especially for static mode seeding. This type of seeding is generally performed within nonprecipitating clouds. Consequently, the lifetime of precipitation cells, as would be observed by radar, is not very useful.

Definition 4 is perhaps the best for static mode seeding. Braham (1960), Schemenauer and Isaac (1984), Cooper and Lawson (1984) and Kochtubajda and Rogers (1984) have all used Definition 4 or a variation of it to describe summer cumuli, most of which were nonprecipitating at selection. However, they were only able to calculate the lifetime after the cloud was selected. This might considerably underestimate the total lifetime. Schemenauer and Isaac (1984) used the term cloud "top" lifetime to help describe some of their assumptions. At the Weather Modification Workshop, Dennis suggested that the word lifetime might be replaced by "life expectancy." Although the words "cloud lifetime" often convey a false meaning to the reader, especially when the definition is not specified, in order to avoid confusion with past work, a more descriptive term has not been adopted for this paper.

The purpose of this paper is to emphasize the importance of cloud lifetime and to describe a few of the measurements that have been made. Although there are many different types of cumuli, the discussion will be restricted to fair-weather summer cumulus clouds, which are usually isolated from each other and not precipitating. This cloud type has been studied extensively and is regarded as one of the simplest cloud types. Certainly, it is relatively easy to make observations within and near such clouds. Discussion will also be limited to points relevant to the companion review paper by Silverman (1986) on static mode seeding.

3.2. Importance

Figure 3.1 shows some of the major cloud physics processes that might occur during the total lifetime (Definition 1) of a typical cumulus. Although the times quoted are used by the weather modification community, they must be considered as approximate and without firm theoretical

TABLE 3.1. Several possible definitions of cloud lifetime.

1. Time from initial condensation to total dissipation
2. Time updraft is above a specified threshold
3. Time liquid water content is above a specified threshold
4. Time with a sector colder than nucleating threshold
5. Time to appearance of first precipitation
6. Time from first precipitation to no precipitation
7. Time a definable turret exists on a cloud mass

or observational foundation. A cloud should probably last longer than 15 to 20 min from initial condensation in order for 1 mm sized particles to develop by the Bergeson–Findeisen process. Schemenauer and Isaac (1984) concluded that the natural cold rain process could be initiated if the cloud-top lifetime (Definition 4) at temperatures colder than −5°C was only a few minutes. If a cloud lasts a sufficiently long time, perhaps 40 to 60 min, then the warm rain process might generate precipitation (>1 mm) sized particles. Accepting the Hallet–Mossop process mechanism as described by Mossop (1978) and the conclusions of Isaac and Schemenauer (1979), then once a warm rain process has matured, the possibility exists for an ice multiplication process to begin. Some workers suggest that the ice multiplication mechanism might start as soon as 20 μm diameter droplets appear. If this hypothesis is proved correct, then the ice multiplication process might "turn on" at the same time as the cold rain process in Fig. 3.1.

For the problem of cloud seeding, Fig. 3.1 illustrates the difficulty of identifying a suitable cumulus cloud for treatment. The cloud must last longer than 15 to 20 min, or no rain will form naturally or artificially. On the other hand, if the cloud lasts sufficiently long, then precipitation will form naturally through the warm or cold rain process. Therefore, a seeding window exists in the range of possible cloud lifetimes. A cloud must be selected that will last longer than 15 min but not too long for rain to form naturally. Of course, such reasoning assumes that the mature natural rainforming process is very efficient such that improvement by static mode seeding is not possible. Most static mode seeding experiments attempt to stimulate precipitation in nonprecipitating clouds.

3.3. Observations

Table 2 summarizes the measurements of summer cumulus lifetimes for conditions in North America. The quoted lifetimes refer to the time that a cloud existed after initial observation at temperatures below −5° to −8°C (Definition 4, Table 3.1). The different assumptions used to calculate lifetime for each location are also summarized in Table 3.2. When reporting cloud lifetimes, it is important to indicate the type of cloud and the physical dimensions of the clouds being studied. For example, cumulus, cumulus congestus and cumulonimbus clouds will

obviously have different lifetimes. The data of Table 3.2 refer mainly to small, isolated continental cumulus congestus not precipitating at selection. However, the early data of Braham (1960, 1981) were much more maritime than continental and ranged from cumulus congestus to cumulonimbus. The data of Table 3.2 are also more descriptive of cold cloud bases near +3°C (Yellowknife, Miles City, Alberta) and +12°C (Thunder Bay) with little data being available for clouds in the southern United States, which might have warmer bases.

When available, the average liquid water content (LWC) measured on the first aircraft cloud penetration is given in Table 3.2. Because of cloud selection criteria and the use of 1 km penetration maxima versus penetration average LWC, the two Miles City datasets have very different first-pass LWC. Cooper and Lawson (1984) basically selected wider clouds with higher LWC than Schemenauer and Isaac (1984). However, the lifetime difference is mainly due to 1) the assumption of exponential instead of linear decay rates, 2) the acceptance of data by Cooper and Lawson as much as 500 m below the initial penetration level 8 min after real or simulated seeding, and 3) the use of 1 km LWC maxima versus penetration average LWC. Using Fig. 13 of Cooper and Lawson and assuming a linear decay rate, a lifetime to 0.1 g m^{-3} (1 km LWC) would be approximately 12 min for data near the initial penetration level in comparison to the 14 min exponential decay rate reported. Schemenauer and Tsonis (1985) reviewed the analysis of Cooper and Lawson and suggested that the evidence for an exponential decay rate was not strong. However, using similar assumptions in the calculation of lifetime, most of the difference between the Miles City values quoted from the two references can be explained.

This author prefers the linear decay model because it provides a definite lifetime, as opposed to a "lingering death" with some liquid water always present, as implied by exponential decay. Visual observations from aircraft clearly show that cumuli can go from vigorous turrets to clear air in 10 to 15 min. In addition, my interpretation

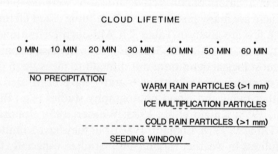

FIG. 3.1. Some of the physical processes leading to precipitation-sized particles (>1 mm) that might occur during the lifetime of a cumulus cloud, with a cloud top at some period colder than −5°C. Lifetime is based upon Definition 1 in Table 3.1 or time from initial condensation. The times are relative and should not be interpreted rigidly.

TABLE 3.2. Summary of lifetimes of summer continental cumulus clouds by location and reference. The two sets of values given for Yellowknife, Thunder Bay and Miles City represent averages for different years.

Location	Number of clouds	LWC* (g m^{-3})	Lifetime** (min)	Details of definition and calculation***	Reference
1. Midwest, Arizona, U.S.	63	—	10–12	A	Braham 1960, 1981
2. Yellowknife, Canada	43	0.4, 0.5	12, >20	B, D, F, H	Schemenauer and Isaac, 1984
3. Thunder Bay, Canada	38	0.9, 1.4	8, 6	B, D, F, H	Schemenauer and Isaac, 1984
4. Miles City, U.S.	21	0.3, 0.7	3, 13	B, D, F, H	Schemenauer and Isaac, 1984
5. Miles City, U.S.	20	1.1	14†	C, E, G, I	Cooper and Lawson, 1984
6. Alberta, Canada	11	≈0.3	12†	C, E, F	Kochtubajda and Rogers, 1984

† Decay constant.
  * Liquid water content on first penetration.
 ** Average or median lifetime.
*** A—Lifetime of cloud portion colder than −5°C.
    B—Lifetime of cloud top colder than about −7°C.
    C—Lifetime of cloud top colder than about −8°C.
    D—Only data within 100 m of initial penetration used.
    E—After 8 min, penetrations made below initial penetration.
    F—Penetration average liquid water content (LWC) used.
    G—Maximum 1 km LWC used.
    H—Average first pass LWC divided by mean LWC rate of change gives lifetime to 0.1 g m^{-3}.
    I—Best fit to ratio of first pass LWC to pass LWC on scatter diagram gives lifetime decay constant.

of recent South African data (Krauss, private communication, 1985) tends to support the linear decay model. However, as indicated in Table 3.2, some authors prefer the exponential model.

Reporting an average or median lifetime also tends to suggest a range of values with a maximum frequency near that lifetime. The real distribution or histogram may show a peak at the lowest or shortest lifetime with lower frequencies at longer lifetimes. Unfortunately, very little data exists describing the nature of such a histogram of lifetimes.

3.4. Conclusions

Lifetime should be measured in any summer cumulus observation program. It would be advantageous if the total cloud lifetime could be measured and not just lifetime after initial detection of a mature cloud. When measurements are being reported, the lifetime definition should be clearly specified, as well as the method of calculating this parameter.

For weather modification purposes, it would be a major step forward if a method could be determined for predicting "cloud life expectancy" from either microphysical or dynamical observations. The literature contains no strong correlation between cloud lifetime and another cloud parameter. Schemenauer and Isaac (1984) did find a moderate correlation with cloud width for Yellowknife, Thunder Bay and Miles City clouds. However, lifetime did not appear related to initial LWC, cloud depth, cloud-base temperature, inside–outside cloud temperature difference, environmental humidity, cloud energy dissipation rate, energy flux or wind shear.

Identifying a cumulus cloud that has a suitable seeding window remains one of the most frustrating problems in weather modification. Many workers, such as Isaac et al. (1982), have used visual clues and a restricted set of physical and microphysical parameters to select clouds. A very specific set of conditions was developed for the HIPLEX-1 Montana experiment (Cooper and Lawson, 1984; Smith et al., 1984) with a number of microphysical and dynamical restrictions to define the ideal candidate. However, both Isaac et al. and Cooper and Lawson reported that the cumulus clouds selected had very limited cloud lifetimes, which resulted in a small effect or no effect on precipitation production due to seeding. Preliminary indications by English and Kochtubajda (1984) show that the Alberta cumulus cloud seeding project, with cloud selection criteria similar to those of Cooper and Lawson, can produce precipitation in clouds lasting longer than 20 min. Because the Alberta and Montana results are different, defining a generalized set of conditions for selecting clouds with a suitable seeding window remains a problem. A statistical dataset for an experiment can then be "contaminated" with clouds whose lifetimes are too short for artificially stimulated rainfall to form. In order to get statistical significance, the dataset must be enlarged considerably.

Continental summer cumuli, which are often selected for weather modification projects that use static mode seeding with an ice nucleant, have typical lifetimes of 10 to 12 min at temperatures lower than −5°C (Table 3.2). This lifetime must be considered marginal for the effective production of artificial rainfall. For example, Braham (1960) found that only half of the clouds lasting longer than 10 to 12 min after initial observation developed a weather radar echo greater than 8 dBZ (<1 mm h^{-1} rain rate). Any weather modification project on summer cumuli should consider whether usable amounts of rainfall can be produced from such small clouds with limited lifetimes.

REFERENCES

Braham, R. R., 1958: Cumulus cloud precipitation as revealed by radar—Arizona 1955. *J. Meteor.,* **15,** 75–83.

——, 1960: Physical properties of clouds. *J. Irrig. Drain. Div., Proc. ASCE,* **86,** 111–119.

——, 1964: What is the role of ice in summer rain showers? *J. Atmos. Sci.,* **21,** 640–645.

——, 1981: Designing cloud seeding experiments for physical understanding. *Bull. Amer. Meteor. Soc.,* **62,** 55–62.

Cooper, W. A., and R. P. Lawson, 1984: Physical interpretation of results from the HIPLEX-1 experiment. *J. Climate Appl. Meteor.,* **23,** 523–540.

Dennis, A. S., 1980: *Weather Modification by Cloud Seeding.* Academic Press, 267 pp.

English, M., and B. Kochtubajda, 1984: Precipitation initiation through cloud seeding. *Proc. Ninth Int. Conf. on Cloud Physics,* Tallinn, U.S.S.R., 707–710.

Houghton, H. G., 1968: On precipitation mechanisms and their artificial modification. *J. Appl. Meteor.,* **7,** 851–859.

Isaac, G. A., and R. S. Schemenauer, 1979: Comments on "Some factors governing ice particle multiplication in cumulus clouds." *J. Atmos. Sci.,* **36,** 2271–2272.

——, J. W. Strapp, R. S. Schemenauer and J. I. MacPherson, 1982: Summer cumulus cloud seeding experiments near Yellowknife and Thunder Bay, Canada. *J. Appl. Meteor.,* **21,** 1266–1285.

Johnson, D. B., and M. J. Dungey, 1978: Microphysical interpretation

of radar first echoes. *Preprints 18th Conf. on Radar Meteorology,* Atlanta, Amer. Meteor. Soc., 117–120.

Kochtubajda, B., and D. C. Rogers, 1984: Seeding effects on liquid water and ice crystal concentrations in Alberta cumulus clouds. *Proc. Ninth Conf. on Weather Modification,* Park City, Amer. Meteor. Soc., 102–103.

Knight, C. A., W. D. Hall and P. M. Roskowski, 1983: Visual cloud histories related to first radar echo formation in northeast Colorado cumulus. *J. Climate Appl. Meteor.,* **22,** 1022–1040.

Ludlam, F. H., 1980: *Clouds and Storms.* Pennsylvania State University Press, 405 pp.

Mason, R. J., 1971: *The Physics of Clouds.* Clarendon Press, 671 pp.

Mossop, S. C., 1978: Some factors governing ice particle multiplication in cumulus clouds. *J. Atmos. Sci.,* **35,** 2033–2037.

Pruppacher, H. R., and J. D. Klett, 1978: *Microphysics of Clouds and Precipitation.* D. Reidel, 714 pp.

Rogers, R. R., 1979: *A Short Course in Cloud Physics.* 2nd ed., Pergamon, 235 pp.

Saunders, P. M., 1965: Some characteristics of tropical marine showers. *J. Atmos. Sci.,* **22,** 167–175.

Schemenauer, R. S., and G. A. Isaac, 1984: The importance of cloud top lifetime in the description of natural cloud characteristics. *J. Climate Appl. Meteor.,* **23,** 267–279.

——, and A. A. Tsonis, 1985: Comments on "Physical interpretation of results from the HIPLEX-1 experiment." *J. Climate Appl. Meteor.,* **24,** 1269–1274.

Silverman, B. A., 1985: Static mode seeding of convective clouds—a review. *Precipitation Enhancement—A Scientific Challenge. Meteor. Monogr.,* No. 43, Amer. Meteor. Soc., 7–24.

Smith, P. L., A. S. Dennis, B. A. Silverman, A. B. Super, E. W. Holroyd III, W. A. Cooper, P. W. Mielke, Jr., K. J. Berry, H. D. Orville and J. R. Miller, Jr., 1984: HIPLEX-1: Experimental design and response variables. *J. Climate Appl. Meteor.,* **23,** 497–512.

Warner, C., 1976: Case studies of seeded Alberta hailstorms. McGill University Rep. MW-86, 72 pp.

CHAPTER 4

Ice Initiation in Natural Clouds

WILLIAM A. COOPER*

University of Wyoming, Laramie, Wyoming

ABSTRACT

Selected concentrations of ice crystal concentrations attributable to nucleation are compiled and summarized. The variability in the observations is discussed, and some conclusions related to natural precipitation formation and to seedability are discussed.

4.1. Introduction

Although ice nuclei (IN) provide the only plausible source for the initial ice that forms in moderately super-cooled clouds, it is awkward that the evidence linking such particles to ice formation is still circumstantial. Usual detectors for IN do not separate nucleation modes, do not preserve "preactivated" nuclei, and do not duplicate important characteristics of natural clouds such as their time scales or levels of supersaturation. Despite many years of effort, even initial ice concentrations in clouds are still not predictable from measured IN concentrations.

The initial ice concentration produced by IN affects the precipitation efficiency and the seedability of clouds because that ice has the longest growth time before the cloud water is depleted by processes such as entrainment and glaciation. The following comments focus on that early stage of ice crystal formation.

4.2. Ice concentrations attributed to nucleation

There are some cloud conditions where secondary processes are unlikely and where it is reasonable to associate the observed ice concentrations with ice nucleation. Included are observations of ice concentrations in regions of laminar flow into stratiform clouds (e.g., Rogers, 1982; Cooper and Vali, 1981; Cooper and Saunders, 1980) or measurements made early in the cloud's lifetime near the tops of growing cumulus clouds (Gagin, 1975; Mossop, 1972; and others). In some other cases the observed ice was attributed to nucleation (e.g., Heymsfield et al., 1979, for undiluted cores of cumulus clouds; Jayaweera and Ohtake, 1973, for stratus clouds), but the observed concentrations in those cases were not given as functions of the formation temperature and so have not been included

here. However, those data do not appear to contradict the trends to be presented.

The data from selected observations are shown in Fig. 4.1. In all these cases, the authors attributed the ice concentrations to primary formation via nucleation. Ice concentrations were averaged over distances of at least 1–2 km and were compared to cloud summit temperatures (for cumulus clouds penetrated within 300 m of cloud top) or to the observation temperature (for cases of laminar flow into stratiform clouds). Most cumulus clouds were selected to be nonprecipitating and growing. Where possible, only first passes through these clouds were used. These precautions minimize possible contributions from secondary ice production processes or from possible influences of the aircraft such as were found by Rangno and Hobbs (1983). In stratiform clouds, most penetrations were at positions where <10 min of ice growth was possible after the air passed through the leading edge of the cloud.

The measurements of Fig. 4.1 show a clear temperature trend, as expected for a nucleation process, and the concentrations are in a range that is reasonable in comparison to conventional measurements of IN concentrations.[1] Some features of these measurements are discussed in the following subsections.

4.2.1. Variance

The scatter in the ice concentrations at any given temperature is large. It is unlikely that this scatter is due to secondary processes, since little riming and no precipitation development occurred in these regions. Although higher variability in cloud properties in a cumulus cloud might be expected to produce more variability in ice concentrations, the scatter is not larger for cumulus cases

---

* Present affiliation: National Center for Atmospheric Research, Boulder, Colorado.

[1]Mossop (1985) has shown a similar plot and reached similar conclusions.

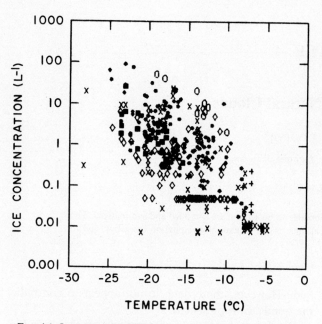

FIG. 4.1. Ice crystal concentration as a function of temperature in cases where the ice concentration can be attributed to nucleation. The observations are (●) Rogers (1982) and Cooper and Vali (1981), both from Wyoming wintertime cap clouds; (○) Cooper and Saunders (1980), wintertime orographic clouds of southwestern Colorado; (×) Gagin (1975), Israel winter cumulus clouds; (◇) Cerni (1982), summertime cumulus clouds of Montana; (■) Cooper (unpublished), cumulus clouds of the Bethlehem area of South Africa; and (+) Mossop (1972), tops of newly formed Australian cumulus clouds. Most of the measured values for concentrations < 0.1 L⁻¹ are upper limits.

served variability at any one location. Figure 4.2 shows the mean concentrations observed in various areas, and it shows that all except the observations from the wintertime storms of southwestern Colorado tend to lie within a factor of 10 of each other. The measurements include summertime and wintertime cases, cumulus and stratiform clouds, five different continents, and widely differing meteorological conditions, yet the averages are surprisingly similar. This consistency is similar to that in IN measurements when similarly averaged (Pruppacher and Klett, 1978). The exception of the Colorado wintertime storms receives some support from the observations of Magono and Lee (1973), who found similarly high ice concentrations near the tops of a widespread snow system.

4.2.3. Comparison to ice nucleus concentrations

Figure 4.3 shows the geometric means of the measurements for 5°C intervals. For reference, the formula suggested by Fletcher (1962) as an approximate representation of IN concentrations is also shown; this formula provides a reasonable (factor of ±10) representation of many later IN measurements, as shown by Pruppacher and Klett (1978). Measurements quoted as being upper limits were included at those upper limits, so the averages in Fig. 4.3 are also upper limits. This figure indicates that the mean concentrations attributable to nucleation are within about

than for stratiform cases. Particularly for the measurements by Rogers (1982) in a wintertime cap cloud, the cloud conditions were quite consistent from case to case, yet there were variations in ice concentration spanning about a factor of 100 at any given temperature. The high variability in the stratiform cases indicates either that the IN population itself has high variability or that the fraction activated and detected as ice varies widely.

The variability in these measured ice concentrations is far greater than the measurement uncertainties. Typical standard deviations associated with the measurements are about ±10% for 2 km averages if the concentration is 1 L⁻¹ and ±3% for concentrations near 10 L⁻¹.

For the composite dataset, a range corresponding to a factor of 6 encompasses about 68% of the measurements, as would a confidence interval spanning one standard deviation. For the individual datasets, the similar confidence limits span ranges corresponding to factors of 3–4. The regression line representing $Y = \log_{10}$ (concentration, L⁻¹) versus $T =$ (temperature, °C) is $Y = -2.35 - 0.135T$, and the correlation between Y and T is -0.65.

4.2.2. Geographic variability

The geographic variability in the mean ice concentration attributable to nucleation is much less than the ob-

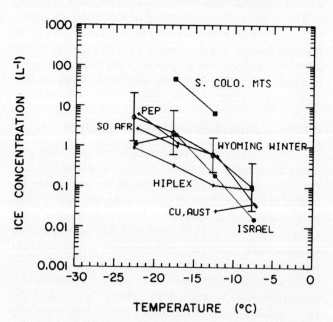

FIG. 4.2. Geometric-mean ice concentrations for the different areas of Fig. 4.1 and also for similar measurements from Spain (Vali and Yoksas, unpublished data from the Precipitation Enhancement Project). Measurements have been averaged over 5°C intervals. All measurements originally quoted as upper limits have been included in these averages at those upper limits. The measurements of Rogers (1982) and Cooper and Vali (1981) have been combined, and the error bars represent the geometric standard deviations for those measurements; these limits are typical for the other measurements as well.

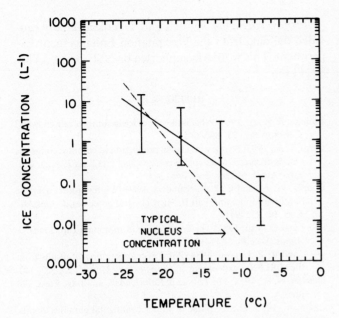

FIG. 4.3. Geometric-mean ice concentrations for the entire dataset, averaged in 5°C intervals. The error bars indicate geometric standard deviations, and the solid line shows the regression line obtained for the logarithm of concentration as a function of temperature. The dashed line is the typical IN spectrum suggested by Fletcher (1962).

an order of magnitude of typical IN concentrations. However, this agreement should not be interpreted as evidence of a cause-and-effect relationship, for two reasons. First, measured IN concentrations were consistently about an order of magnitude below ice crystal concentrations in the cases where corresponding measurements were available (Gagin, 1975; Rogers, 1982; Cooper and Vali, 1981; Cooper and Saunders, 1980; Cerni, 1982). Second, the IN measurements did not incorporate realistic separation of nucleation modes, and in some of the cases shown in Fig. 4.1, the nucleation modes detected by the IN detectors could not be the same as those active in the clouds. (Cooper and Vali, for example, argued that the nucleation process in the wintertime cap clouds had to be either a contact or a condensation–freezing process.)

4.2.4. Cumulus versus stratified clouds

The different areas do not differ significantly in mean ice concentration, except that the observations from the Colorado mountains seem anomalously high. The difference between stratiform and cumuliform clouds, however, may be significant. The highest ice concentrations were observed in stratiform cloud systems, while the lowest were observed in cumuliform clouds. This is not the expected trend in ice nucleus concentrations, since all of the stratiform clouds were observed in wintertime while most of the cumulus clouds (except those from Israel and Spain) were summertime clouds. Since aerosol particle concentrations are generally higher in summertime than in wintertime, due to enhanced surface sources and faster

gas-to-particle conversion rates, IN might be expected to follow this same trend. Also, cumulus clouds usually have their sources of air in the boundary layer, and hence, such air might be expected to have higher IN concentrations associated with surface particle sources, while many stratiform clouds form in laminar flow of air at altitudes well above the surface. Droplet concentrations in cumulus clouds are generally higher than those of stratiform clouds, and other indications are of a higher total aerosol content in the air forming the cumulus clouds.

The higher concentrations of ice measured in stratiform clouds may arise from the longer times available for nucleation and growth, the different nucleation modes that can act, or the possibility that preactivated IN may be present in the air flowing into wintertime stratiform cases.

4.3. Implications for natural efficiency and seedability

Figures 4.1 and 4.2 suggest that the typical concentration of ice crystals that is formed via nucleation does not vary widely with geographic location. If that is true, other cloud characteristics (such as lifetime, liquid water content, and the rate of secondary ice formation) are likely to be more important causes of geographic variations in seedability.

Many clouds occur in all these areas in which ice concentrations are less than 1 L^{-1}, and more than half the cases had ice concentrations of <10 L^{-1} for all temperatures for which there were data ($-25°$ to $-5°C$). At temperatures warmer than $-20°C$, most of the ice concentrations were thus below conventional estimates (e.g., Ludlam, 1955) of the ice concentration needed for efficient precipitation formation.

Secondary ice formation can affect the precipitation in two ways. First, as secondary particles appear, they will grow at the expense of the supercooled water; even if they appear late in the lifetime of the cloud, they will still compete with the continued growth of the ice crystals formed earlier. Second, if secondary processes produce ice within a few minutes after the first ice appears, the initial ice concentration that provides embryos for precipitation formation may be increased above the values of Fig. 4.3. Secondary production may be this fast in cases where coalescence growth produces raindrops by the time the cloud reaches the freezing level, since freezing of such raindrops avoids the much longer time (>15 min) required for growth of graupel from ice crystals and causes the ice concentration to increase more rapidly than if no coalescence growth occurs (Hallett et al., 1978; Koenig, 1963; Braham, 1964). In such cases, the primary ice concentration has little effect on the precipitation efficiency because secondary processes cause important increases in ice concentration only a few minutes after ice first appears. As Silverman (1986) suggests, such clouds may be poor candidates for seeding.

In cases where secondary production processes only appear late in the lifetime of the cloud, in association with precipitation formation, their primary effect is not to change the concentration of large ice particles that compete for the available liquid water but rather to deplete the liquid water available for such growth. Such secondary ice may not be a serious competitor to seeding, since seeding might still affect the early ice concentration leading to growth of graupel.

The ice concentrations needed for efficient precipitation development are not well known and depend strongly on the available liquid water content, lifetime of the clouds, and precipitation mechanisms. Ludlam (1955) estimated about 20 L^{-1} to be a favorable concentration for wintertime orographic clouds, and Rokicki and Young (1978) calculated that concentrations needed for fastest precipitation *initiation* were 100–1000 L^{-1} (for parcels ascending adiabatically).

For cumulus clouds in which precipitation forms via riming of ice particles, an estimate of the desirable ice concentrations may be made by comparing the characteristic time describing the natural decay of the liquid water content to that characterizing depletion of the liquid water content by growing graupel particles. If the natural decay is already determined by depletion due to growing graupel particles, an increase in the graupel concentration would be undesirable since the available water would be partitioned among more particles but the total mass in the precipitation would remain about the same. If the decay is determined primarily by entrainment or glaciation to small ice particles, then increasing the graupel concentration to a magnitude that permits the graupel to compete with these other sources of decay would be desirable. In clouds such as those discussed by Cooper and Lawson (1984), which had lifetimes < 1000 s and which only rarely had the potential to grow graupel larger than 3 mm in diameter, the graupel concentration that these arguments favor is about 0.5 L^{-1}. Graupel concentrations of 10 L^{-1}, for 2 mm graupel, cause depletion of the available cloud water in characteristic (exponential decay) times of about 1 min, so this concentration is surely much too high. Since the natural ice concentration formed near the tops of a cumulus cloud is a likely source for the graupel particles that form in a single isolated cell, Figs. 4.1 and 4.2 indicate that the seeding opportunity in such clouds may be restricted to clouds with cloud-top temperatures warmer than about −15°C.

Acknowledgments. G. Vali and T. Yoksas kindly provided the data from the Precipitation Enhancement Experiment. This work was supported by NSF Grant ATM-8211134.

REFERENCES

Braham, R. R., Jr., 1964: What is the role of ice in summer rain showers? *J. Atmos. Sci.,* **21,** 640–645.

Cerni, T. A., 1982: Primary ice crystal production in cumulus congestus clouds of eastern Montana. *Preprints Conf. on Cloud Physics,* Chicago, Amer. Meteor. Soc., 346–349.

Cooper, W. A., and C. P. R. Saunders, 1980: Winter storms over the San Juan Mountains. Part II: Microphysical processes. *J. Appl. Meteor.,* **19,** 927–941.

——, and G. Vali, 1981: The origin of ice in mountain cap clouds. *J. Atmos. Sci.,* **38,** 1244–1259.

——, and R. P. Lawson, 1984: Physical interpretation of results from the HIPLEX-1 experiment. *J. Climate Appl. Meteor.,* **23,** 523–540.

Fletcher, N. H., 1962: *The Physics of Rainclouds.* Cambridge Press, 390 pp.

Gagin, A., 1975: The ice phase in winter continental cumulus clouds. *J. Atmos. Sci.,* **32,** 1604–1614.

Hallett, J., R. I. Sax, D. Lamb and A. S. Ramachandra Murty, 1978: Aircraft measurements of ice in Florida cumuli. *Quart. J. Roy. Meteor. Soc.,* **104,** 631–651.

Heymsfield, A. J., C. A. Knight and J. E. Dye, 1979: Ice initiation in unmixed updraft cores in northeast Colorado cumulus congestus clouds. *J. Atmos. Sci.,* **36,** 2216–2229.

Jayaweera, K. O. L. F., and T. Ohtake, 1973: Concentrations of ice crystals in Arctic stratus clouds. *J. Rech. Atmos.,* **7,** 199–207.

Koenig, L. R., 1963: The glaciating behavior of small cumulonimbus clouds. *J. Atmos. Sci.,* **20,** 29–47.

Ludlam, F. H., 1955: Artificial snowfall from mountain clouds. *Tellus,* **7,** 277–290.

Magono, C., and C. W. Lee, 1973: The vertical structure of snow clouds, as revealed by "snow crystal sondes," Part II. *J. Meteor. Soc. Japan,* **51,** 176–190.

Mossop, S. C., 1972: The role of ice nucleus measurements in studies of ice particle formation in natural clouds. *J. Rech. Atmos.,* **6,** 377–389.

——, 1985: The origin and concentration of ice crystals in clouds. *Bull. Amer. Meteor. Soc.,* **66,** 264–273.

Pruppacher, H. R., and J. D. Klett, 1978: *Microphysics of Clouds and Precipitation.* Reidel, 714 pp., cf. p. 242.

Rangno, A. L., and P. V. Hobbs, 1983: Production of ice particles in clouds due to aircraft penetrations. *J. Climate Appl. Meteor.,* **22,** 214–232.

Rogers, D. C., 1982: Field and laboratory studies of ice nucleation in winter orographic clouds. Ph.D. dissertation, University of Wyoming, 161 pp.

Rokicki, M. L., and K. C. Young, 1978: The initiation of precipitation in updrafts. *J. Climate Appl. Meteor.,* **17,** 745–754.

Silverman, B. A., 1986: Static mode seeding of convective clouds—a review. *Precipitation Enhancement—A Scientific Challenge, Meteor. Monogr.,* No. 43, Amer. Meteor. Soc., 7–24.

CHAPTER 5

Aggregates as Embryos in Seeded Clouds

ANDREW J. HEYMSFIELD

National Center for Atmospheric Research, Boulder, Colorado

ABSTRACT

The growth of ice particles through aggregation is investigated for seeded clouds using currently available field data and a numerical particle-growth model. Observations indicate that the aggregation process is fairly common, even when moderate liquid water contents, $\sim$0.5 g m^{-3}, are available for particle growth through accretion. The modeling study suggests that certain temperature ranges are especially conducive to aggregate formation.

5.1. Introduction

Early proponents of weather modification reasoned that if a convective cloud were seeded with concentrations of ice particles sufficient for the amount of liquid water available in it, the cloud's natural precipitation mechanism might be increased. Until recently little attention has been paid to the various processes by which the seeded crystals might grow—especially to the role of aggregation. The present study reports investigations of the various processes by which particles grow in seeded clouds, using available field data and a numerical particle growth model and paying particular attention to the role of aggregation.

5.2. Observations of precipitation embryos in seeded clouds

Holroyd and Jiusto (1971) reported one of the first investigations of snowfall from a heavily seeded cloud in which the dominant precipitation form was snow crystal aggregates. In this instance the heavy seeding amplified the preexisting process of snow crystal aggregation, although this was not the intention. They reported a total ice crystal concentration of about 300 L^{-1} in the seeded region and a mean diameter of 200 μm for the individual crystals composing the aggregates.

In 1980, Hobbs and Politovich reported investigations of the effects of different modes of artificial seeding on the structure of clouds in the area of Miles City, Montana. The maximum ice crystal concentrations in the seeded clouds were about 300 L^{-1}, and only one of the clouds contained aggregates. The cloud containing the aggregates had been heavily seeded with dry ice close to its top. Hobbs and Politovich used a replicator and an optical ice crystal counter in their investigations. In more recent studies, two-dimensional (2D) imaging probes have been used to distinguish among individual crystals, aggregates, and graupel. Cooper et al. (1982a,b), using 2D probes to in-

vestigate seeded clouds in the Montana area, reported that the primary precipitation particle type in many of the seeded clouds was, in fact, the ice crystal aggregate. The concentrations of ice crystals measured by Cooper et al. 2 min after seeding were comparable to those reported by Hobbs and Politovich.

Investigations of seeded clouds in other geographical areas have also produced some evidence of aggregation. Strapp et al. (1979), studying clouds around Yellowknife and Thunder Bay, Canada, found that, in the seeded regions of the clouds, the largest ice particles grew at rates faster than could be explained by deposition/accretion alone, indicating that aggregation of ice particles might account for the enhanced rates of growth. Unfortunately, there were no accompanying measurements of the ice particle habits that might have answered the question. Later investigations of seeded clouds in the same geographical areas were reported by Isaac et al. (1982); these studies, in which 2D imaging probes were used to measure ice crystal habits, revealed that particle shapes were often irregular and suggested that they may have been aggregates. The concentrations were approximately 50 L^{-1}.

Marwitz and Stewart (1981) investigated ice crystal habits that developed over a period of 111 min in a seeded convective cloud associated with a wintertime storm over the Sierra Nevada. Abundant liquid water was available throughout the period, and the particles initially appeared to be rimed; however, in the later stages, the particles appeared less heavily rimed and aggregates were observed.

5.3. Aggregate formation interpreted from measurements and modeling studies

The most comprehensive dataset currently available for evaluating the conditions in which aggregates develop in seeded clouds comes from the High Plains Experiment (HIPLEX) case studies in Montana (Cooper et al., 1982a,b; Cooper and Lawson, 1984). Table 5.1 gives a

34   AGGREGATES AS EMBRYOS IN SEEDED CLOUDS

TABLE 5.1. Summary of HIPLEX 1979 and 1980 measurements applicable to aggregation growth in seeded clouds.

Cloud number	Parameter (max values)	Observations relative to seed time					Cloud-top temperature at seed time (°C)	General comments
		−2½ min	+2 min	+5 min	+8 min	+11 min		
2	W (m s^{-1})	11	2	0	0	i	14.0	Initial precipitation aggregates
	LWC (g m^{-3})	2.3	0.8	0.3	some			
	Ice conc. (L^{-1})	0.3	700	700	>10			
	Comments	High LWC Negative buoyancy at cloud top	Cloud eroding	Signs of aggregation Aggregates of columns	Large aggregates Rimed columns			Entrainment important
4	W (m s^{-1})	7	2	2	<0	i	15.5	No graupel formed
	LWC (g m^{-3})	0.7	0.1	0.1	0.1			
	Ice conc. (L^{-1})	<1.0	40	100	20			
	Comments	No organized updraft	Seed plume 1 km wide	Crystals have light rime Ice spread over 3 km	Small region of LW Ice spread over 4 km	5 mm aggregates		Precipitation aggregates
6	W (m s^{-1})	6	<0	<0	<0	i	∼ −15	Very weak precipitation development
	LWC (g m^{-3})	1	0.5	0.4	0.2			
	Ice conc. (L^{-1})	0.03	300	>40	200			
	Comments	Cloud decaying rapidly	Ice plume 0.3 km wide		Seed plume Ice content <0.1 g m^{-3} Columns	Aggregates		Aggregates were very low density
8	W (m s^{-1})	4	<0				−13.0	No precipitation
	LWC (g m^{-3})	1.2	0.2					
	Ice conc. (L^{-1})	≈0	200	Cloud descended below aircraft				
	Comments	Cloud negative buoyant Cloud top −8°C descending	Ice plume 0.3 km wide Cloud top −7°C					
9	W (m s^{-1})	14	>0	8	4	i	−14.0	Precipitation was graupel which probably grew by accretion and aggregates
	LWC (g m^{-3})	1.3	0.9	0.3	0.2			
	Ice conc. (L^{-1})	—	400	>400	300			
	Comments		Ice plume 0.4 km wide	Cloud decaying High column concentration Some plates lightly rimed	Cloud decayed Plume throughout cloud			
10	W (m s^{-1})	12	<0	<0	<0	i	11.0	Precipitation developed through accretion
	LWC (g m^{-3})	1.8	>1.0	∼1.0	(−1/0)			
	Ice conc. (L^{-1})	0.02	10	3				
	Comments	High turbulence Water drops	Low conc. seeded crystals Cloud descending	Cloud is descending	Cloud below aircraft	Graupel uniform size		
12	W (m s^{-1})	8	2	0	2	i	−17.9	Weak precipitation heavily over-seeded (LW depleted)
	LWC (g m^{-3})	0.7	0.4	0.2	0			
	Ice conc. (L^{-1})	0	>1000	>650	∼100			
	Comments		Depletion important Highest ice conc. seen Ice water content 0.3 g m^{-3}	Ice throughout cloud		Crystals riming, aggregates	LWC increased up to 0.5 g m^{-3}	Aggregation important

TABLE 5.1. (*Continued*)

Cloud number	Parameter (max values)	Observations relative to seed time					Cloud-top temperature at seed time (°C)	General comments
		−2½ min	+2 min	+5 min	+8 min	+11 min		
13	W (m s⁻¹)	<2	2	2	2	i	−12.8	Precipitation mostly large aggregates
	LWC (g m⁻³)	1	0.4	<0.1	~1			
	Ice conc. (L⁻¹)	<0.4	800	60	400			
	Comments	Cloud descending	Cloud collapsing		Columns Aggregates of columns Little riming Ice content 0.1 g m⁻³		Depletion probably inhibited growth	
14	W (m s⁻¹)	5	2	0	2	i	−15.0	Cloud initially had little LW, but maintained for long period
	LWC (g m⁻³)	0.6	0.5	0.4	0.2			
	Ice conc. (L⁻¹)	5.8	>0	400	30			
	Comments	Ice already present	Obvious seed plume	Little turbulence		Precipitation graupel		Precipitation developed through accretion
15	W (m s⁻¹)	7	4	2	4	i	−11.0	Natural precipitation were graupel
	LWC (g m⁻³)	1	0.5	0	0.1			
	Ice conc. (L⁻¹)	0.3	40	300	40			
	Comments		Natural ice present			Many needles Some large graupel		Seeding produced mostly rimed needles, crystals
19	W (m s⁻¹)	14	6	4	4	i	−9.5	Precipitation, a few large aggregates, few small graupel
	LWC (g m⁻³)	2.2	0.6*	0.2*	0.05			
	Ice conc. (L⁻¹)	0	300	800	300			
	Comments	Cloud top only −9.5	Little LW depletion	Cloud decayed Crystals-columns Ice content 0.1 to 0.3 g m⁻³	Depletion probably important			Insufficient LW for growth
20	W (m s⁻¹)	7	10	10	6	i	~ −17	Riming of crystals at −13°C into graupel contributed to rainfall
	LWC (g m⁻³)	1.1	1.1	0.2*	~0			
	Ice conc. (L⁻¹)	low	200	800	60			
	Comments	Graupel 2–3 mm present	Strong updrafts	Plates	Plume throughout cloud Updrafts decayed Long columns Aggregates			

* In main body of plume.
i Insufficient data.

summary of these results, showing values for various parameters at the −5°C level in the seeded clouds. The values are given as a function of the time from seeding with dry ice at or near the cloud tops, at temperatures from about −10° to −12°C. As shown in the table, within a few minutes of seeding, the maximum updrafts (W) had decreased dramatically and the tops had descended, indicating that the clouds were decaying. The decreases in the liquid water content (LWC) following seeding, also shown in the table, are further indications of decaying clouds. However, in two instances noted in the table under general comments, the seeding itself was thought to account for the depletion of the liquid water available for accretional growth. The peak ice crystal concentrations (ice conc.) shown in the table are often much higher than those usually observed in natural clouds and, in fact, much higher than the perceived optimum 10 L⁻¹ desired in seeding operations. Characteristics of hydrometeor development in the seeded clouds relevant to the formation and growth of aggregates are also noted in Table 5.1.

Considerable information on aggregation is available from observational studies in naturally precipitating clouds. In a particularly relevant study, Rogers (1974) reported that most of the aggregates that he examined contained one crystal that was substantially larger than any others. Rogers suggested that the aggregates began as large crystals whose erratic pattern of fall, characterized by extensive horizontal excursions, may have dramatically increased the volume swept out by the crystal and, hence, the possibility of aggregation.

Modeling studies have also clarified some aspects of the aggregation process in natural clouds. Over a period, dt, the change in frequency, F, of collisions between two particles—either two crystals, an aggregate and a crystal, or two aggregates—can be represented as

$$\frac{dF}{dt} = \frac{\pi}{4}(D_1 + D_2)^2 C_1 C_2 |V_1 - V_2| E, \qquad (5.1)$$

where D_1 represents the diameter of particle 1 and D_2 of particle 2, C the concentration, V the terminal velocity, and E the collection efficiency. Passarelli (1978) used this basic equation to develop an analytical model of snowflake aggregation. He showed the existence of a snow-size spectral equilibrium that is due to the counteracting effects of vapor deposition and aggregation. Passarelli and Srivastava (1979) used this equation to develop an approach for calculating aggregation growth through stochastic collection, accounting for the fact that snowflakes of the same mass can have a spectrum of terminal velocities. This approach indicated that aggregates can grow more rapidly than previously thought. Modeling studies have not yet considered combined growth through diffusion, accretion, and aggregation.

5.4. The role of aggregates in seeded clouds

The following discussion on the role of aggregates in seeded clouds derives from the information given in Table 5.1 and from the other studies previously discussed, and, except where noted, the ideas represent my views.

A. *The time required for the observed onset of aggregation in seeded clouds is 5 to 8 min.*

B. *Over this period, mechanisms must be operative to allow crystals to achieve a range of terminal velocities.* From Eq. (5.1), note that differential terminal velocities are required to facilitate particle collisions. A range of terminal velocities can be achieved through differential crystal growth rates, resulting, for example, from 1) subtle differences in the geometry of the individual crystals, 2) a slight variation in growth temperatures, or 3) differences in the vapor density from the center to the edge of the seeding curtain, particularly in clouds seeded with dry ice.

C. *Most of the precipitation that develops in convective clouds that decay shortly after being seeded is apparently produced by aggregates.*

D. *From C, it follows that the aggregation process can be important in enhancing precipitation efficiency.*

E. *Aggregation is unimportant in precipitation development when the liquid water content exceeds several tenths of a gram per cubic meter over a period at least 10 min following seeding.*

The experimental data given in Table 5.1 and the conclusions drawn from them lead to questions as to the role of aggregates in seeded clouds.

A. Are there certain temperatures at which aggregate production is faster than at others?

B. If single crystals do, in fact, aggregate, can the aggregates then grow into graupel? This question is important because graupel have higher terminal velocities than aggregates and are therefore more likely to produce rain at the surface, thus increasing the effects of seeding.

C. What liquid water content is required for single crystals to grow into graupel before they can aggregate? This question is also important for the reason given in B.

5.5. Particle growth model

The particle growth model described by Heymsfield (1982) has been used as the basis for considering these questions. In drawing conclusions from this model it must be kept in mind that many approximations have been used to represent the growth of particles, and thus the results can only provide some insight into the growth processes. As reported, the model includes accretional and diffusional growth of planar crystals and aggregates. It has recently been extended to consider the growth of columnar and needle crystals as well (Parrish and Heymsfield, 1985). In its application here, the growth of crystals through aggregation will also be considered. I believe that this study is the first treatment of combined aggregational and accretional growth.

The model as applied to the present study considers growth to take place at constant temperatures ranging from $-6°$ to $-16°C$ over a period of 12 min at a pressure of 400 mb. Ice particles are initialized in the model as 10 μm spheres in concentrations of 100 L^{-1} at temperatures of $-6°$, $-10°$, $-12°$, and $-16°C$. Their growth over the 12 min period is considered in 5 s time steps. The LWC is initialized with values of 0.25 or 1.0 g m^{-3}. The LWC decreases with time according to the mean decay rate of 832 s reported by Cooper and Lawson (1984) for the HIPLEX clouds:

$$LWC = LWC_0 e^{-t/832}, \qquad (5.2)$$

where LWC_0 represents the initialized LWC and t the time in seconds from crystal nucleation. The efficiency at which droplets are collected by the growing crystals is calculated from considerations of the characteristic crystal shapes, dimensions, and terminal velocities. The form of the droplet size distribution is specified from the LWC, the latter based upon a parameterization from measurements in a Colorado storm (Heymsfield, 1982). The mean droplet diameter is about 12 μm for a LWC of 0.25 g m^{-3} and 15 μm for a LWC of 1.0 g m^{-3}.

At each time step in the model, single values are predicted for the crystal dimension, and a terminal velocity is predicted for given growth conditions. In a seeded cloud, it is reasonable to assume that the terminal velocities of crystals nucleated at the same time and place will vary for reasons previously discussed. In this study crystal velocities are assumed to vary about a single mean value, $\bar{V}$, with a normal (Gaussian) distribution and a standard deviation, σ, of $\pm$12.5% of $\bar{V}$. In this instance the standard deviation is 50% of that reported for aggregates by Passarelli and Srivastava (1979); however, review of available ice-crystal terminal velocity data suggests that the value selected is reasonable. A spectrum for crystal concentration versus terminal velocity has been obtained using the assumed velocity distribution form and variance, and an example is given in Fig. 5.1. In this figure and in the calculations, concentrations are given for $\pm 2\sigma$ of the calculated terminal velocity, $0.75\bar{V}$ to $1.25\bar{V}$, in increments of 1 cm s^{-1}. The concentration–velocity spectrum has been adjusted to contain the correct total crystal concentration.

Aggregate production and growth are calculated according to the number of collisions between particles given by Eq. (5.1). In applying the equation, we take the time step between collisions to be 1 min. The collision (and collection) efficiency is assumed to be 1.0, somewhat lower than that found by Passarelli (1978) but still quite reasonable. Aggregates are produced by collisions between crystals of all velocities from $0.75\bar{V}$ to $1.25\bar{V}$; the diameter of two colliding crystals is given by $1.5D$, where D is the crystal diameter. The velocity variance spectrum for aggregates is the same as that given for crystals. Aggregates grow by diffusion and through collisions with crystals and other aggregates. Collisions between aggregates and crystals lead to a particle diameter $D_a + 0.5D$, where D_a is the aggregate diameter. Collisions between aggregates lead to a particle diameter of $1.5D_a$. As particles collide, concentrations of those particles are correspondingly reduced.

The following simplifying assumptions have been used to facilitate these calculations: 1) throughout the calculations crystal concentrations are 100 L^{-1} (with time, concentrations are actually reduced by turbulence and diffusion); 2) relative humidities are assumed not to fall below 100%; 3) crystals grown into graupel do not collide with other crystals to become aggregates and are not collected by aggregates (my personal experience suggests that graupel do not aggregate); and 4) aggregates do not grow by accretion. In calculations at 1.0 g m^{-3}, this last assumption may lead to an overestimate of the concentration of aggregates because some may develop into graupel. Note also that the collection efficiencies of the droplets are based upon theoretical studies and thus are only approximate at best. In addition, the growth and riming characteristics of crystals developing into graupel and the influence of changes in temperature during such growth are only moderately well understood, and thus the calculations should be regarded as approximate.

5.6. Results of model calculations

5.6.1. Growth without aggregation

To illustrate several potentially important aspects of the aggregation process, the first model calculations are of the growth of single crystals by diffusion and accretion without aggregation. Figures 5.2a and 5.2b show the maximum dimensions and terminal velocities of the crystals as a function of time. The following points are relevant: 1) crystals grow much faster at $-6°$ and $-16°$C than they do at $-10°$ and $-12°$C, 2) crystal terminal velocities increase more slowly at $-6°$ and $-16°$C than they do at $-10°$ and $-12°$C, and 3) crystals growing with an initial LWC of 0.25 or 1.0 g m^{-3} at $-6°$ and $-16°$C take a long time to grow into graupel while those at $-10°$ and $-12°$C grow quickly.

The importance of points 1 and 2 is illustrated below using the results from the Appendix, where the probability of crystal collisions is given by

$$P = KD^2\bar{V}, \tag{5.3}$$

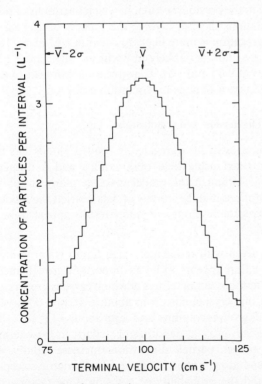

FIG. 5.1. Example of the technique used to generate a spectrum of particle concentration versus terminal velocity for a particle that has a calculated terminal velocity of 100 cm s^{-1} and a concentration of 100 L^{-1}.

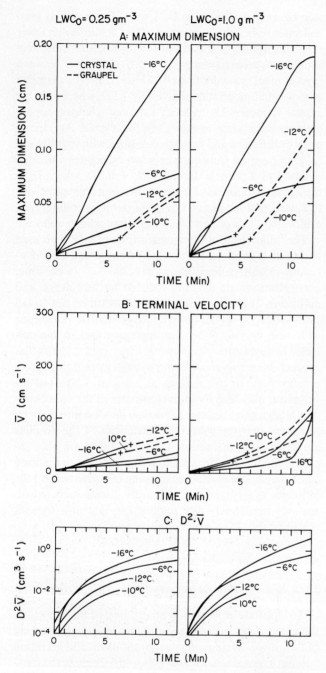

FIG. 5.2. Calculations from the model at constant growth temperatures that do not incorporate the effects of aggregation. Left panels, LWC = 0.25 g m^{-3}; right panels, LWC = 1.0 g m^{-3}. (a) Particle dimension (maximum along the *a*- or *c*-axis); solid line indicates that the particle is a crystal, dashed line indicates it has grown into a (spherical) graupel. (b) Terminal velocity. (c) $D^2 \cdot \bar{V}$, a term related to the probability of crystal collisions.

where K has the same value for all growth temperatures and liquid water contents, D is the crystal diameter, and $\bar{V}$ is the terminal velocity. This equation can be used to compare the probability of crystal collisions for different growth temperatures and liquid water contents, if the

crystal concentrations are assumed to be the same in each case. Plots of $D^2\bar{V}$ versus growth time are given in Fig. 2c, where the data are shown up until the time crystals become graupel. This figure indicates that particles at −6° and −16°C have a mugh higher probability of colliding (to become aggregates) than those at −10° and −12°C.

The importance of the final point, point 3, arises from the assumption that graupel do not aggregate. As is shown in Fig. 2c, if the LWC = 0.25 g m^3, the process of aggregation may continue for a considerable time at any given temperature. This is also true of particles growing at −6° and −16°C, if the LWC is increased to as high as 1.0 g m^{-3}.

5.6.2. Growth with aggregation

The results of the model calculations of aggregation are summarized in Figs. 5.3 and 5.4, which give plots of particle concentrations and diameters at particular times during the 12 min growth period. Concentrations are calculated in 200 μm diameter ranges. The figures also show the nature of the particles—whether they are crystals or graupel (solid bars) or whether they are aggregates (open bars)—as well as the ratio of ice water content in crystals (*C*) or graupel (*G*) compared to that in aggregates (*A*). A large ratio, of C/A or $G/A \gg 1$, indicates that accretional growth is dominant, while a ratio of $\ll 1$ indicates aggregational growth is dominant. These model calculations involve aggregation, but the results emphasize the points made in the previous section, in which the results of model calculations did not involve the process of aggregation: 1) aggregation is more likely to occur at −6° and −16°C, 2) aggregation is unlikely to occur at −10° and −12°C, and 3) at −6° and −16°C aggregation can lead to rapid development of large precipitation embryos.

5.7. Discussion and conclusions

The growth of aggregates in seeded clouds has been investigated using field observations and a numerical model. Although the model involves many simplifying assumptions and the scarcity of data restricts the modeling of aggregate development, some results appear to be relevant.

A. Even with moderately high liquid water contents, aggregation appears to be an important growth mechanism at those temperatures at which crystals grow rapidly along one crystal axis: from about −4° to −6°C, where crystals grow as columns, and from about −13° to −18°C, where they grow as dendrites or stellars. In these temperature ranges, particle dimensions increase rapidly, with correspondingly rapid increases in volumetric sweep-out rate and the probability of collision. One factor which increases the possibility that crystals will develop into ag-

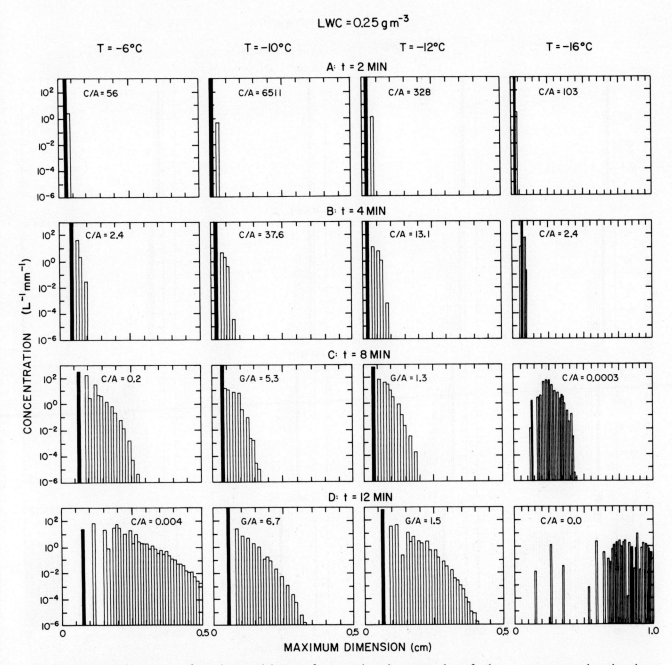

FIG. 5.3. Calculated size spectrum of crystals, graupel that grow from crystals, and aggregates shown for the same temperature along the columns and at the same time for different temperatures along rows. Solid shaded bars indicate the crystal or graupel concentrations, unshaded bars indicate aggregate concentrations. Ice water content in crystals (C) or graupel (G) is compared to that in aggregates (A) in the form of ratios. Liquid water content = 0.25 g m^{-3}. Note the scale change for the calculations at T = −16°C.

gregates is the relative inefficiency at which these crystals rime because of their large cross-sectional area and correspondingly low terminal velocities. It is unlikely that graupel or particles that are heavily rimed will aggregate. It is interesting to point out that the HIPLEX measurements taken at −6°C (see Table 5.1) did indicate many aggregates. As asymmetrical particles fall to lower levels, it is likely that they will continue to aggregate and will not, therefore, be restricted to the given temperature levels.

B. Aggregation appears to be a fairly unimportant growth mechanism at temperatures at which crystals grow relatively uniformly along the crystal axes: from about −8° to −12°C, where crystals grow as columns or fairly thick plates. Crystal dimensions increase relatively slowly and the probability of collision also increases slowly. Symmetric particles rime relatively efficiently because they have a small cross-sectional area and, consequently, a fairly high terminal velocity. They are, therefore, likely

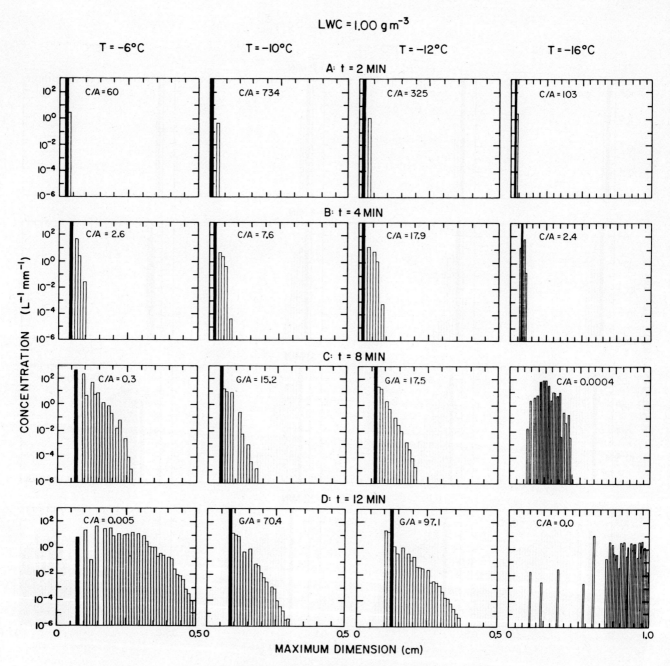

FIG. 5.4. As in Fig. 3 except for LWC = 1.0 g m⁻³.

to grow into graupel with only modest liquid water content and are not likely to aggregate, even if they fall into one of the favorable temperature ranges.

In closing, it is important to point out that to understand more fully the development and importance of aggregates, additional measurements are needed in seeded clouds.

Acknowledgments. The author wishes to thank Nancy Knight and Marie Boyko for their review of this paper and Frances Huth for typing the manuscript.

APPENDIX

Discussion of the Probability of Crystal Collisions

Equation (5.1) describes the probability of collisions between two particles in terms of their characteristics. In the particle growth model for a given time, the crystal diameter has a single value, D, so that in Eq. (5.1) $D_1 = D_2 = D$. From the assumed normal distribution of terminal velocities about the single value that is calculated at a given time, the number concentration versus velocity distribution can be written as

$$N = \frac{N_T}{\sqrt{2\pi}} \exp\left\{-\frac{1}{2}[(V - \bar{V})/\sigma]^2\right\}, \qquad (5.4)$$

where N_T is the total crystal concentration, N the concentration of particles of velocity V, $\bar{V}$ the calculated terminal velocity, and σ the standard deviation, assumed in the calculations to have a value of $0.125\bar{V}$. Consider now particles in concentrations of N_1 and N_2 that have grown during a certain period. Solving for V_1 and V_2 from Eq. (5.4), then

$$V_1 = \bar{V} + 0.125\,\bar{V}\{2\ln[N_T/(\sqrt{2\pi}\,N_1)]\}^{0.5}, \qquad (5.5)$$

$$V_2 = \bar{V} + 0.125\,\bar{V}\{2\ln[N_T/(\sqrt{2\pi}\,N_2)]\}^{0.5}, \qquad (5.6)$$

where σ is replaced by $0.125\bar{V}$, so that

$$V_1 - V_2 = \alpha\bar{V}, \qquad (5.7)$$

where α is a proportionality constant. Since N_T is the same for all growth conditions used in the model, the value of α is constant for all cases.

From Eq. (5.1) and the preceding discussion, the probability of collisions between crystals in concentrations N_1 and N_2 is given by

$$P = KD^2\bar{V}, \qquad (5.8)$$

where K is a constant that has the same value for all model runs. The probability of collisions between two particles from one run of the model can be compared, in a relative sense, to that from another model run from the term $D^2\bar{V}$ in Eq. (5.8).

REFERENCES

Cooper, W. A., and R. P. Lawson, 1984: Physical interpretation of results from the HIPLEX-1 experiment. *J. Climate Appl. Meteor.,* **23,** 523–540.

——, —— and T. A. Cerni, 1982a: *Cloud Physics Investigations by the University of Wyoming in HIPLEX 1979.* Department of Atmospheric Science, University of Wyoming, 301 pp.

——, ——, —— and A. R. Rodi, 1982b: *Cloud Physics Investigations by the University of Wyoming in HIPLEX 1980.* Rep. No. ASs140, University of Wyoming, 256 pp.

Heymsfield, A. J., 1982: A comparative study of the rates of development of potential graupel and hail embryos in High Plains storms. *J. Atmos. Sci.,* **39,** 2867–2897.

Hobbs, P. V., and M. K. Politovich, 1980: The structure of summer convective clouds in eastern Montana. II: The effects of artificial seeding. *J. Appl. Meteor.,* **19,** 664–675.

Holroyd, E. W., III, and J. E. Jiusto, 1971: Snowfall from a heavily seeded cloud. *J. Appl. Meteor.,* **10,** 266–269.

Issac, G. A., J. W. Strapp, R. S. Schemenauer and J. I. MacPherson, 1982: Summer cumulus cloud seeding experiments near Yellowknife and Thunder Bay, Canada. *J. Appl. Meteor.,* **21,** 1266–1285.

Marwitz, J. D., and R. E. Stewart, 1981: Some seeding signatures in Sierra storms. *J. Appl. Meteor.,* **20,** 1129–1144.

Parrish, J. L., and A. J. Heymsfield, 1985: A user guide to particle growth and trajectory models. NCAR Tech. Note 259 plus Instruction Aid, 69 pp.

Passarelli, R. E., Jr., 1978: Theoretical and observational study of snow size spectra and snowflake aggregation efficiencies. *J. Atmos. Sci.,* **35,** 882–889.

——, and R. C. Srivastava, 1979: A new aspect of snowflake aggregation theory. *J. Atmos. Sci.,* **36,** 484–493.

Rogers, P. C., 1974: The aggregation of natural ice crystals. Master's thesis, Department of Atmospheric Resources, University of Wyoming, Rep. No. AR110, 35 pp.

Strapp, J. W., H. G. Leighton and G. A. Isaac, 1979: A comparison of model calculations of ice crystal growth with observations following silver iodide seeding. *Atmos.-Ocean,* **17,** 234–252.

CHAPTER 6

A Review of Dynamic-Mode Seeding of Summer Cumuli

HAROLD D. ORVILLE

South Dakota School of Mines and Technology, Rapid City, South Dakota

ABSTRACT

This paper reviews the field experiments and theoretical studies relating to the ice-phase seeding of summer convective clouds for the purpose of affecting their dynamic evolution and precipitation production. The review reports on studies of both tropical and extratropical clouds, citing the physical evidence for microphysical and dynamic changes and reviewing the numerical modeling efforts in support of the field experiments. The statistical evidence is also reviewed. A critique and discussion of the results is given, and many questions related to these dynamic-mode seeding hypotheses are posed. Strategies for attacking the many unsolved problems are presented briefly.

6.1. Introduction

6.1.1. Historical background

The spectacular results sometimes achieved by cloud seeding with ice-forming nucleants were first reported by Kraus and Squires (1947) for cumuliform clouds. The drop of 136 kg of dry ice into one cloud in a seemingly uniform field of clouds resulted in the "explosive" growth of the seeded cloud and more than 12 mm of rain over 130 km^2 (nearly 1300 acre-feet or 1600 kT of water). Figure 6.1a–c from Smith (1974) shows the results. More tests followed; rain was produced from the seeding of individual clouds with both dry ice and silver iodide (Squires and Smith, 1949; Bethwaite et al., 1966). However, the dynamic effects on the growth of clouds was not emphasized in the early years, although greater persistence of the seeded clouds was noted. Perhaps the use of small amounts of silver iodide, only 20 g per cloud in one of the field experiments, precluded such external dynamic effects as enhanced cloud growth from being produced.

It was in the 1960s in the Caribbean and in Arizona that the dynamic stimulation of the cloud was focused upon as one result of the cloud seeding (Simpson et al., 1965; Weinstein and MacCready, 1969). The potential for increasing rain via this process was also stressed, but a direct relationship of the increased cloud-top height to rain production was difficult to document. Some evidence was produced by radar studies.

While the rainfall from individual clouds was easily affected, given the proper atmospheric conditions, the effect on the rainfall over a larger area was less clear. Experiments in South Dakota, North Dakota, and Florida were established (Dennis and Koscielski, 1969; Dennis et al., 1975; Woodley and Sax, 1976). The design of the Florida Area Cumulus Experiment (FACE) was based

upon a dynamic seeding hypothesis, while the Dakota experiments used a "seedability" concept developed in FACE to stratify data for statistical analysis. Subsequently, other field experiments have been reanalyzed with the "dynamic seeding potential" of the clouds as a test variable (e.g., McCarthy, 1972).

More recently, these concepts have been used in planning projects in South Africa and in Illinois and in analyzing the cloud seeding potential of clouds in the Ivory Coast region. But before we go on, let's stop to consider more precisely what the "dynamic-mode" seeding concept is and examine its evolution over the past 20 years or so.

6.1.2. Various scenarios for dynamic seeding

The concept of dynamic seeding in its simplest form (circa 1965) was that massive ice-phase cloud seeding of a cloud turret at the −5° to −10°C level would rapidly convert all or most of the supercooled liquid to ice, releasing large amounts of latent heat (of fusion) to the cloudy atmosphere and causing enhanced growth of the cloud in both depth and breadth, usually in a two-stage process. It was also hypothesized that more rain than would have occurred naturally would result.

In addition to the latent heat of fusion, many papers considered an additional heat release as the cloud adjusts to saturation with respect to ice instead of liquid (Woodley, 1964; Simpson, 1967; Weinstein, 1970). This latter heat release has been calculated by Saunders (1957), MacCready and Skutt (1967), Orville and Hubbard (1973), and Fukuta (1973). The "heat release" due to saturation adjustment at the time of freezing has been shown to be a *negative* contribution if amounts of supercooled water greater than about 1.5 g kg^{-1} are to be frozen rapidly in these warmer regions of the cloud. This is caused by the

FIG. 6.1. (a) 5 February 1947. Inland cumulus clouds before dry ice was dropped; (b) cloud growth 13 min after seeding; and (c) the fully developed anvil 40 min after seeding. (From Smith, 1974.)

FIG. 6.1. (Continued)

requirement that the cloud stay at saturation with respect to ice; the heating caused by the freezing of large amounts of liquid water has a tendency to *sub*saturate the air with respect to ice. Consequently, the newly formed ice must sublimate and cool the air to maintain 100% relative humidity with respect to ice. The subsequent ascent of the parcel follows an ice adiabat, which releases more heat to the parcel than if the parcel had followed a water-saturated adiabat, at least to $-30°C$ or so. (The changing latent heat of condensation accounts for a decreasing difference at the colder temperatures.) Of the three "heating" effects identified, the latent heat of fusion and the ice adiabatic ascent of the parcel are positive, while the saturation adjustment is sometimes negative. In any event, the *initial* heat release is not the commonly assumed 0.33°C per gram of liquid frozen *plus* an added amount due to depositional heating, but normally 0.33°C per gram of liquid frozen *minus* a temperature increment due to sublimational cooling (in the "warm" regions of the cloud and with a few to several grams of supercooled liquid water being frozen).[1]

A fourth effect is also identified in the MacCready and Skutt and Fukuta papers. The amount of ice existing after freezing is normally different from the amount of liquid before freezing, leading to a changed loading effect on the parcel. This effect is generally much smaller than the three temperature effects mentioned above. In an extreme case of freezing 15 g kg^{-1} of supercooled liquid at $-5°C$ and at 525 hPa, 14.1 g kg^{-1} of ice results. The decrease of 0.9 g kg^{-1} of condensate is equivalent to nearly a 0.25°C temperature excess effect on parcel buoyancy (the actual temperature change being nearly 2.5°C).

A final point regarding buoyancy effects should be mentioned here. Koenig and Murray (1983) have discussed the relative sizes of individual forces on a cloud parcel and the size of the net force, which is, in general, much smaller than the individual thermal, vapor, and condensate terms. A slight rearrangement of these forces can change the size and even the sign of the net force, leading to the implication that dynamic effects can possibly be caused by rearranging these forces (for example, by increasing or decreasing the fall speeds of particles, as might occur if different growth modes of the ice particles are enhanced or different size distributions of the liquid particles are affected). Orville and Chen (1982) have also analyzed quantitatively the effects of the latent heat of

---

[1] See Orville et al. (1984) for conditions when the depositional heating during saturation adjustment is very important for the temperature increases and vertical motions inside a cloud.

fusion and precipitation loading caused by ice-phase seeding. The effects can be much more complicated than envisioned by the early practitioners of "dynamic-mode seeding."

Whatever the heat input and temperature change, a few tenths of a degree temperature increase is sometimes crucial to the survival or growth of a cloud turret, which is beset by loading forces, entrainment, neighboring cloud turret circulations, etc. These effects tend to reduce the buoyancy of the cloud and halt its growth or cause it to dissipate. However, if the cloud receives an added boost by the early and rapid freezing of the liquid, several effects may occur to change precipitation from the cloud. *One,* the cloud may rise higher in the atmosphere and condense or deposit more vapor onto particles that may subsequently be involved in the precipitation process. *Two,* the additional ice particles may not become involved in precipitation, but may be blown out an anvil and have no effect on precipitation. *Three,* even if the initial ice particles are blown out the anvil, the cloud may become so invigorated that more vapor is drawn into the cloud and greater condensation/deposition rates occur, leading to more precipitation. *Four,* the invigorated updraft may exhaust more cloud condensate into the anvil than is formed at lower levels, leading to less precipitation. *Five,* the enhanced midlevel inflow to the cloud may lead to a cutoff turret and little or no change in precipitation (Simpson and Dennis, 1974; Simpson, 1980). *Six,* the increased midlevel inflow may mix with the precipitation and cloud particles, cause evaporation or sublimation, and, acting in concert with the enhanced precipitation loading, form a stronger downdraft than would have occurred in the unseeded cloud (Simpson, 1980; Woodley et al., 1982b). This strong downdraft then diverges at the earth's surface and causes increased convergence in the inflowing boundary-layer air leading to new, more vigorous cell development and enhanced precipitation in the area. *Seven,* the ice particles formed by the seeding may reach precipitation size earlier than the ice particles in an unseeded cloud and sweep out the supercooled liquid streaming upward, releasing latent heat of fusion and resulting in some of the same scenarios outlined above but in a delayed sense. Or *eight,* the early "artificial" precipitation may prematurely, or at a nonoptimum time, sweep out the supercooled liquid feeding the central storm updraft, leading to a diminished storm circulation and decreased precipitation (Orville and Chen, 1982).

Table 6.1 gives a present day exposition of the dynamic seeding hypothesis (Woodley et al., 1982b) as viewed for tropical clouds.

For extratropical clouds, the emphasis on rapid freezing was downplayed (Weinstein and MacCready, 1969; Matthews, 1981; Dennis, 1980), and effects other than cloud-top growth were emphasized at times. Glaciation at a lower level of the cloud was desired, but the mechanisms

for this to occur were thought to be early development of precipitation and accretional sweepout of the supercooled liquid, leading to the effects hypothesized in item seven.

The present picture is one of considerable complexity and of great time and space dependency. The seeding agents should be placed in locations and at times when effective results can be expected. The quantity and spread of the ice-forming agents are also important considerations. Normally, the desired action is to produce about 100 ice crystals per liter of air at the $-5°C$ level, although early formation of precipitation to sweep out supercooled liquid could require less seeding material.

Certainly, the microphysical effects of the ice-phase seeding must be detected if dynamic effects are to ensue. Consequently, I next review the physical evidence for microphysical effects in projects attempting to produce dynamic effects.

6.2. Physical evidence for microphysical and dynamical effects

The projects attempting to alter the dynamics of clouds, or that have been evaluated and found to suggest dynamic effects, have been conducted in both tropical and extratropical regions and with maritime and continental-type cumuliform clouds. The evidence for the occurrence of the "proper" microphysical characteristics and for the changes following seeding are now given for these two regions.

6.2.1. Tropical clouds

Microphysical and other measurements. Several papers report on efforts to measure the cloud liquid, number and size of ice particles, temperatures, updrafts, and radar echo characteristics in tropical clouds (Ruskin, 1967; Simpson, 1967; Woodley, 1970; Sax, 1976; Sax et al., 1979). One objective of these efforts was to determine if adequate amounts of supercooled liquid and few enough ice crystals were present to make dynamic seeding a viable procedure. Measurements before and after seeding were generally attempted in order to detect the seeding effect.

"First generation" experiments. The studies of the Caribbean clouds in Project Stormfury provide much of the evidence for microphysical effects in tropical clouds seeded to produce a dynamic effect. Ruskin (1967) reports on measurements inside a cloud turret developing on 5 August 1965 over the sea 480 km south of Puerto Rico. Sixteen pyrotechnic units, dropped at 100 m intervals and at about 7.5 km height, provided about 1.7 kg of silver iodide in their 6 km fall before burnout. Five minutes after the drop, four cloud physics instrumented aircraft penetrated the cloud at various elevations, perpendicular to the plane of the flare drops. Other seemingly similar turrets in the cloud line were also sampled. Assumptions

TABLE 6.1. Summary of hypothesized dynamic seeding chain of events*

Stage I: Initial vertical tower growth

1. Rapid glaciation of the updraft regions of supercooled convective tower(s) by silver iodide pyrotechnic seeding.

2. Invigoration of the updrafts through buoyancy increase produced by the release of latent heats of fusion and perhaps deposition; the latter may or may not contribute as the cloud air approaches saturation relative to ice.

3. Pressure falls beneath the actively growing tower due to upward acceleration and upper level warming followed by increased inflow at mid to low levels (surface to 6 km) which fuels the initial stage of cloud growth.

   Stage I may last 10-20 min, sometimes longer.

Stage II: Horizontal cloud expansion, secondary growth

4. Enhanced downdrafts below the invigorated seeded tower as the precipitation and evaporatively cooled air moves downward.

5. Convergence at the interface between the downdraft and the ambient low-level flow, instigating tower ascent fed by the warm, moist inflow.

6. Growth of secondary towers (which, in turn, might be seeded).

7. Horizontal enlargement of the cloud by joining of the feeder towers, leading to wider protected updraft(s), augmented condensation, water content, rainfall.

   Stage II may last 30-50 min.

Stage III: Interaction with neighboring clouds

8. Seeding of secondary towers in the parent cloud results in their growth, followed by expansion and intensification of the downdraft area which then moves outward to interact with outflows from neighboring clouds (which might also have been seeded). This increases the convergence on a larger scale, deepens the moist layer and results in new cloud growth and merger in the convergent regions between the cloud systems. These new towers are normally seeded as well.

9. The increased seeding-induced growth and merger of clouds on the mesoscale coupled with sinking in their near environments results in a mesoscale region of warming (50 km on a side). The resulting thermally direct mesoscale circulation provides additional low-level mass and moisture convergence to fuel new cloud development and perhaps to prolong the lives of the older cloud systems. Further, under certain conditions, the upward branch of the mesoscale circulations may become saturated and produce a period of stable (non-convective) rainfall.

Stage IV: Increased area rainfall

10. Seeding increases rainfall over the floating target by:
    a. enhancing the growth of the directly treated cloud towers.

    b. inducing additional cloud growth and larger cloud systems through the mechanism of downdraft interactions.

    c. indirectly increasing the efficiency of cloud elements as they grow in the more moist environment provided by the larger cloud systems; and

    d. augmenting the supply of available moisture through the enhancement of the thermally-direct mesoscale circulation.

11. Seeding increases rainfall over the total target by:
    a. obtaining more rain from the available moisture than would have occurred naturally, and/or by

    b. enhancing the moisture supply to the target.

---

*From Woodley et al. (1982b).

about the degree of spread of the AgI and rough calculations indicated that about 180 ice nuclei active at $-5°C$ should have been produced per liter of air, although these numbers can be disputed with as little as we know about seeding delivery and dispersion and AgI activity.

The results of Ruskin's study combined with Simpson's companion article (Simpson, 1967) regarding photographic and radar evidence of changes in the seeded and unseeded turrets gave evidence of sufficient liquid water $(2.0–2.4 \text{ g m}^{-3})$ and little ice (less than 10% coverage of the particle replicator films) before seeding, and impressive changes after seeding (Johnson–Williams measured liquid water content decreased to a few tenths of a gram per cubic meter, small and large raindrops decreased in number, and the percentage of ice increased to 80% or more). The seeded cell grew to 10.8 km while other clouds in the vicinity topped out at about 8.1 km.

Temperature changes in the seeded and adjacent unseeded regions of the cell led to some consternation. Temperature increases of $1.5°C$ in the heavily iced seeded region and of $2.0°C$ in an adjacent region occurred. Simpson postulated an increased updraft in previously inactive clouds to explain the temperature change in the central unseeded portion. In the seeded region, she calculated that freezing of the liquid water could produce $1.2°C$ change *plus* an additional $0.5°C$ from the adjustment from water to ice saturation. The preceding discussions of the heating effects (in section 6.1.2) indicate that a heating of about $1.2°C$ would occur minus $0.5°C$ due to sublimational cooling for a net heating of $0.7°C$, if freezing occurred instantaneously and isobarically (Orville and Hubbard, 1973).

However, these discussions of temperature changes and (implicitly) vertical motions treat a very difficult observational problem. Examination of the pertinent motion, temperature, and water mass equations shows that the changes would be more easily detected in a Lagrangian framework (i.e., following the seeded air parcel) than in an Eulerian framework [i.e., at a constant height in the cloud (Orville et al., 1975)]. This is due to the fact that it is extremely difficult to measure the nonlinear advective terms in the Eulerian form of the equations ($w\partial T/\partial z$, or $w\partial\theta/\partial z$, etc., w, T, θ, being vertical velocity, temperature, and potential temperatures, respectively). These terms are just as large as most of the "physical effect" terms in the equations (the sum of the terms being equal to the Lagrangian time derivative). The 1965 Stormfury cloud-seeding experiment evaluations appear to have concentrated on measurements at the seeding level, without attempting to estimate the nonlinear effects from the four airplane sampling patterns, which would be of dubious value, anyway, because of the large height distances between aircraft. Nevertheless, that was a pioneering effort in coordinating aircraft operations and measurements and comparing results with model output. Future multilevel

pass data should be analyzed with results of the newer, time-dependent cloud modeling studies in mind.

Woodley (1970) reported on a field experiment in south Florida during May 1968 in which 19 clouds were studied, 14 seeded with silver iodide pyrotechnics and 5 unseeded. The seeded clouds grew an average 3.2 km higher than the unseeded clouds. Estimates of rainfall increases were made from 10 cm radar data. An average increase of 100 to 150 acre-feet 40 min after seeding was indicated (more than doubling the natural amounts). Woodley reasoned that "the seeded clouds were larger and more lasting and processed more moisture than their unseeded counterparts, which accounted for the increase in precipitation." He also suggested that the subsequent natural precipitation processes were enhanced.

Valiant attempts had been made to detail the effects of dynamic-mode seeding in tropical clouds. Much useful information and great stimulation was given to cloud physics from these experiments. Many cloud seeding experiments around the world began to be analyzed for dynamic effects.

FACE-1 and FACE-2 continued the Florida-based efforts and led to a "second generation" attempt to document the microphysical and dynamic changes in heavily seeded clouds.

"Second generation" experiments. The 1970s found the emphasis on tropical cloud seeding shifting to those clouds forming over the Florida peninsula and an attempt to affect areas of 10 000 km^2 and more. Even though the emphasis was areawide, it was still important to demonstrate that the target clouds were affected by the cloud seeding.

Results of FACE '73 (part of FACE-1, which extended from 1970 to 1976) revealed no significant icing effects on the treated clouds (Sax, 1974), due primarily to a failure in the ice crystal instrument to detect ice crystals smaller than 200 μm (in maximum dimension). However, cloud physics sampling of clouds in FACE '75 and FACE '76 left little doubt that the seeded clouds were glaciated by the seeding (Sax, 1976; Sax et al., 1979; Sax and Keller, 1980; Hallett, 1981). The production of crystalline ice and graupel occurred more abundantly in the seeded clouds and more clearly in the 1976 than the 1975 field experiment. This last effect was attributed as most likely due to new, more efficient seeding flares introduced in the 1976 experiment. Figure 6.2, adapted from Sax et al. (1979), shows the effects in the seeded and nonseeded clouds regarding ice crystal production.

The Sax and Keller (1980) paper documents very neatly the development of precipitation in the unseeded Florida clouds in 1975 and 1976 and shows an active ice multiplication process at times. There appeared to be a progressive sequence of cloud water, rain water, graupel, and crystalline ice detected near the $-10°C$ level in the sampled 1976 clouds. These authors detected a rapid micro-

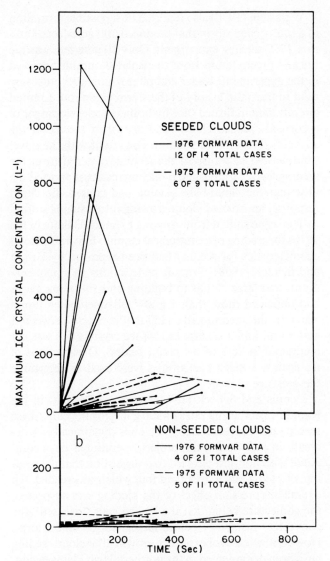

FIG. 6.2. (a) Evolution of maximum ice crystal concentration with time in clouds seeded during the FACE '75 and FACE '76 programs; cases not shown contained no crystals upon repeat penetration. (b) Evolution of maximum ice crystal concentration with time in naturally developing (unseeded) Florida cumuli; all cases not shown contained no crystals during any penetrations. (Adapted from Sax et al., 1979.)

physical evolution of the clouds in some cases. "A vigorously growing, sharply defined, "boiling" cloud element containing an abundant quantity of supercooled water near the −10°C level can change, within the span of 3–5 min, to a collapsing, wispy appearing, diffuse precipitating mass with hardly any supercooled water remaining in the form of cloud droplets."

Other data on the Florida clouds showed several climatological features of the Florida target clouds in FACE '75 (Sax et al., 1979). The 39 growing clouds that were sampled had ample liquid water contents (peak and mean values of 5 and 4 g m^{-3}), adequate vigor (15 and 8 m s^{-1} peak and mean updrafts), and an average width of 2 km, characteristics that indicate good cloud seeding potential.

Simpson and Dennis (1974) and Simpson et al. (1980) discuss the effects of cloud merger, caused either naturally or by cloud seeding. A comparison of cloud photographs and radar PPI images in the 1974 article shows how individual cells grow and merge, one of the cells having been seeded. More than a factor of 10 increase in precipitation was thought to have occurred. These authors point out that going from a single cloud experiment to a wide area project makes the evaluation more difficult by a factor of 10 or more. Cloud interactions and mesoscale systems increase the natural variability and allow for propagating seeding effects.

Cunning et al. (1982) presented a very thorough study of cloud evolution and merger in south Florida. They used radar, visual cloud, and raingage data to document the merger of two convective systems on 19 August 1975. New cloud development and differential motion between convective elements were the two main causes of merger. This last effect also shows up in modeling studies, which superimpose mesoscale convergence on the cloud scale (Chen and Orville, 1980).

A paper by Cunning et al. (1979) shows an explosive growth case from FACE '75. It was hypothesized that an unseeded cell was affected by seeded cells around it and grew at an accelerated rate. Doppler radar analysis provided the radar reflectivities and indicated a growth rate of the cloud of 3.5 km in only 2 min.

Although the sampling of individual clouds provided strong evidence for a microphysical and dynamic effect from the artificial seeding of clouds, it was also recognized that similar effects happen to unseeded clouds growing through natural ice regions in the atmosphere.

Cunning and DeMaria (1981), responding to Simpson (1980) regarding dynamic seeding, gave evidence for pressure responses to dynamic-mode seeding. They argued that the rapid growth of the seeded convective tower on one FACE day (25 August 1975) induced a surface low-pressure region (0.35 hPa decrease) by hydrostatic and/or nonhydrostatic means, prior to the development of precipitation-induced downdrafts at the surface. Simpson and Cooper (1981) in their reply emphasized that the seeded clouds were not isolated and that downdrafts from other regions may have caused the explosive growth of the observed cloud.

The dynamic seeding of clouds in the tropics was becoming a more complicated and complex exercise, with several outcomes of the seeding recognized, but difficult to pin down in most cases.

6.2.2. Extratropical clouds

Scientists at Pennsylvania State University and with Meteorology Research, Incorporated (MRI) were working on the buoyancy concept for enhancing cloud development in the southwestern part of the United States at about the same time as the Stormfury work was progressing in the Caribbean and Florida. Consequently, the clouds were of a different microphysical makeup, presumably with cooler cloud bases and less active coalescence processes.

Weinstein and MacCready (1969) reported on dynamic seeding of isolated cumulus clouds near Flagstaff, Arizona. Their cloud seeding technique was considerably different from that used in the tropical tests. Seeding was done from an aircraft circling in an updraft, 300 m or so below cloud base, such that 120 to 240 g of AgI were added to each seeded cloud over a period of about 20 min. Pairs of clouds were selected, one seeded and one not, during a test day. In July and August 1967, 21 clouds were studied, 11 seeded and 10 unseeded; 9 days had paired clouds. Only the first cloud was treated randomly; the treatment of the second cloud depended on the actions taken with respect to the first.

This study concentrated on those clouds that were neither too small nor too large, those that would top out in the $-8°$ to $-25°C$ layer. A one-dimensional, steady-state cloud model was used to help select cloud sizes that would respond most favorably to seeding. Measurements focused on cloud-top height, rainfall amount (depth), and rain duration.

Results were impressive. The average cloud-top height of seeded clouds was 10.2 km; unseeded clouds, 7.8 km. The duration of the rain averaged 10 additional minutes, and the total depth of rain increased 2.00 mm (to 4.63 mm) over the unseeded cases. The average radius of the clouds was 1.3 km; broadening of the clouds was not measured. The initial echo was lower in the unseeded clouds than in the seeded ones, in contrast to results reported by Dennis and Koscielski (1972) from South Dakota seeded and unseeded clouds. There was some evidence that clouds seeded with two generators responded more clearly than those seeded with only one.

Two other papers give evidence of cloud microphysical and dynamic effects in the Arizona clouds. Weinstein and Takeuchi (1970) report on a case in which two pyrotechnics were dropped into a supercooled cumulus cloud near Flagstaff, each one releasing 25 g of AgI in a 3-km fall. A thin turret grew 1.35 km above the main cloud mass. Aircraft sampling of the cloud was done 9 min before seeding and 5 and 8 min after seeding. Liquid water contents of greater than 1.0 g m^{-3} were found before seeding; large numbers of small ice crystals were found after seeding. The observed temperature change after the seeding ($0.75°C$) was found to be consistent with MacCready and Skutt's (1967) calculations, but note the difficulties of interpretation of temperature measurements mentioned in section 6.2.1. The authors note that many more ice crystals were observed in the tower than predicted by nucleation tests of the pyrotechnics.

Weinstein (1970) also reported on a possible formation of a multicelled storm that produced 100 acre-feet of rain in a 1969 seeding experiment. Only 10 acre-feet was the normal precipitation from the isolated cumuli observed in the experiment. Radar and photographic records were used to trace the history of the storm evolution. Limited aircraft data indicated "ice hydrometeor development of a correctly seeded cloud" and arrival of the ice at the aircraft's level ($-9°C$) at a time consistent with the travel time needed to transport the seeding material from the seeding level. Steady-state modeling results supported the idea that the cloud-base seeding had caused the cloud response, an isolated cloud growing into a major storm.

The scene shifted from Arizona to South Dakota in the 1970s for seeding of extratropical cumuli. Favorable cloud characteristics for seeding had been reported by Hirsch and Schock (1968). Aircraft penetrations in small, medium, and large clouds in building and subsiding stages had indicated more than 1 g m^{-3} of supercooled liquid water in the medium-size clouds (diameters between 1 and 3 km) and a general lack of ice crystals (ice was encountered in 26% of the cloud passes). Observations of ice against a black background were used to determine the presence of ice.

Dennis and Schock (1971) hypothesized that, in the Rapid Project, using rates of 300 g h^{-1} for the cloud-base seeding may have affected the clouds dynamically. A decrease in hail during the project operational days compared with non-test days led to a suggestion that the seeding affected a larger area than that which was seeded. An overall suppression effect of the seeding was suspected, the conjecture being that the seeded cells "released" the convective instability before larger storms could form. However, no cases of explosive growth after cloud seeding have been documented in the northern High Plains region.

Other cloud physics measurements have been made in Illinois and Texas related to dynamic seeding. During a 1978 field season in Illinois, Ackerman and Johnson (1979) report on results of 237 aircraft passes through clouds on seven days in June. Average maximum liquid water contents (JW-instrument) were greater than 1.25 g m^{-3}, and average maximum updrafts $5-14 \text{ m s}^{-1}$. More than 75% of the cloud regions with updrafts greater than 1.5 m s^{-1} had water contents greater than 1 g m^{-3}. In addition, an active ice process was observed; ice multiplication processes associated with large drops were measured in clouds growing past the $-10°C$ level. This led the authors to conclude that seeding to produce a more efficient ice process was probably not appropriate, but that seeding to produce more glaciation and invigorate the clouds probably was.

The measurements in Texas took into account the organization of radar echoes (Chen et al., 1979). These authors concluded that more intense rainfall was associated with rain cells oriented perpendicular to the wind shear and that attention should be paid to the mesoscale features of the cloud systems. They stated that "efforts to organize groups of clouds into more productive cloud systems should recognize the preferred mode and spacing existing at the time and work toward strengthening the natural tendencies for organization." More will be said about this mesoscale connection below.

Cloud models of these dynamic and microphysical processes gave additional evidence regarding dynamic seeding effects and will be reviewed next.

6.3. Numerical modeling efforts

6.3.1. Tropical clouds

The development and application of a one-dimensional, steady-state cloud model (1DSS) was a distinctive part of the tropical cloud dynamic seeding experiments (Simpson et al., 1965; Simpson et al., 1967; Simpson and Wiggert, 1969, 1971). The model was developed from spherical vortex and starting plume theories (Levine, 1959; Turner, 1962), as was a similar one by Squires and Turner (1962). The model simulated the rise of a parcel, accounting for entrainment, condensate loading, thermal and vapor buoyancy effects, and the formation of various types of cloud water and ice contents. Freezing of the liquid water was accomplished at temperatures lower than $-25°C$ or so (some models at $-40°C$) in the unseeded cases and in the interval $-4°$ to $-8°C$ in the seeded runs. The additional growth of the seeded model cloud over the unseeded model cloud gave rise to the concept of "dynamic seedability," normally defined as this height difference and denoted "S" in Simpson's work.

The model was developed to help explain the observations obtained in the Caribbean field experiments. Only the first phase of the growth (the vertical extension) and not the subsequent broadening of the cloud could be simulated by the model.

An impressive result of the use of the model is given in Simpson et al. (1967) and reproduced here as Fig. 6.3. The results show the estimated seeding effect E versus the (predicted) seedability S; E equals zero for a perfect prediction for an unseeded cloud and equals S for a perfect prediction for a seeded cloud.

The randomized test provided 14 seeded clouds and 9 control (unseeded) clouds. A few of the seeded clouds grew "explosively." Average growth was about 1.6 km after seeding. The highly significant results (statistically) were achieved in only one test season, illustrating the value of a good predictive model to a statistical experiment.

The 1965 field data were used to improve the model (Simpson and Wiggert, 1969). It was further developed, especially regarding precipitation fallout, and applied to data from a 1968 randomized experiment (Simpson and

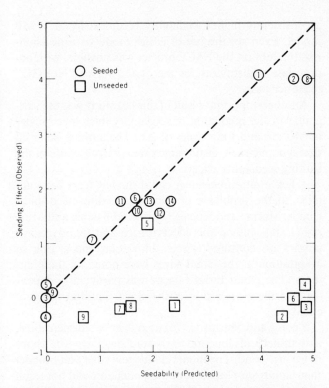

FIG. 6.3. Seedability versus seeding effect for the 14 seeded (circles) and 9 control (squares) clouds studied in 1965. Note that seeded clouds lie mainly along a straight line with slope 1 (seeding effect is close to seedability), while control clouds lie mainly along a straight horizontal line (showing little or no seeding effect regardless of magnitude of seedability). Units of each axis are in kilometers. (From Simpson et al., 1967.)

Wiggert, 1971). The model was denoted EMB, for Experimental Meteorology Branch model. This study produced a prediction of rainfall and rainfall increases due to dynamic seeding, although the actual rainfall was greatly underestimated. [Later work (Silverman et al., 1976) has indicated that the 1DSS models are not really suitable for direct rainfall predictions because of their poor simulation of the rainfall process. Used indirectly, though, through curves of radar height versus rainfall amounts, they might be skillful. In this case, the cloud-top height predicted by the model replaces the radar echo height to then predict the rainfall amount.]

Another one-dimensional model was developed to simulate the dynamic seeding of tropical clouds (Cotton, 1972; Cotton and Boulanger, 1975). Features of the entraining jet model of Squires and Turner (1962) and of the bubble model of Simpson and Wiggert (1969) were combined. However, more sophisticated ice microphysical processes were used by Cotton, with 21 categories of pristine and rimed ice crystals simulated. This microphysical treatment illustrated the importance of supercooled rainwater for producing rapid freezing and glaciation of the cloud. Also, a comparison of seedability as predicted by this model with the seedability of the EMB model showed a consistently higher seedability predicted by the EMB model. Even so, the use of either model as a decision-making tool on the FACE project was possible, with appropriate adjustments for the "GO" versus "NO GO" reference base.

Another important result of this last study was the variability in the prediction of seedability over space scales of 110 km and time scales of 2 h. The authors believed that synoptic-scale disturbances were a likely cause of the shifting seedability effects.

A few multidimensional cloud models have been applied to the problems of dynamic seeding of tropical clouds. Murray and Koenig (1972) and Koenig and Murray (1976) studied the effects of ice and liquid microphysics on cumulus towers. Important effects due to evaporation at the cloud edges were noted by these authors. The cloud turret's decay was particularly dependent on the evaporation of the cloud liquid at the cool cloud cap.

Koenig and Murray (1976) used their two-dimensional, axisymmetrical cloud model to simulate massive infusions of ice in a cloud. The model cloud grew taller and broader than clouds with less ice, but the simulations did not result in more rainfall, perhaps due to the continual supply of ice to simulate seeding instead of an instantaneous pulsed increase.

Levy and Cotton (1984) used a three-dimensional cloud model to analyze the effects of cloud seeding on Florida clouds. Their interest was in trying to see how large releases of latent heat in middle cloud levels would affect the cloud system and pressure patterns in the cloud and subcloud layer. Seeding was simulated by increasing the ice crystal concentration to 100 L^{-1} at the $-10°C$ level and above for 10 min after the tower reached that temperature. The results were examined with respect to dynamic responses and the communication of the effects to the subcloud layer. The authors found that the glaciation caused vertical motion changes by as much as 2.5 m s^{-1}, but only weak responses in the subcloud layer and no additional precipitation. Horizontal responses at the level of seeding were much stronger than the vertical responses lower in the cloud.

A thesis by W. T. Chen (1982) also considered the heavy seeding of a Florida-type cloud using the two-dimensional, time-dependent cloud model of Orville and associates referenced below. This model was modified to account for the warmer cloud bases and more efficient coalescence processes of tropical clouds, compared with the extratropical clouds normally simulated in the model. The sounding used was the same as that used in the study of Levy and Cotton (1984), which resulted in the clouds reported on by Cunning and DeMaria (1981). Comparisons with the Cunning and DeMaria (1981) cloud outlines were favorable. Results of seeding these model clouds showed about a 3 m s^{-1} increase in vertical velocity and enhanced

precipitation processes early in the life cycle of the cloud, but decreases in the later stages.

6.3.2. Extratropical clouds

One-dimensional models. Just as for the tropical studies, the one-dimensional models were the first to be developed and applied to the dynamic seeding experiments of extratropical clouds. Weinstein and Davis (1968) developed a model at Pennsylvania State University, which has had wide distribution and application. It was a steady-state model of an entraining jet or plume and was applied in field trials in Pennsylvania and Arizona in the 1960s. The Weinstein and MacCready paper (1969) shows correlations of 0.88 between predicted and observed cloud-top heights for both seeded and unseeded clouds. These authors also used the model to predict rainfall and rain duration—items of more dubious predictability from one-dimensional models. However, the correlation coefficient for the rainfall was 0.89 for 21 test clouds.

Building on this work, Weinstein (1972) applied the model to construct a dynamic seeding climatology of the western half of the United States, in terms of the additional rain predicted by the seeded version of the model.

In addition, Weinstein (1970) extended his modeling work to include a one-dimensional, time-dependent model. His results showed some important aspects of the time dependence, such as the fact that the simulated seeding could occur at a time in the life stage of the cloud to reduce rainfall but increase cloud top. However, the model was not used extensively, and no further results have been reported.

The Weinstein–Davis 1DSS model was adapted by Hirsch (1971) to the South Dakota projects. That model, with the addition of the microphysics in Wisner et al. (1972), has been used by the Bureau of Reclamation for several years on its environmental time-share network. The program is called GPCM, Great Plains Cumulus Model. Whereas Weinstein simulated the seeding of clouds by freezing all of the liquid at $-8°C$ (or $-15°C$) and natural clouds at $-25°C$, Hirsch froze the supercooled water over a 20°C interval ($-20°$ to $-40°C$ in unseeded clouds and $-5°$ to $-25°C$ in seeded clouds). The change is exponential, slow at first and rapid as the final freezing temperature is approached.

The GPCM has been applied to various areas of the Midwest. Results from soundings in South Dakota and North Dakota indicate relatively small dynamic seeding potential, 400–800 m on an average. Figure 6.4 shows results from the spring/summer of 1971–72 using western North Dakota soundings. The negligible seedability of the shallow and deep clouds is a characteristic result of these models. The low end of the curve in Fig. 6.4 includes cases in which the freezing temperature is not reached by the model cloud, and the high end includes the cases in

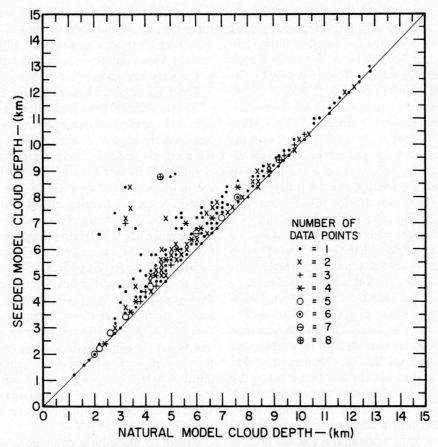

FIG. 6.4. Graph showing natural model cloud depths versus seeded model cloud depths. Various symbols are used at points where multiple data points overlapped. The average increase in model depth was 770 m. Rawinsonde and radar data for the 1971–72 summer seasons in western North Dakota were used to provide input to the cloud model. (From Miller et al., 1975.)

which both seeded and unseeded clouds reach the tropopause, with no additional growth possible.

The experiments in South Dakota and North Dakota gave evidence of positive seeding effects on rain during those days on which a one-dimensional cloud model predicted some dynamic seedability (Dennis and Koscielski, 1969; Dennis et al., 1975). Little extra cloud height was expected, but some invigoration of the updraft was postulated (Chang, 1976).

Matthews (1981) has used the cloud model to determine the climatology of dynamic seedability in the three field sites of HIPLEX:[2] Big Spring, Texas; Goodland, Kansas; and Miles City, Montana. Substantial seedability was found in all three sites, but with more opportunities existing in the southern and northern High Plains. A dynamic modification potential (i.e., a height change in seeded model clouds relative to the unseeded model clouds) was found in approximately 50% of the soundings

analyzed. Satellite data were used to classify the days and determine if certain types of cloud convection were present. Another one-dimensional parcel model was used by Matthews and Silverman (1980) to predict the effects of mesoscale lifting on convection. These effects were often more significant than the effects of cloud seeding and would tend to mask the seeding effects in a field experiment.

Wiggert et al. (1982) applied the FACE one-dimensional cloud model to conditions in Illinois to analyze the "dynamic seed-ability" of the region. In addition, they sampled the microphysical properties of the clouds. Using 10 summers of data, the median maximum seedability was 2–2.5 km. Prefrontal clouds appeared to be more susceptible to seeding having more liquid, less ice, and more vigor than the postfrontal clouds.

Multidimensional cloud modeling. Multidimensional, time-dependent cloud models can be used for the simulation of precipitation processes. These models simulate the formation, evolution, and fallout of precipitation from one level in the atmosphere to another. Their use for testing the dynamic-mode seeding techniques in extratropical

---

[2] High Plains Experiment—A Bureau of Reclamation field experiment in the late 1970s and early 1980s.

clouds has been sparse, although the seeding routines in two-dimensional, time-dependent (2DT) models have become relatively sophisticated compared to those techniques used in one- and three-dimensional models. Papers with pertinent results are those by Hsie et al. (1980), Orville and Chen (1982), and Kopp et al. (1983).

A feature of these 2DT models is that a conservation equation for the seeding agent is added to the basic set of equations and interactions of the seeding agent with the cloud and precipitation fields and motion fields are allowed for. Consequently, various cloud seeding methods can be tested: cloud-base seeding, cloud-top flare drops, ground seeding, solid carbon dioxide (CO_2) drops, and others. In addition, because all of the dependent variables are coupled, the physical effects of the seeding can be isolated, effects having to do with the latent heat of fusion release or the loading of precipitation. The several results relating to dynamic-mode seeding are described next.

The paper by Hsie et al. (1980) simulated the AgI seeding on three different sounding days, two from Miles City, Montana, and one from St. Louis. One of the Montana cases and the St. Louis case were situations with relatively large precipitation amounts; the other Montana case was a light raining cloud, but was the one that most of the seeding variations were tested on, such as seeding amount, location, and timing. The results indicated a rather small time window (about 6 min) in which to seed and relatively small sensitivity to seeding amount; this last fact is quite possibly due to the microphysical simulations which solve for mass of the cloud ice, but not the number concentrations. Rain and graupel/hail amounts were increased in the light rain case and the St. Louis warm base cloud case and decreased slightly in the large rain Montana case.

Although primarily a microphysical seeding simulation, the study of CO_2 seeding (Kopp et al., 1983) is mentioned here because the final effect of the seeding was due to a dynamic interaction of two cloud cells, in addition to the microphysical effect of earlier precipitation formation. The seeded cell reacted more vigorously to the cell merger because of the precipitation fallout from the first cell, which enhanced the boundary-layer convergence and the intensity of the merged cells.

Earlier work by Orville et al. (1980) had indicated that merged cells produced nearly twice as much precipitation, but that merger was relatively infrequent and depended on cloud cell spacing, intensity (buoyancy), and timing (or life stage) of the cells. Turpeinen (1982) saw no effect from timing or intensity on merger in his three-dimensional simulation of clouds on one day in GATE. No calculation was made by Turpeinen of rain increase from the mergers.

Other evidence of the effects of mergers and cloud interactions is given in Orville and Chen (1982). Their study was designed to quantitatively separate the effects of latent heat of fusion and precipitation loading due to AgI seed-

ing. A series of cloud model runs, with specific effects turned on or off and then the model results subtracted from each other, gave quantitative differences in vertical velocity, cloud liquid water mixing ratios, rain, etc., due to the early formation of ice by the seeding.

The dynamic effects were many: increases and decreases of vertical velocity, enhanced fallout, and new cell development in convergent outflows, among others. The stimulation of the seeded cell via seeding came about through the accretion and associated freezing of the supercooled cloud liquid by precipitating ice, not by the freezing of the individual cloud droplets by the seeding material. These accretional freezing effects had been evident in Koenig (1966), Cotton (1972), and Wisner et al. (1972). Orville and Chen (1982) referred to these effects as *indirect* freezing, and the freezing of cloud water by the seeding agent as a *direct* freezing effect.

The net results of the heavy seeding was to stimulate the first cell in a series of cells, but to decrease the overall precipitation of the storm because of the premature sweepout of the supercooled water in feeder clouds. One link in the dynamic-mode seeding chain of events had been broken, leading to opposite effects than those usually postulated. However, only one cell was seeded; in an operational or research project all cells may have been seeded and different results (of unknown sign at this time) would have occurred.

Regarding the various dynamic effects, these authors showed that in comparison to the unseeded cell, the peak domain-averaged kinetic energy increased by 100% if loading of all condensate was turned off, a 50% increase occurred if only precipitation loading was eliminated, and a 15% decrease occurred if the latent heat of fusion was omitted. Another important result was the illustration that microphysical changes could have significant effects on the total storm precipitation, given the same initial dynamic and thermodynamic conditions.

These authors found that, in the one HIPLEX model case studied, substantial accretional freezing occurred about 8 min after seeding. The release of latent heat of fusion caused a temperature increase of about 1°C and resulted in a 2.5 m s^{-1} vertical velocity increase compared with the unseeded case (about a 10% change, but this effect is certainly dependent on the sounding). The redistribution of the loading effect by the earlier precipitation formation resulted in significant changes in cloud cell interactions (see also Koenig and Murray, 1983).

Before leaving these cloud modeling discussions, it should be mentioned that important microphysical modeling of the rate of glaciation and the amount of ice needed to form efficient precipitation processes in various situations has been carried out by Jiusto (1973) and Lamb et al. (1981). Lamb et al. present new observational data from Florida cloud samples "that the primary microphysical role of seeding is the creation of many small ice particles that substitute for the secondary ice splinters of naturally induced glaciation. The aerodynamic capture of the splinters by the supercooled rain leads to the formation of new graupel particles and the rapid release of fusional heat." A relatively narrow time window was calculated for the heat effects from seeding-induced glaciation. More about these processes is given in a companion paper by Hallett (1981).

6.4. Statistical evidence

Statistical evidence relating to a single cloud seeding effect is evident in Fig. 6.3 from Simpson et al. (1967). Dennis (1980) lists a few other projects involving isolated clouds that appeared to give convincing statistical evidence (Bethwaite et al., 1966; Panel, 1973; Dennis et al., 1975; Weinstein and MacCready, 1969; McNaughton, 1973). The physical effects have been so clear that little doubt exists about man's ability to affect cloud growth and precipitation in some situations.

A problem arises, though, when larger area effects are tested in field experiments. Most of the statistical evidence comes from exploratory analyses, which are suggestive of rainfall increases on days with predicted seedability (i.e., a height increase due to ice-phase cloud seeding). However, because the analysis is exploratory and done after the fact, statistical significance is not possible.

There are at least two randomized experiments that attempted to test the dynamic seeding hypothesis over a broad area, the FACE and North Dakota projects. Both of these experiments have been reviewed by the Statistical Task Force of the Weather Modification Advisory Board (WMAB) (1978). Neither FACE-1 nor the North Dakota Pilot Project passed the criteria of a strong statistical test. Operational features and the exploratory nature of the projects precluded either project from qualifying as a "confirmatory" test of the dynamic seeding hypothesis.

However, FACE-2 was set up as a confirmatory experiment; results have just recently been published (Barnston et al., 1983; Woodley et al., 1982a, 1983; Meitin et al., 1984). FACE-1 results had indicated small differences in seed (S) and no-seed (NS) rainfalls in the 0–2 h period after seeding, substantial differences in the 2–5 h period after treatment with the S rainfall peaking 1.5 h after the NS rainfall, and small differences again after 5–6 h (Woodley et al., 1982b). Point estimates of effects were 1.30 and 1.43 in the floating target (FT) and total target (TT) areas. These and other rainfall data led the investigators to set up three levels of confirmation involving various combinations of seed, no-seed rainfall ratios in the 0–6 h period after treatment and double ratios of seed, no-seed rain in the 2–5 h period and 0–2 h period after treatment. Unfortunately, none of these ratios or double ratios achieved statistical significance, so the weakest level of confirmation has not been realized in FACE-2.

The authors reason that the failure may be due to a number of factors: "1) an unknown and possibly intermittent seeding effect; 2) inadequate predictor equations; and 3) a limited sample size."

Replication of the FACE-1 rainfall analyses was accomplished in FACE-2. A clear difference in the FACE-2 data when compared with the FACE-1 results was the earlier rainfall in the seeded versus the unseeded cases. In addition, the linear analysis in FACE-2 using the FACE-1 predictor variables (prewetness, model predicted rainfall, mean vector wind speed, and large square rainfall) yielded much smaller point estimate of treatment effects (1.06 and 1.09 in the floating target and the total target) and 95% confidence limits that bracketed 1.00, the no-effect value.

A single, large rainfall event on a no-seed day (29 July 1978) is blamed for much of the failure to verify the FACE-1 results, although the timing effect would still fail even if that one day were thrown out. In addition, Smith and Miller (1984) show that on a log-probability scale, the FACE-2 rainfall amounts are approximately lognormal, with the extreme day "in line" with a lognormal distribution.

Later exploratory analysis concerning extra-area effects has indicated that the mean S rainfall was greater than the mean NS rainfall in and downwind of the target, but less upwind of the target (Meitin et al., 1984). The rainfall over an extended area was estimated by radar, raingage, and satellite techniques. The rainfall differences were within 200 km of the FACE target area and within 8 h after treatment. The results do not have strong inferential support.

In extratropical regions, the WMAB (1978) analysis of the North Dakota Pilot Project (Dennis et al., 1975) cited several operational problems as weakening the results and causing possible bias. Many personnel knew the seeding decision, and meteorologists had to make a decision during operations whether to increase the seeding dose for hail suppression.

The Statistical Task Force of the Board concluded that the overall differences in rain for seeded and unseeded days were negligible and not significant. However, when the seeded and unseeded days were stratified as to whether the day was characterized by dynamic seedability (as defined above), then the rainfall differences approached significance. Days with low 500-mb temperatures ($-15°C$ to $-20°C$) gave less rainfall on seeded days. Results on hail were inconclusive.

Of the projects that have been analyzed after the fact, Project Whitetop was probably the largest and longest running. Original statistical analyses indicated a decrease in both rainfall and radar echo cover, with the largest decreases appearing 5 h after seeding (Flueck, 1971). Statistical support for the rain decrease was weak, for the radar echo cover moderate to strong.

The stratifications of the results indicated that on days with maximum echo tops below about 6 km (−10°C), these seeding effects were negative, but with little statistical support. With echo tops above 12 km (about −40°C), the seeding effects were strongly negative with strong statistical support. With echo tops between these extremes, the target–control differences were +38% to +57% in echo coverage with weak statistical support, and +68% to +100% in rain amount with strong statistical support. A recent paper by Ackerman and Sun (1985) indicates that this is just the cloud-top range in which dynamic seeding effects are strongest in Illinois. The authors used one-dimensional cloud models and aircraft observations of clouds in their study.

Braham (1979) speculates that some of the tall clouds (over 12 km) in Whitetop may have been caused by the seeding and were essentially overseeded, sweeping up all of the seeding material in the subcloud layer. Clouds of intermediate size were able to use the seeding material beneficially and were given moderate to low dosage rates, the clouds being isolated to some degree from the seeding material. Finally, because the area seeded was very large, clouds of all types and in all stages of "seedability" were seeded, leading to counterproductive results.

Braham concluded that Project Whitetop should be regarded more as an exploratory experiment, raising more questions than it answered. It should probably be repeated, he stated, from a cloud physics viewpoint, but ethically it is difficult to justify, in his opinion, rerunning an experiment that gave evidence of rainfall decrease in a rain sensitive area.

The statistical reanalysis of the Rapid Project (Dennis and Koscielski, 1969; Chang, 1976; Dennis et al., 1976) indicated that days with positive ΔH predicted (increased height) showed little seeding effect, but that on days in which Δw_{max} (of about 2 m s^{-1} or 10% increase) was predicted by a one-dimensional cloud model, strong positive seeding effects were indicated. Dennis et al. (1976) concluded that one effect of the cloud seeding was to intensify the cloud circulation *and* the necessary low-level convergence to permit the conversion of the clouds to functioning rain showers.

6.5. Critique and discussion of results

This review of dynamic-mode seeding has been a revelation to me; the several documented instances of dynamic effects of seeding individual clouds serve to reinforce a personal experience of cloud seeding with dry ice that appeared to cause rainfall from the only clouds seeded in a rather uniform cloud field. On the other hand, I am impressed with the immense difficulty of translating success involving individual clouds to success over an area of 10^4 km^2 or more, as in FACE and the NDPP, and doing so in economically significant amounts of water

increase. I am also impressed by the differences in approach and seeding techniques used in the tropical and extratropical regions.

Following is a critique of the various types of evidence and then a series of questions with discussion that arise from this review.

6.5.1. Critique

Physical evidence. Not having participated actively in the field experiments designed to test the dynamic-mode seeding responses, I am reluctant to criticize the operational methods, experimental design, and observational data obtained. However, certain advances made in the past few years indicate items that would have strengthened the early experiments. Foremost among these improvements are the new cloud physics aircraft instrumentation and associated computerized analysis methods and programs, along with digitized conventional and Doppler weather radar. Equipment to test the hypotheses was marginal in the early experiments.

In addition, the coverage of experimental units was probably inadequate in both space and time. Even now coverage problems are acute. Aircraft sample a tiny (but important) fraction of the cloud volume–time interval; radars sample only the precipitation stages of clouds. The cycling of the radar antenna is normally too long to follow the development of seeded and unseeded cells, particularly those radars used on the early experiments. Doppler radars were not available then for weather research. Consequently, the microphysical evolution and dynamic stimulation of clouds were difficult to document and are items of active research today in summer cumuli seeding operations.

There was also an inadequate theoretical knowledge of how clouds should react to seeding, the various rates and methods used. The many complexities of the cloud microphysical–dynamical interactions were not predictable in the steady-state (or bubble or plume) cloud models. Inadequacies still exist, and will always exist to some extent, but the strong time dependency and the cloud–cloud and cloud–environment interactions are becoming clarified as satellite, radar, and modeling studies proceed.

Finally, regarding field experiments, the inadequate knowledge of seeding material characteristics and the transport and diffusion of the seeding material have cast doubt on some results of the field studies. Too much emphasis has been placed on how many grams of seeding material were used and the rate of delivery, and not enough on the effectiveness and timeliness of the agent in forming ice crystals. The work of Finnegan and his associates (DeMott et al., 1983, and papers at Park City, Utah, AMS Weather Modification Conference, 1984) are clarifying some of these principles and helping to reconcile results in past projects that seemed to be in conflict.

The transport of the seeding material to individual

clouds is not easy, but methods have been developed over the years and appear effective. To seed large areas for dynamic effects is more difficult, and the optimum methods have probably not been found yet. Even after effective transport is accomplished, the "right" amount of diffusion cannot be guaranteed. It appears that the clouds diffuse and transport the agent in different proportions, depending on the growth stage of the cloud. Vigorous clouds transport and diffuse the seeding agent in thin ribbons, less active clouds in broader regions (Linkletter and Warburton, 1977). Certainly, more effort is needed to determine how to provide the proper amount of seeding material at the proper time and in the proper location. Probably all experiments have suffered dilution of their effects due to nonoptimum placement and quantity of seeding material.

Numerical models. This critique of numerical models applies only to their use as "test beds" for the dynamic-mode seeding concepts and is not a comprehensive criticism of cloud models. In hindsight, probably too much attention has been paid to one-dimensional, steady-state cloud models to answer questions about a highly time-dependent process. The application of the 1DSS models to simulate precipitation processes was beyond the models' capabilities, and, consequently, the use of the models to predict changes in precipitation due to ice-phase cloud seeding is rarely justified. One exception, as suggested above, might be the application of this type of model when good radar top versus precipitation curves are available and the 1DSS models are used to predict cloud top. This method has yet to be tested and suffers from our inability to specify the proper updraft radius, which is crucial for the 1DSS model results. Warner (1970) has detailed other criticisms of these models with reply by Simpson (1971), Cotton (1971), Weinstein (1971), and Warner (1971).

All models suffer from inadequate simulation of the microphysical processes, to a greater or lesser extent. The extreme complexities of the ice processes make it impractical to include all facets in coupled microphysical–dynamical, multidimensional, time-dependent models, so simplifications have to be made. The importance of crystal habit, particle density and terminal velocity, aggregation, accretion, ice nucleation, coalescence, and many other processes is still under active investigation, some of which was started because of the cloud seeding experiments. The early experiments could not adequately account for these processes in the theories and operational methods. A modicum of hope was relied upon then, and even now, that the various effects were occurring as postulated.

Related to these weaknesses in microphysical modeling is the simplistic modeling of the seeding process. The change of liquid to ice at predetermined temperature criteria is oversimplified, but is commonly done in the one-dimensional models. The icing of the cloud depends on many other items as well, such as updraft (condensation rate), ice nuclei amount and type, nuclei and crystal dispersion, liquid sweepout rate by the larger ice particles, etc. Improvement in seeding routines is made when the number of ice crystals is increased in time-dependent models. Better yet is the inclusion of equations to treat the seeding agents in the models. Only then can the time dependency of the seeding processes be examined and the importance, or even possibility, of freezing be determined. Of great importance in such models is when the freezing is effected and what influence this has on the model cloud development.

The dynamic-mode field experiments have lacked numerical modeling support over the entire scale of the experiment. Individual cloud elements are seeded and several scales of interactions are expected. Cloud models with 100 m or so grid intervals are needed to track the seeding agent and simulate the cloud-scale responses. Midlevel inversions require small grid intervals so that enough grid points are available to faithfully represent the atmospheric sounding. These inversions are important for inhibiting early convection and for allowing the atmosphere to store up energy for the later deep convection to occur. The dynamic-mode seeding concept depends, at times, on the ability of seeded clouds to break through the inversion, while unseeded clouds cannot. The computer resources required to simulate these conditions are formidable.

Coarser grids may be adequate for the downdraft interactions in the boundary layer, but much larger domains are then needed to include the cloud-scale effects on the mesoscale. Nested grids may help in future studies (Clark and Farley, 1984).

The cloud and precipitation interactions with the boundary layer require that active lower boundary surfaces be modeled. Heating and evaporation rates at the earth's surface should be included in the models that attempt to understand the dynamic effects of seeding and cloud interactions. In addition, mesoscale convergence–divergence values are important in some instances and need to be simulated in the cloud-scale models. Past experiments have lacked such modeling support, but future experiments would have available such models.

Statistical evidence. The statistical methods of some of the projects have been adequately reviewed and critiqued in Volume II of the WMAB (1978) report. The Statistical Task Force pointed out that randomization and "blindness" regarding the seeding decision are essential for research experiments, factors not always successfully accomplished in past experiments. Most of the experiments were exploratory in nature and, consequently, could not lead to definitive results. The task force discussed the wisdom of parallel explorations, after-the-fact analysis, the need for establishing better covariates, and the transition to a confirmatory experiment, among other important statistical considerations. Needless to say, there were many statistical deficiencies in the studies reviewed above.

What needs to be done for an adequate statistical experiment is difficult to accomplish and has rarely been accomplished in any existing, or prior, field experiment. The importance of specifying an adequate hypothesis and obtaining the measurements needed to test the hypothesis should not be underemphasized. The efforts in FACE-2 (Woodley et al., 1983) and in HIPLEX (Smith et al., 1984) are monumental. More work along such lines is needed for the dynamic-mode seeding concept.

This critique then leads to a number of questions concerning dynamic-mode seeding and related discussions.

6.5.2. Discussion and questions

1) *With the different rates of seeding in the different areas of the country, how can the dynamic response of the clouds to the seeding be as similar as they sometimes are?*

The Flagstaff and Florida cloud seeding methods were vastly different, primarily in rate of release. Both projects may have produced 100 or more ice crystals per liter active at $-5°$ to $-10°C$. But, in Arizona, the seeding was conducted from below cloud base with the aircraft generators releasing 120 to 240 g in 20 min, while in some of the Florida tests, ten to twenty 50 g flares per cloud were dropped into cloud turrets at about the $-10°C$ level in less than a minute, so there was a substantial difference in the amount of seeding material per unit time added to the clouds. Nevertheless, the indicated mean height changes of seeded clouds in both locations were in the 1 to 2 km range. How could the microphysics or dynamics of the storms and the thermodynamics of the environment be so different that they reacted in similar ways to two very different seeding rates? Some careful modeling and field experimentation might offer some clues. (One point to consider is that the models used to predict cloud top are steady state and have no way to react to the seeding rates.) Could the two seeding methods be interchanged and be just as effective? Finally, how many ice nuclei are enough, and at what rate must they be supplied in the two regions?

These questions and comments lead to the next basic question—a major one in the cloud physics sense.

2) *What are the dynamic and microphysical responses to the ice-phase seeding and the rate of this seeding?*

Can we document these responses and learn which ones will occur and at what place and at what time? Can we determine a seeding rate threshold for a dynamic effect?

These questions lead to a whole host of other questions concerning the cloud growth changes in both vertical and horizontal extent; the change in vertical velocity magnitude and profile (maximum value high or low in-cloud); the stimulation of inflow and downdrafts by the augmented updraft and the precipitation fallout; the frequency of explosive growth and its importance for a positive seeding effect; the influence of the dynamic-mode

seeding effects on precipitation development and on the rain and hail processes; the sensitivity of the responses to the atmosphere's thermodynamic and dynamic structure, and the ice-phase seeding rate and location; the pressure response of the seeded clouds; and on and on.

The answers to these questions are crucial to an understanding of the seeding effects and the construction of a viable hypothesis and appropriate experimental designs. Measurements reported in other sections of this review have given partial answers to some of these questions (such as cloud-top height changes) and have led to field experimental designs. The field experiments have, in turn, led to some of the other questions above for which we have incomplete data and theory at this time. For example, the focus on downdraft augmentation, new cell development, and cell merger directs attention to radar studies (Turpeinen and Yau, 1981; Westcott, 1984) and cloud modeling studies (Chen and Orville, 1980; Tripoli and Cotton, 1980; Orville et al., 1980; Turpeinen, 1982; Orville, 1984) to quantify the effects. Whether the results will lead to better field designs is yet to be determined. Some of the numerical studies that simulate seeding and time-dependent precipitation processes would indicate earlier rain from FACE seeded clouds, in agreement with FACE-2 results, but in disagreement with FACE-1 results.

A comment concerning the complexity of the hypotheses is in order here. The hypothesis illustrated in Table 6.1 is very complex. Such hypotheses are inherently unstable, as the failure of any one link will cause the succeeding links to fail or be suspect. An extremely thorough knowledge of the microphysical and dynamic interactions on the cloud and the mesoscale is necessary, in this instance, to design a viable experiment.

3) *Even though the seeding effects are convincing on individual clouds, why are the areawide effects so difficult to detect?*

The obvious answer to such a question is the compounding of effects that occur when an entire field of clouds is involved. Even if concentrated, directed seeding is used in place of broadcast seeding, the many interactions of clouds with one another via transfer of particles, anvil production, midlevel subsidence, and low-level outflow, etc., are so varied and so little understood it seems amazing that we have progressed as far as we have in detailing the dynamic effects of glaciogenic seeding.

4) *Can we identify those cases that give negative effects?*

It seems crucial that we be able to determine those cloud situations that would have no effect or would likely produce less rain if seeded. Limited modeling results show plausible scenarios for such effects, such as the premature sweepout of supercooled liquid from feeder cells. But this implies a very good measurement and modeling capability for operational use. The suggestive statistical results give simpler answers: seed only intermediate-size clouds and hope that the seeding does not stimulate too many of the clouds to the large category ($\geqslant 12$ km).

We need to learn more about the negative effect cases. After all, successful understanding of the precipitation processes and successful operational cloud seeding imply an ability to specify what will happen to a cloud or a field of clouds in any given situation. This is as true for precipitation forecasting as it is for cloud modification. As cloud physicists and meteorologists, we fail in one sense if we cannot make progress in clarifying the responses of clouds to various natural and artificial stimuli—we can and we have. The use of artificial stimuli gives us a way to prove to the rest of the scientific community that we know our science; this technique should be focused upon in the future.

5) *What are the interactions of clouds with the mesoscale, including the effects of stimulated cloud growth and of precipitation fallout on extra cell development, enhanced convergence, and the like?*

These are crucial questions that the companion paper by Fritsch will cover to some degree. The solutions will contribute to an understanding of the perplexing timing difference that occurs on seeded days compared with nonseeded days in FACE, different in the two phases of FACE. The timing differences are obvious in cloud modeling experiments, but the effect on the mesoscale of seeding many clouds and possibly enhancing merger are not known.

6) *Can we model the effects of ice-phase seeding for dynamic effects on both cloud scale and mesoscale and the interaction between the two scales?*

As indicated in the preceding sections on numerical models, some modeling effort has been devoted to these problems. However, much more effort is needed. More cloud physicists need to devote their attention to precipitation physics problems and the purposeful modification of the processes.

A major problem relates to Question 5; i.e., How can the details of the cloud-scale precipitation modeling be incorporated in mesoscale models utilizing grid intervals of 40 to 100 km? This might ultimately involve two-way interactions among the models. Until now, the downscale interaction has been simulated (Chang and Orville, 1973; Chen and Orville, 1980; Tripoli and Cotton, 1980) as has at least one model on the upscale influence of cloud seeding for dynamic effects (Fritsch and Chappell, 1981). Nevertheless, much more needs to be done, which will have far-reaching implications regarding precipitation forecasting, acid rain deposition, and other aspects of air pollution, as well as on weather modification.

7) *Do we have the tools and techniques to answer the questions posed above?*

Cotton will be considering these items in his paper. Certainly, the status of cloud physics aircraft instrumen-

tation, conventional and Doppler radar capabilities, ground measuring systems for wind and thermal profiles, statistical techniques, new seeding agents, and new, larger, faster computers brighten the prospects for attacking the problems in a more complete fashion than before.

8) *Can we learn about ice-phase seeding for dynamic effects on operational projects?*

State operational projects in North Dakota and Illinois are being planned, or are already planned, which hypothesize certain dynamic effects from the cloud seeding. Additional microphysical and cloud dynamic studies are being supported on these projects by NOAA and offer hope of clarifying some of the many issues raised by this type of cloud seeding.

6.6. Strategy

In order to decide on strategies for seeding clouds and for purposes of designing hypotheses, it is useful to distinguish, as well as we can, between the dynamic effects of seeding and the microphysical effects. I propose that possible dynamic effects include the following: 1) enhanced growth of the seeded cloud in both the vertical and horizontal, particularly if the horizontal expansion is caused by new cell growth; 2) enhanced vertical velocity within the seeded cloud cells; 3) enhanced midlevel inflow and stronger downdrafts; 4) increased numbers of cells formed by increased convergence in the boundary layer; 5) increased rain from individual stimulated cloud cells; 6) increased duration of rain; and 7) increased area of rain from the dynamically stimulated cells and region.

Possible microphysical effects would include the following: 1) creation of precipitation and radar echoes in previously observed nonechoing clouds without any of the above dynamic effects; 2) increased numbers of precipitating cells independent of the boundary-layer flow; 3) the clearing of cloudy regions via precipitation fallout; 4) the decrease in size of graupel or hail particles with little change in storm flow; and 5) the increase in rain duration, rain area, and rain amount from the seeded cells.

For dynamic-mode seeding, I see a need for a strong modeling effort at the cloud scale and mesoscale using data from previous field experiments to validate the models. Various seeding strategies should be tested and the primary signals noted. We should attempt to discover new seeding scenarios and validate old ones, if possible, with time-dependent models of the cloud and precipitation processes.

At the same time, field measurements of crucial cloud characteristics should be attempted. Chemical tracer experiments can help in determining airflow in clouds and the transport of seeding material. Not much research field effort is occurring; consequently, it would be wise to make measurements on the federal/state operational weather modification programs.

It is important that precipitation simulation/mesoscale models be developed to predict precipitation over an area. Better covariates are needed for the field experiments and might be developed in such work. Coordination of conventional radar, satellite, and modeling efforts is needed to improve both observations and modeling on the mesoscale. Cooperative efforts with the STORM[3] program and other multiagency projects should be explored.

We need to determine a number of field experimental designs, including hypotheses, physical measurements, numerical models, and statistical methods required to evaluate the projects. We should concentrate on the whole spectrum of events: precipitation increase, redistribution, and decrease. Cloud seeding research experiments, concentrating on cumulus clouds, should be reestablished.

On any and all cloud or mesoscale research field experiments, we should attempt to learn as much as possible about the natural precipitation processes, applying seeding methods, when appropriate, to test our understanding of the processes. We should look for natural seeding situations in the data and analyze for glaciation effects. Enough long-term support must be provided if definite answers are to result for this most difficult, but most rewarding, task of successfully developing ice-phase seeding of cumulus clouds.

Acknowledgments. I thank Ms. Carol Vande Bossche, Mrs. Sandra Palmer, and Mrs. Joie Robinson for their help in typing and preparing the manuscript.

REFERENCES

Ackerman, B., and D. B. Johnson, 1979: Physical assessment of seeding opportunities in the U.S. midwest. *Preprints Seventh Conf. on Inadvertent and Planned Weather Modification,* Banff, Amer. Meteor. Soc., 152–153.

——, and R-Y. Sun, 1985: Investigation of two 1-D cloud models. *J. Climate Appl. Meteor.,* **24,** 617–628.

Barnston, A. G., W. L. Woodley, J. A. Flueck and M. H. Brown, 1983: The Florida Area Cumulus Experiment's second phase (FACE-2). Part I: The experimental design, implementation and basic data. *J. Climate Appl. Meteor.,* **22,** 1504–1528.

Bethwaite, F. D., E. J. Smith, J. A. Warburton and K. J. Heffernan, 1966: Effects of seeding isolated cumulus clouds with silver iodide. *J. Appl. Meteor.,* **5,** 513–520.

Braham, R., Jr., 1979: Field experimentation in weather modification. *J. Amer. Stat. Assoc.,* **74,** 57–104.

Chang, L. P., 1976: Reevaluation of Rapid Project data. M.S. thesis, Department of Meteorology, South Dakota School of Mines and Technology, 58 pp.

Chang, S., and H. D. Orville, 1973: Large-scale convergence in a numerical cloud model. *J. Atmos. Sci.,* **30,** 947–950.

Chen, C. H., and H. D. Orville, 1980: Effects of mesoscale convergence on cloud convection. *J. Appl. Meteor.,* **19,** 256–274.

Chen, P. C., M. E. Humbert, T. B. Smith, D. M. Takeuchi and G. J.

Mulvey, 1979: Radar echo organization in the mesoscale environment and a design concept for mesoscale cloud seeding experiments. *Preprints Seventh Conf. on Inadvertent and Planned Weather Modification,* Banff, Amer. Meteor. Soc., 171–172.

Chen, W. T., 1982: Numerical simulation of cloud seeding effects on a maritime tropical cloud. M.S. thesis, Department of Meteorology, South Dakota School of Mines and Technology, 111 pp.

Clark, T. L., and R. D. Farley, 1984: Severe downslope windstorm calculations in two and three spatial dimensions using anelastic interactive grid nesting: A possible mechanism for gustiness. *J. Atmos. Sci.,* **41,** 329–350.

Cotton, W. R., 1971: Comments on "Steady-state one-dimensional models of cumulus convection." *J. Atmos. Sci.,* **28,** 647–648.

——, 1972: Numerical simulation of precipitation development in a supercooled cumuli—Part II. *Mon. Wea. Rev.,* **100,** 764–784.

——, and A. Boulanger, 1975: On the variability of "dynamic seedability" as a function of time and location over south Florida: Part I. Spatial variability. *J. Appl. Meteor.,* **14,** 710–717.

Cunning, J. B., and M. DeMaria, 1981: Comments on "Downdrafts as linkages in dynamic cumulus seeding effects." *J. Appl. Meteor.,* **20,** 1081–1084.

——, R. I. Sax and R. L. Holle, 1979: Morphology of seeded clouds as determined from triple-Doppler radar—a case study. *Preprints Seventh Conf. on Inadvertent and Planned Weather Modification,* Banff, Amer. Meteor. Soc., 142–143.

——, R. L. Holle, P. T. Gannon and A. I. Watson, 1982: Convective evolution and merger in the FACE experimental area: Mesoscale convection and boundary layer interactions. *J. Appl. Meteor.,* **21,** 953–977.

DeMott, P. J., W. C. Finnegan and L. O. Grant, 1983: An application of chemical kinetic theory and methodology to characterize the ice nucleating properties of aerosols used for weather modification. *J. Climate Appl. Meteor.,* **22,** 1190–1203.

Dennis, A. S., 1980: *Weather Modification by Cloud Seeding.* Academic Press, 267 pp.

——, and A. Koscielski, 1969: Results of a randomized cloud seeding experiment in South Dakota. *J. Appl. Meteor.,* **8,** 556–565.

——, and M. R. Schock, 1971: Evidence of dynamic effects in cloud seeding experiments in South Dakota. *J. Appl. Meteor.,* **10,** 1180–1184.

——, and A. Koscielski, 1972: Height and temperature of first echoes in unseeded and seeded convective clouds in South Dakota. *J. Appl. Meteor.,* **11,** 994–1000.

——, D. E. Cain, J. H. Hirsch and P. L. Smith, Jr., 1975: Analysis of radar observations of a randomized cloud seeding experiment. *J. Appl. Meteor.,* **14,** 897–908.

——, J. H. Hirsch and L. P. Chang, 1976: The role of low-level convergence in controlling convective rainfall and its possible modification by seeding. *Preprints Int. Conf. on Weather Modification,* Boulder, 49–54.

Flueck, J. A., 1971: Statistical analyses of the ground level precipitation data. Project Whitetop, Part V. Department of Geophysical Sciences, University of Chicago.

Fritsch, J. M., and C. F. Chappell, 1981: Preliminary numerical tests of the modification of mesoscale convective systems. *J. Appl. Meteor.,* **20,** 910–921.

Fukuta, N., 1973: Thermodynamics of cloud glaciation. *J. Atmos. Sci.,* **30,** 1645–1649.

Hallett, J., 1981: Ice crystal evolution in Florida summer cumuli following AgI seeding. *Preprints Eighth Conf. on Inadvertent and Planned Weather Modification,* Reno, Amer. Meteor. Soc., 114–115.

Hirsch, J. H., 1971: Computer modeling of cumulus clouds during Project Cloud Catcher. Rep. 71-7, Institute of Atmospheric Sciences, South Dakota School of Mines and Technology, 61 pp.

——, and C. L. Schock, 1968: Cumulus cloud characteristics over western South Dakota. *J. Appl. Meteor.,* **7,** 882–885.

---

[3] Storm-scale Operational and Research Meteorology.

Hsie, E. Y., R. D. Farley and H. D. Orville, 1980: Numerical simulation of ice-phase convective cloud seeding. *J. Appl. Meteor.,* **19,** 950–977.

Jiusto, J. E., 1973: Seeding requirements for rapid glaciation or simulation of a mixed phase cloud. *WMO/IAMAP Scientific Conf. on Weather Modification,* Tashkent, WMO/IAMAP 169–178.

Koenig, L. R., 1966: Numerical test of the validity of the drop-freezing/splintering hypothesis of cloud glaciation. *J. Atmos. Sci.,* **23,** 726–740.

——, and F. W. Murray, 1976: Ice-bearing cumulus cloud evolution: Numerical simulation and general comparison against observations. *J. Appl. Meteor.,* **15,** 747–762.

——, and ——, 1983: Theoretical experiments on cumulus dynamics. *J. Atmos. Sci.,* **40,** 1241–1256.

Kopp, F. J., H. D. Orville, R. D. Farley and J. H. Hirsch, 1983: Numerical simulation of dry ice cloud seeding experiments. *J. Climate Appl. Meteor.,* **22,** 1542–1556.

Kraus, E. B., and P. Squires, 1947: Experiments on the stimulation of clouds to produce rain. *Nature,* **159,** 489–491.

Lamb, D., R. I. Sax and J. Hallett, 1981: Mechanistic limitations to the release of latent heat during the natural and artificial glaciation of deep convective clouds. *Quart. J. Roy. Meteor. Soc.,* **107,** 935–954.

Levine, J., 1959: Spherical vortex theory of bubble-like motion in cumulus clouds. *J. Meteor.,* **16,** 653–662.

Levy, G., and W. R. Cotton, 1984: A numerical investigation of mechanisms linking glaciation of the ice-phase to the boundary layer. *J. Climate Appl. Meteor.,* **23,** 1505–1519.

Linkletter, G. O., and J. A. Warburton, 1977: An assessment of NHRE hail suppression seeding technology based on silver analysis. *J. Appl. Meteor.,* **16,** 1332–1348.

McCarthy, J., 1972: Computer model determination of convective cloud seeded growth using Project Whitetop data. *J. Appl. Meteor.,* **11,** 818–822.

MacCready, P. B., and R. F. Skutt, 1967: Cloud buoyancy increase due to seeding. *J. Appl. Meteor.,* **6,** 207–210.

McNaughton, D. L., 1973: Seeding single cumulus clouds in Rhodesia with silver iodide: 1968–69. *J. Wea. Modif.,* **5**(1) 88–102.

Matthews, D. A., 1981: Natural variability of thermodynamic features affecting convective cloud growth and dynamic seeding: A comparative summary of three High Plains sites from 1975 to 1977. *J. Appl. Meteor.,* **20,** 971–996.

——, and B. A. Silverman, 1980: Sensitivity of convective cloud growth to mesoscale lifting: A numerical analysis of mesoscale convective triggering. *Mon. Wea. Rev.,* **108,** 1056–1064.

Meitin, J. G., W. L. Woodley and J. A. Flueck, 1984: Exploration of extended-area treatment effects in FACE-2 using satellite imagery. *J. Climate Appl. Meteor.,* **23,** 63–83.

Miller, J. R., Jr., A. S. Dennis, D. E. Cain and J. H. Hirsch, 1975: Precipitation management potential in western North Dakota as revealed by radar echoes and cloud model studies. Rep. 75-4, Institute of Atmospheric Sciences, South Dakota School of Mines and Technology, 163 pp.

Murray, F. W., and L. R. Koenig, 1972: Numerical experiment on the relation between microphysics and dynamics in cumulus convection. *Mon. Wea. Rev.,* **100,** 717–732.

Orville, H. D., 1984: Comments on "Cloud interactions and merging on day 261 of GATE." *Mon. Wea. Rev.,* **112,** 387–388.

——, and K. G. Hubbard, 1973: On the freezing of liquid water in a cloud. *J. Appl. Meteor.,* **12,** 671–676.

——, and J. M. Chen, 1982: Effects of cloud seeding, latent heat of fusion, and condensate loading on cloud dynamics and precipitation evolution: A numerical study. *J. Atmos. Sci.,* **39,** 2807–2827.

——, F. J. Kopp and C. G. Myers, 1975: The dynamics and thermodynamics of precipitation loading. *Pure Appl. Geophys.,* **113,** 983–1004.

——, Y. H. Kuo, R. D. Farley and C. S. Hwang, 1980: Numerical simulation of cloud interactions. *J. Rech. Atmos.,* **14**(3–4), 499–516.

——, R. D. Farley and J. H. Hirsch, 1984: Some surprising results from simulated seeding of stratiform-type clouds. *J. Climate Appl. Meteor.,* **23,** 1585–1600.

Panel on Weather and Climate Modification (T. F. Malone, Chairman), 1973: *Weather and Climate Modification: Problems and Progress.* National Academy of Science, Washington, DC, 258 pp.

Ruskin, R. E., 1967: Measurements of water-ice budget changes at −5°C in AgI-seeded tropical cumulus. *J. Appl. Meteor.,* **6,** 72–81.

Saunders, P. M., 1957: The thermodynamics of saturated air: A contribution to the classical theory. *Quart. J. Roy. Meteor. Soc.,* **83,** 342–350.

Sax, R. I., 1974: On the microphysical differences between populations of seeded vs. non-seeded Florida cumuli. *Preprints Fourth Conf. on Weather Modification,* Ft. Lauderdale, Amer. Meteor. Soc., 65–68.

——, 1976: Microphysical response of Florida cumuli to AgI seeding. *Second WMO Scientific Conf. on Weather Modification,* Boulder, WMO 109–116.

——, and V. W. Keller, 1980: Water-ice and water-updraft relationships near −10°C within populations of Florida cumuli. *J. Appl. Meteor.,* **19,** 505–514.

——, J. Thomas and M. Bonebrake, 1979: Ice evolution within seeded and non-seeded Florida cumuli. *J. Appl. Meteor.,* **18,** 203–214.

Silverman, B. A., D. A. Matthews, L. D. Nelson, H. D. Orville, F. J. Kopp and R. D. Farley, 1976: Comparisons of cloud model predictions: A case study analysis of one- and two-dimensional models. *Preprints Int. Conf. on Cloud Physics,* Boulder, Amer. Meteor. Soc., 343–348.

Simpson, J., 1967: Photographic and radar study of the Stormfury 5 August 1965 seeded cloud. *J. Appl. Meteor.,* **6,** 82–87.

——, 1971: On cumulus entrainment and one-dimensional models. *J. Atmos. Sci.,* **28,** 449–455.

——, 1980: Downdrafts as linkages in dynamic cumulus seeding effects. *J. Appl. Meteor.,* **19,** 477–487.

——, and V. Wiggert, 1969: Models of precipitating cumulus towers. *Mon. Wea. Rev.,* **97,** 471–489.

——, and ——, 1971: 1968 Florida seeding experiment: Numerical model results. *Mon. Wea. Rev.,* **102,** 115–139.

——, and A. S. Dennis, 1974: Cumulus clouds and their modification. *Weather and Climate Modification,* Wiley and Sons, 842 pp.

——, and H. J. Cooper, 1981: Reply to Cunning and DeMaria. *J. Appl. Meteor.,* **20,** 1085–1088.

——, R. H. Simpson, D. A. Andrews and M. A. Eaton, 1965: Experimental cumulus dynamics. *Rev. Geophys.,* **3,** 387–396.

——, G. W. Brier and R. H. Simpson, 1967: Stormfury cumulus seeding experiment 1965: Statistical analysis and main results. *J. Atmos. Sci.,* **24,** 508–521.

——, N. E. Westcott, R. J. Clerman and R. A. Pielke, 1980: On cumulus mergers. *Arch. Meteor. Geophys. Bioklim.,* **A29,** 1–40.

Smith, E. J., 1974: Cloud seeding in Australia. *Weather and Climate Modification,* Wiley and Sons, 842 pp.

Smith, P. L., and J. R. Miller, Jr., 1984: Comment on "The Florida Area Cumulus Experiment's second phase (FACE-2). Part II: Replicated and confirmatory analyses." *J. Climate Appl. Meteor.,* **23,** 1484–1485.

Smith, P. L., and E. J. Smith, 1949: The artificial stimulation of precipitation by means of dry ice. *Aust. J. Sci. Res.,* **A2,** p. 232.

——, and J. S. Turner, 1962: An entraining jet model for cumulonimbus updraughts. *Tellus,* **14,** 422–434.

Tripoli, G. J., and W. R. Cotton, 1980: A numerical investigation of several factors contributing to the observed variable intensity of deep convection over south Florida. *J. Appl. Meteor.,* **19,** 1037–1063.

Turner, J. S., 1962: The starting plume in neutral surroundings. *J. Fluid Mech.,* **13,** 356–368.

Turpeninen, O., 1982: Cloud interactions and merging on day 261 of GATE. *Mon. Wea. Rev.,* **110,** 1238–1254.

——, and M. K. Yau, 1981: Comparisons of results from a three-dimensional cloud model with statistics of radar echoes on day 261 of GATE. *Mon. Wea. Rev.,* **109,** 1495–1511.

Warner, J., 1970: On steady-state one-dimensional models of cumulus convection. *J. Atmos. Sci.,* **27,** 1035–1040.

——, 1971: Reply to "On steady-state one-dimensional models of cumulus convection." *J. Atmos. Sci.,* **28,** 651–652.

Weather Modification Advisory Board, 1978: The management of weather resources. Vol. II: The role of statistics in weather resources management. Report to the Secretary of Commerce, Washington, DC, 94 pp.

Weinstein, A. I., 1970: A numerical model of cumulus dynamics and microphysics. *J. Atmos. Sci.,* **27,** 246–255.

——, 1971: Comments on "Steady-state one-dimensional models of cumulus convection." *J. Atmos. Sci.,* **28,** 648–651.

——, 1972: Ice-phase seeding potential for cumulus cloud modification in the western United States. *J. Appl. Meteor.,* **11,** 202–210.

——, and L. G. Davis, 1968: A parameterized numerical model of cumulus convection. Rep. No. 11, Dept. of Meteorology, Pennsylvania State University, NSF Grant GA-777.

——, and P. B. MacCready, 1969: An isolated cumulus cloud modification project. *J. Appl. Meteor.,* **8,** 936–947.

——, and D. M. Takeuchi, 1970: Observations of ice crystals in a cumulus cloud seeded by vertical-fall pyrotechnic. *J. Appl. Meteor.,* **9,** 265–268.

Westcott, N., 1984: A historical perspective on cloud mergers. *Bull. Amer. Meteor. Soc.,* **65,** 219–226.

Wiggert, V., R. I. Sax and R. L. Holle, 1982: On the modification potential of Illinois summertime convective clouds with comparisons to Florida and FACE observations. *J. Appl. Meteor.,* **21,** 1293–1322.

Wisner, C., H. D. Orville and C. Myers, 1972: A numerical model of a hail-bearing cloud. *J. Atmos. Sci.,* **29,** 1160–1181.

Woodley, W. L., 1964: Cloud growth relative to seeding. *J. Rech. Atmos.,* **1,** 73–79.

——, 1970: Precipitation results from a pyrotechnic cumulus seeding experiment. *J. Appl. Meteor.,* **9,** 242–257.

——, and R. I. Sax, 1976: The Florida Area Cumulus Experiment: Rationale, design, procedures, results, and future course. NOAA Tech Rep. ERL 354-WMP06. 206 pp.

——, J. A. Fleuck, R. Biondini, R. I. Sax, J. Simpson and A. Gagin, 1982a: Clarification of confirmation in the FACE-2 experiment. *Bull. Amer. Meteor. Soc.,* **63,** 273–276.

——, J. Jordan, A. Barnston, J. Simpson, R. Biondini and J. Flueck, 1982b: Rainfall results of the Florida Area Cumulus Experiment, 1970–76. *J. Appl. Meteor.,* **21,** 139–164.

——, A. G. Barnston, J. A. Flueck and R. Biondini, 1983: The Florida Area Cumulus Experiment's second phase (FACE-2). Part II: Replicated and confirmatory analyses. *J. Climate Appl. Meteor.,* **22,** 1529–1540.

CHAPTER 7

Evaluation of "Static" and "Dynamic" Seeding Concepts through Analyses of Israeli II and FACE-2 Experiments

ABRAHAM GAGIN

Department of Atmospheric Sciences, Hebrew University of Jerusalem, Israel

ABSTRACT

In static-mode seeding two assumptions are usually made: a deficiency in concentrations of natural ice crystals is the reason for delay, or even failure, of precipitation formation in certain cloud conditions; and, moderate increases in ice crystal concentrations, obtained by glaciogenic seeding of such clouds, will result in rainfall enhancement either by making the already existing process of rain formation more effective or by inducing precipitation formation in clouds that otherwise would not have precipitated naturally.

The basic assumption behind seeding for dynamic effects is that increased cloud buoyancy, achieved through conversion of supercooled water to ice by seeding, will cause an increase in cloud depth, which in turn will result in stronger rainfall intensities, areas and durations.

These basic assumptions are examined in terms of physical and statistical analyses of data from Israeli II (a static-mode seeding project) and FACE-2 (a dynamic-mode seeding project).

7.1. Introduction

In designing cloud seeding experiments, due consideration should be given to the need for obtaining data that will allow the testing of the basic physical hypotheses underlying the experiments. While it is important to record the effects of seeding on rainfall at the ground, the statistical evaluation of this parameter alone cannot constitute an acceptable result of a successful seeding effect.

In attempting to provide substantiation of the underlying physical hypotheses, it is possible to adopt one of two general approaches:

(i) That which requires each step in the chain of events leading to precipitation formation, in both the treated and nontreated clouds, to be specified in advance within the framework of a detailed conceptual model and later to be verified by quantitative observations.

(ii) That which defines the conditions that are conducive to positive seeding effects on the basis of a general conceptual model, resting on previous studies of cloud microphysics and dynamics of the clouds and cloud systems involved, and analyzes the results for their physical plausibility within stratifications of the experimental data.

The first of those two approaches is used to establish the physical basis for rainfall enhancement techniques and hence is very basic, expensive and time consuming. The second is obviously more risky. It requires, however, less in the way of facilities and highly skilled human resources and, under favorable conditions, may provide quicker answers at a reduced cost. The crucial factor in the latter approach is the complexity of making sound physical hypotheses on the basis of circumstantial scientific evidence only. It would be appropriate to point out that the second approach cannot be a substitute for the first, but, rather, it should be used to facilitate the transfer of knowledge gained in the conduct of the basic studies by the first approach to similar rainfall regimes.

HIPLEX-I (Smith et al., 1984) is the only experiment for rainfall enhancement thus far to be designed and conducted according to the first of these approaches. Unfortunately, it was terminated prematurely and hence did not completely fulfill its objectives. Nevertheless, it provided very important quantitative data on the early stages of precipitation formation. The present report utilizes data obtained in two experiments (Israeli II and FACE-2), both of which used the second approach in an attempt to check the validity of some of the basic hypotheses underlying cloud seeding techniques for increasing rainfall either by producing microphysical effects in the treated clouds or by triggering dynamic changes in them.

7.2. Evaluation of the basic postulates of "static" seeding as inferred from analyses of Israeli II

7.2.1. General remarks

Cloud seeding, aimed at rainfall enhancement through the production of static effects on cloud microstructure, was carried out in two consecutive, long-term, randomized experiments referred to as Israeli I and II. Daily rainfall, averaged over the entire target area, was found to be increased, under seeding, by about 15% in Israeli I (Gabriel, 1970; Gagin and Neumann, 1974) and by about 13% in Israeli II (Gagin and Neumann, 1981). These results were significant at less than 5%. In subareas of the target, roughly the same distance from the seeding line, positive effects of 24% and 18% were found. In these latter cases the statistical significance levels were less than 3%.

At the outset of each of these experiments (Neumann et al., 1967), the following two assumptions were made:

• The so-called ice-crystal mechanism is the most efficient precipitation-forming process for the given regional meteorological conditions. It is assumed that there are periods when a deficiency of natural ice nuclei results in a delay, or even failure, of precipitation initiation.

• AgI smoke was selected as the seeding agent. This implied a hope for achieving rain stimulation by ice-crystal formation, either through making the existing process of rain formation more effective or through inducing precipitation formation in clouds that otherwise would not have precipitated naturally.

The cumulus clouds treated in the Israeli experiments were organized along cold fronts, in postfrontal bands, or in Bénard-cell patterns. Physical studies accompanying the Israeli experiments (summarized by Gagin, 1981) showed that winter Eastern Mediterranean clouds are significantly continental in nature. These clouds are totally deficient of large drops (>100 μm) at all altitudes, a fact indicating the absence of an efficient collision coalescence process (Gagin and Neumann, 1974). Furthermore, it was found (Gagin, 1975) that in all probability precipitation elements in these clouds form by the combined processes of ice-crystal nucleation and growth by vapor deposition, followed by accretion. The resulting graupel particles were found to have low-density porous structures, at least initially. Their concentrations were found to be roughly an order of magnitude less than that of the ice crystals, just below the tops of clouds having summit temperatures warmer than −21°C. Both ice crystals and graupel concentrations were found to have a clear dependence on cloud summit temperature. Graupel concentrations fell below the detection limit of our instruments (10^{-3} L^{-1}) for clouds with top temperatures warmer than −11°C. More recent studies (Gagin, 1980), utilizing a vertically pointing radar and a disdrometer, found that raindrop concentrations on the ground (roughly 1 km below the 0°C isotherm) depend on cloud-top temperatures in a manner resembling that of the dependence of graupel concentration on summit temperatures.

Routine observations of cloud-top heights and temperatures were made to permit us to 1) test some of our hypotheses with regard to the physics of precipitation processes and the effect of seeding on them, 2) break down the overall results in order to detect whether there is some physically systematic set of results that would provide

physical plausibility to the statistical results, and 3) give some indication of the relative efficiency of cloud seeding on clouds of different properties.

The determination of cloud-top height and temperature distribution was done as follows. The project's C-band radar, located near Tel-Aviv, was operated continuously on experimental days to yield echo-height distributions in both control and target areas. These areas are roughly at the same distance (80–125 km) from the radar. Echo-height distributions were obtained once every hour as long as there were echoes in the experimental area. The RHI scan mode was used manually to determine the height of every echo. Comparisons of radar echo top-heights with aircraft, or vertically pointing X-band radar observations on the same clouds, show that the inaccuracies involved in the radar determination are within ±0.5 km.

Daily echo-height distributions, accumulated from the hourly data, were transformed into temperature distributions through the use of nearby radiosonde observations. These distributions were shown to have different characteristics on days with different rainfall amounts (Gagin and Neumann, 1974; Gagin, 1981). Days with larger rainfall amounts were found to be characterized by clouds with higher (colder) clouds. Cloud-base temperatures remained essentially the same on all days.

The modal value of these distributions is considered the most representative characteristic of the cloud populations. In recognition of the inherent inaccuracies of radar echo-height measurements, the grouping of days having similar modal values was done with interval values of about 1 km or a temperature interval of about 6°C.

One of the salient results of Israeli II (Gagin and Neumann, 1981) is that the treated clouds responded to seeding in a manner that depended systematically on cloud-top temperature. On days when the modal values of the cloud-top temperature distributions were −15° to −21°C, maximum positive effects of about 46% were observed, a result statistically significant at less than 1%. On the other days, when the modal values of cloud-top temperatures were either warmer than −10°C or colder than −21°C, the effects of seeding were found to be either nil or insignificant. These experimental results were anticipated by theoretical studies that actually predicted these effects (Neumann et al., 1967; Gagin and Neumann, 1974; Gagin and Steinhorn, 1974). While both the direct results of the physical studies (Gagin, 1975) and the results of the statistical analyses of the seeding experiments as summarized above provide a reasonably acceptable corroboration of the first of the above-stated assumptions, the following summary of recent results will attempt to relate some findings in support of the second of these assumptions.

The present study provides further clues for understanding how our static-mode seeding resulted in rainfall increases. It uses recording raingage data to study seeding effects on the daily values of duration, intensity and num-

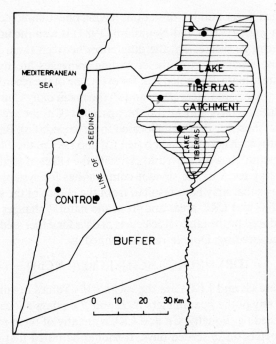

FIG. 7.1. Map showing the location of the recording raingages, denoted by full circles (4 in the Control and 6 in the LKC). The hatched area defines the LKC. Also shown is the line of aircraft seeding.

ber of rain periods. Definite and statistically significant results relating seeding to these factors will provide support to the validity of the second of the assumptions stated above. Thus, an increase in the number of rain events following seeding can be interpreted as a result of rain initiation in cloud that would not have precipitated naturally. An increase in rain duration, beyond that resulting from the increased number of rain events, is an indication that seeding affected the efficiency of already-precipitating clouds.

7.2.2. The effect of seeding on the components of rainfall

A summary of the design of Israeli II. Since this study makes use of data from Israeli II, it is appropriate to sum-

marize some of the salient features of the design of this experiment. The Israeli II experiment was conducted during the six winter rainfall seasons of 1969–75. Its purpose was to examine the possibility of enhancing rainfall, by static cloud seeding, over the catchment areas of Lake Kinneret (Lake Tiberias). The core area of the target was known as the LKC (Fig. 7.1). The test hypotheses, the criteria for defining the conditions suitable for seeding, and the details of the design of Israeli II are given in Gagin and Neumann (1981). The total number of experiment days was 388. Of these, 209 were randomly allocated to seeding; the rest were unseeded. Allocations were carried out, at probability of 0.5 in both cases, in such a manner that they were independent from day to day.

A table of random allocation of dates, without blocking, was made before the beginning of each season by the project statistician. This list was known to all involved in the operations of the experiment, but not to the rainfall observers of the Israeli Meteorological Service, who were totally independent of the experiment and therefore can be regarded as "blind" to the seeding operations. However, it should be noted that the experiment design required that, in every case where a day allocated to seeding was not actually seeded (e.g., because no suitable clouds developed, or for any other reason), the days were counted as seeded days. While this requirement obviously resulted in a reduction of the overall positive result of the experiment, it was considered an efficient way to remove any selection bias.

We reserved the Mediterranean coastal area west of the target as a control area (C). In Israel, on all days of rain, the winds at cloud-base level and above always have a westerly component.

Data reduction and processing of measurements taken by recording raingages. Figure 7.2 shows a typical chart of recorded rainfall in this experiment. The various rain events on this day are numbered consecutively. The integrated duration of rainfall for all of these 14 rain events is 164 minutes, and the total amount of rainfall precipitated at this station on this day is 14.5 mm. Since most, if not all, rainfall in Israel is produced by cumuliform

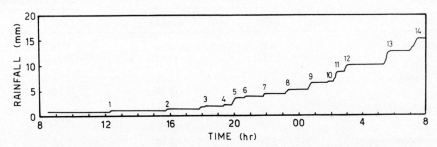

FIG. 7.2. An example of a chart of a recording raingage, depicting the determination of the daily number of rain events (P), the daily duration of rain (D), and the total amount of daily rainfall (R), in Jerusalem on 14 March 1984. On that day, $P = 14$, $D = 2.74$ h, and $R = 14.5$ mm.

clouds, the distinction between various rain events is fairly straightforward and rather simple.

In the analyses described below, data from all 388 days of Israeli II are used. Control area rainfall characteristics have been determined from four rain-recording stations. For the LKC we have used six such stations. Their locations are shown in Fig. 7.1. In view of the exceptionally high correlations of daily rainfall between stations within any one of the experimental areas and between the control and the LKC (Gagin and Neumann, 1981), and considering the magnitude of the effort required for data reduction for 388 days, it was felt that a relatively small number of stations would suffice to provide reliable daily means of rain properties for the two areas in question, i.e., the Control and the LKC. As stated above, the LKC was the core target of Israeli II; it was therefore decided to restrict this study to an evaluation of the results in the LKC only.

It is appropriate to note that while there are some inaccuracies in reading rain duration from strip-chart records, the application of ratios minimizes their impact in the calculation of seeding effects. This is also true in the case of the systematic differences between the rain components of the target and the controls, which are expressed as double ratios.

7.2.3. Results

Overall statistical analyses. Table 7.1 gives the daily mean value of rainfall (R) in millimeters, its duration (D) in hours, and the number of rain events (P) in the Control (C) and LKC under conditions of seeding and no-seeding. While the formal results of the Israeli experiments are based on days allocated to seeding independently of whether the clouds on these days were seeded, we also included in Table 7.1 the means of these three parameters on days allocated to seeding and actually seeded. For comparison, the daily mean rainfall, as obtained from a

much larger number of stations having only nonrecording raingages (Gagin and Neumann, 1981) is also included. As can be seen easily, the differences between them and those obtained from the recording raingages are really quite minute. It is noted that for every one of the recording raingage parameters (R, D and P) the mean daily Control values on days allocated to be unseeded (Cn) are greater than those for the days allocated to be seeded (Cs). Based on the high correlation between Cn and LKCn, the "double ratio" has been used to estimate the effect of seeding on any test variable since it compensates for systematic effects that may have resulted from the choice of the specific C and LKC areas and from the natural changes introduced by the random selection of days for either seeding or no-seeding. Double ratio is defined as

$$(DR)_j = [(LKCs)_j/(Cs)_j] \cdot [(Cn)_j/(LKCn)_j] \quad (7.1)$$

where Cs and LKCs are the mean daily values obtained for any of the j parameters R, D and P on days allocated to seeding. Equally, Cn and LKCn are any of these j parameters on unseeded days. It should be noted that the DR should equal unity if seeding had no effect and if the sample is sufficiently large. Table 7.2 gives the DRs calculated for these various rain parameters. The one-sided randomization significance level of the DR (Gabriel and Feder, 1969) was obtained from 1000 random permutations of the data. The significance of the observed DR was estimated by the percentage of cases in which the computer permutations gave a DR equal to or greater than that observed in the actual seeding experiments. It is seen from Table 7.2 that the excess of the DR over 1, in the case of R and D, is at least twice as large as the SE. This fact is reflected in the significance levels. The levels of significance for seeding effects on both the daily recorded rainfall, of about 25%, and the mean daily duration of rain, of about 18%, are highly significant, i.e., at the levels of 0.6% and 1.8%, respectively.

TABLE 7.1. Mean daily values of rainfall parameters for the control and LKC areas obtained from recording raingages on seeded and unseeded days. Values in parentheses are for days allocated for seeding and actually seeded. Also given are values of mean rainfall from nonrecording gages read once daily.

Parameter	Mean seeded		Mean unseeded	
	LKC	Control	LKC	Control
R, Rainfall (mm/24 h)	8.82 (10.32)	8.36 (9.51)	7.24 (7.24)	8.57 (8.57)
D, Duration of rain (h)	2.21 (2.59)	1.58 (1.80)	2.18 (2.18)	1.83 (1.83)
P, Number of rain events	5.61 (6.54)	4.63 (5.34)	5.09 (5.09)	4.72 (4.72)
Rainfall from nonrecording gages (mm/24 h)	8.89 (10.39)	8.30 (9.54)	7.32 (7.32)	8.05 (8.05)
Number of experiment days Number of days actually seeded	209 (174)	209 (174)	179 (179)	179 (179)

TABLE 7.2. Double ratios [DR, Eq. (1)], standard errors (SE) and randomization significance pertaining to all three rain parameters of the LKC with Control. The numbers in parentheses refer to days allocated to seeding and actually seeded.

Parameter	Mean seeded		Mean unseeded		DR	SE	Randomization significance level (%)
	LKC	Control	LKC	Control			
R, Rainfall (mm/24 h)	8.82 (10.32)	8.36 (9.51)	7.24 (7.24)	8.57 (8.57)	1.25 (1.28)	0.09 (0.10)	0.6 (0.1)
D, Duration of rain (h)	2.21 (2.59)	1.58 (1.80)	2.18 (2.18)	1.83 (1.83)	1.18 (1.21)	0.08 (0.09)	1.8 (0.4)
P, Number of rain periods	5.61 (6.54)	4.63 (5.34)	5.09 (5.09)	4.72 (4.72)	1.13 (1.14)	0.06 (0.06)	2.3 (1.9)

The effect of seeding on the total daily mean number of rain events, though positive, does not seem to be significant. The question of why the apparently positive effect of seeding on the overall daily mean number of rain events does not appear to be statistically significant will be dealt with in the following section.

Stratification of the data by cloud-top temperatures. Table 7.3 gives the results for daily mean rainfall as extracted from the recording raingages in the LKC and Control areas for several stratifications based on the daily modal cloud-top temperatures. In addition to the randomization test of significance, Table 7.3 gives the significance levels computed by the nonparametric Wilcoxon–Mann–Whitney (WMW) test. The DRs were calculated according to Eq. (7.1). The SEs were computed from Eq. (7.2), which is a modified version of that used by Gabriel and Feder (1969) and defined in more detail in the appendix of Gagin and Neumann (1981), i.e.,

$$(SE)_j = \left\{ 4 \sum_i [(LKC_j)_i / \sum_e (LKC_j)_e - (C_j)_i / \sum_e (C_j)_e]^2 \right\}^{1/2} \quad (7.2)$$

where $(LKC_j)_i$ and $(C_j)_i$ are the mean rain parameters R, D and P on day i, and similarly $(LKC_j)_e$ and $(C_j)_e$ are the mean rain parameters on day e, and the summations are over all experimental days, seeded and unseeded.

Not surprisingly, Table 7.3 reveals the same pattern of results as given in Gagin and Neumann (1981). In the present study, however, the effects of seeding on daily rainfall seem somewhat larger, but the significances are somewhat less. It should be noted that these results pertain to the LKC versus Control areas, whereas earlier results related to the total "North" target area. Thus, we now find overall positive effects in the LKC, under seeding, of about 53% and 45% on days when the modal values of daily cloud-top temperatures are in the range of −15° to −21°C and −11° to −21°C, respectively. The corresponding WMW significance values are 3.0% and 1%. As before, days with modal cloud-top values either warmer than −11°C, or colder than −21°C, exhibit insignificant seeding effects.

Table 7.4 gives the seeding results on rain duration in the LKC for various stratifications of cloud-top temperature. While the overall effect of seeding on the daily duration of rain was an increase of 18%, significant at 1.8% (Table 7.2), we find that in the groups of days when the modal cloud-top temperatures are either in the range of −15° to −21°C or −11° to −21°C the effects are larger and more significant. Here the positive effects on daily duration of rain are 53% and 47%, significant at 0.1% and

TABLE 7.3. Daily mean rainfall in the LKC and Control areas for seeded (S) and unseeded (US) days, and seeding effect on daily mean rainfall expressed in the form of double ratios and their significance. Based upon recording gages.

Temperature interval (°C)	Number of days		Mean seeded		Mean unseeded		DR/SE	SE	Significance level (%)	
	S	US	LKC	Control	LKC	Control			Randomization	WMW
T < −26	49	38	10.66	10.24	9.87	10.97	1.158	.159	18.2	19.1
T < −21	70	52	11.14	10.26	10.55	10.92	1.124	.126	16.3	14.4
−26 < T < −21	21	14	12.25	10.30	12.41	10.78	1.033	.122	43.9	38.1
−11 < T	27	21	9.85	8.06	7.12	7.87	1.347	.301	19.1	23.3
−15 < T	45	38	9.51	7.82	6.88	7.85	1.388	.189	5.5	7.0
−26 < T < −11	59	53	11.37	9.80	10.24	11.88	1.346	.130	1.3	1.0
−26 < T < −15	41	36	12.41	10.94	10.31	11.96	1.316	.161	5.1	4.0
−21 < T < −11	38	39	10.77	9.64	9.45	12.27	1.450	.166	1.3	1.0
−21 < T < −15	20	22	12.57	11.61	8.97	12.70	1.533	.236	5.3	3.0

TABLE 7.4. Daily average duration of rain in the LKC and Control areas for seeded (S) and unseeded (US) days, and seeding effect expressed in the form of double ratios (DR) and their significance (see text).

Temperature interval (°C)	Number of days		Mean seeded		Mean unseeded		DR	SE	Significance level (%)	
	S	US	LKC	Control	LKC	Control			Randomization	WMW
$T < -26$	49	38	2.30	1.83	2.84	2.22	0.986	.163	52.4	44.2
$T < -21$	70	52	2.57	1.91	2.83	2.12	1.009	.129	49.1	36.4
$-26 < T < -21$	21	14	3.20	2.10	2.82	1.85	1.000	.202	48.8	24.5
$-11 < T$	27	21	2.67	1.85	1.61	1.26	1.126	.213	30.7	22.6
$-15 < T$	45	38	2.46	1.66	2.27	1.92	1.247	.156	8.3	17.2
$-26 < T < -11$	59	53	2.92	1.82	2.90	2.40	1.327	.122	0.6	0.2
$-26 < T < -15$	41	36	3.27	2.01	2.80	2.22	1.289	.147	5.1	0.2
$-21 < T < -11$	38	39	2.70	1.66	2.93	2.60	1.469	.153	0.6	0.2
$-21 < T < -15$	20	22	3.33	1.92	2.79	2.46	1.531	.208	2.2	0.1

0.2%, respectively. The pattern found for the daily rainfall is repeated. On days when the cloud-top modal values are either warmer than $-11°C$ or colder than $-21°C$, the effects are either not significant or nil.

Whereas we found that the overall 25% increase in daily rainfall in the LKC was associated with a somewhat smaller (18%) increase in rain duration (Table 7.2), in the case of days with modal cloud-top temperatures between $-15°$ and $-21°C$ and between $-11°$ and $-21°C$, the increases in rainfall are brought about by similar increases in the duration of rainfall, i.e., the 53% increase in rainfall is obtained by a 53% increase in duration, while the 45% increase in rainfall is obtained by a 47% increase in duration.

It is interesting to point out that the 18% increase in daily rain duration can be explained in a rather simple, straightforward way. Thus, if one divides the mean daily rain duration of the unseeded days in the LKC, 2.18 hours, by the mean daily number of rain periods, 5.61, the mean duration per rain period is found to be 0.39 hours or 23.3 minutes.

Radar measurements on cumuliform cells in the same area indicated that a duration of 23 minutes is typical of cumulus clouds with top heights of about 5.5 km (or top temperatures of about $-18°C$). This top height is very close to the mode of the clouds treated in Israel. The overall increase of 18% in daily rain duration can, therefore, be interpreted as being due to an increase of $0.18 \times 23.3 = 4.2$ minutes. Assuming that the duration of precipitation of any single cloud is determined roughly by the time it takes for an ice crystal to grow to the size of a raindrop and emerge from cloud base, then we can calculate the additional duration of rain due to seeding by estimating the time required for an artificially nucleated crystal at $-5°C$ to reach the $-11°C$ level, where the first natural embryos of rain elements form. The corresponding height interval between the $-5°$ and $-11°C$ isotherms is about 1000 m (assuming a lapse rate of $6°C$ km^{-1}). In the early stages of the growth of an ice crystal, its fall velocity is negligible in comparison to that of an updraft of about 4 m s^{-1}, typical of clouds with tops of about 5 km (Gagin and Steinhorn, 1974). Thus the additional growth time is roughly 1000/4 sec, or about 4.2 minutes. This reasonable agreement, although obtained by a fairly simplistic method, is quite striking.

In Table 7.2 we showed that the overall effect of seeding on the number of rain periods was a positive 13%. The significance level of this result is also found to be quite high, i.e., 2.3%. Table 7.5, which shows the effects of seeding on the daily mean number of rain periods as parti-

TABLE 7.5. Daily average number of rain periods in the LKC and Control areas for seeded (S) and unseeded (US) days, and the seeding effect expressed in the form of double ratios and their significance (see text).

Temperature interval (°C)	Number of days		Mean seeded		Mean unseeded		DR	SE	Significance level (%)	
	S	US	LKC	Control	LKC	Control			Randomization	WMW
$T < -26$	49	38	5.88	5.42	6.93	6.30	0.986	0.111	54.0	34.2
$T < -21$	70	52	6.89	5.64	7.30	6.37	1.065	0.095	25.6	13.4
$-26 < T < -21$	21	14	9.25	6.13	8.33	6.54	1.184	0.169	18.9	10.0
$-11 < T$	27	21	5.73	4.60	3.53	3.59	1.268	0.128	4.9	4.3
$-15 < T$	45	38	5.95	4.82	4.84	4.67	1.188	0.124	8.7	8.7
$-26 < T < -11$	59	53	8.08	6.06	6.50	5.84	1.197	0.097	2.8	0.5
$-26 < T < -15$	41	36	8.88	6.46	6.51	5.76	1.216	0.117	5.9	0.2
$-21 < T < -11$	38	39	7.44	6.02	5.84	5.58	1.179	0.115	8.1	1.5
$-21 < T < -15$	20	22	8.49	6.80	5.36	5.26	1.226	0.154	11.2	0.3

tioned by the daily modal top temperature, reveals a remarkable result. By scanning the last two columns of Table 7.5, pertaining to the significance levels of the various stratifications of the data according to cloud-top temperatures, it can be seen that, unlike the cases of rainfall and duration, where the positive and significant effects are confined to clouds colder than $-11°C$, seeding also seems to produce a positive (27%) and significant (4.9%) effect on the number of rain periods on days when the modal cloud-top temperatures are warmer than $-11°C$.

Since a rain period indicates the passage of a rain cloud over the rain recording station, these results confirm the postulate regarding the effect of "static" seeding in the initiation of rain in clouds that are on the verge of producing rain (i.e., those with tops warmer than $-11°C$) but are deficient in the number of natural ice crystals. The introduction of a seeding agent, such as AgI (having nucleation threshold temperatures of about $-5°$ to $-8°C$), initiates ice crystals earlier and at lower elevations in the clouds. This results in the initiation of rain in clouds that otherwise would not have precipitated naturally. That this effect is nonexistent in deeper clouds with colder top temperatures is obviously due to the ability of these clouds to initiate rain naturally. Such differential effects are, in all probability, the reason for the absence of an effect, on this parameter, on days with modal cloud tops colder than $-21°C$, when the vast majority of the clouds are colder than $-11°C$.

An important question, from both the cloud physics and hydrological points of view, is the effect of seeding on rainfall intensity. Measurement of the intensity of each rain event directly from raingage charts can result in large inaccuracies due to the inability to measure the slope of the curve delineating the accumulation of rainfall on the chart. A rough approximation of the effect of seeding on "rain intensity" was obtained by using the ratios of daily recorded rainfall, divided by the daily duration of rain. Daily means for all stations in the LKC and the Control areas and for all days of Israeli II are given in Table 7.6. This table also gives the DRs and their SEs as well as the corresponding significance levels obtained by both the randomization and the WMW significance tests. This analysis suggests that rainfall intensity, as defined above, was not affected by seeding. The DR has, practically, a value of unity and the significance tests suggest inconclusiveness.

Throughout this report we have examined the effects of seeding, through means, DRs and significance levels, both for all days allocated to seeding and for days allocated to seeding and on which seeding was actually carried out. Not surprisingly, in all partitions and for all three rain parameters the results suggest larger and more significant effects when the data are restricted to days on which seeding took place. We realize, and have consistently so stated, that the results using all days allocated to seeding (regardless of whether seeding was carried out) must constitute the "formal" results of our experiments, since the restricted dataset may be biased for unknown reasons. On the other hand, any real seeding effects should be strongest and most consistent when the data are restricted to "actually seeded" days. The fact that our results follow such a pattern has strengthened our confidence that we correctly understand essential features of the meteorological physics operating in our experiments.

7.2.4. Downwind displacement of seeding effects

In both Israeli I and II, cloud seeding was performed by airborne patrol seeding at cloud-base altitudes along lines upwind of the target areas. Figure 7.3 displays the transport and dispersion pattern under typical wind conditions, calculated by assuming that the seeding aircraft are moving point sources emitting plumes that diffuse according to GDP models (Gagin and Aroyo, 1985). It can be seen that the areas between 20 and 50 km are exposed to a concentration $\geqslant 10\ L^{-1}$ of AgI particles, active at $-15°C$, for more than 50% of the time. Areas closer to or further from the seeding line have significantly reduced exposures to such concentrations. While this seeding method does not affect the whole target area uniformly, it has the great advantage of providing time for the dispersal and dilution, required for "static" effects, to take place in the free atmosphere, prior to activation of AgI particles in the clouds.

In Israeli I, the seeding line was offshore. The area of maximum effect occurred at distances of 20 to 50 km downwind from the seeding line. In Israeli II, we shifted the line of seeding eastward, i.e., inland, to affect the eastern catchment area of Lake Kinneret. This resulted in a corresponding shift of the area of maximum effect eastward by a distance comparable to the shift of the line of seeding.

TABLE 7.6. Recording raingage data pertaining to "average rainfall intensity" as defined in the text. The means, the DR and its SE are given with the aforementioned tests of significance. The numbers in parentheses refer to days allocated to seeding and actually seeded.

| Parameter | Mean seeded | | Mean unseeded | | DR | SE | Significance (%) | |
	LKC	Control	LKC	Control			Randomization	WMW
Rainfall intensity (mm h^{-1})	3.50 (3.64)	5.22 (5.32)	3.14 (3.14)	4.36 (4.36)	0.93 (0.95)	.080 (.071)	78.4 (71.2)	43.9 (41.8)

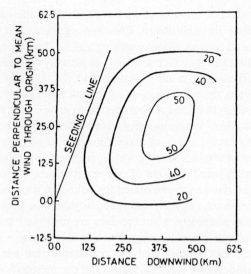

FIG. 7.3. The transport and dispersion pattern of AgI particles emitted from an aircraft flying in a patrol mode along the seeding line. Isopleths denote percentage of time when AgI particle concentration was ≥ 10 L^{-1}.

Thus, the statistical results pertaining to the areal distribution of seeding effects are physically plausible since they reflect the diffusion pattern of the seeding materials produced by a seeding technique aimed at producing static effects in clouds.

7.2.5. Concluding remarks—"static" mode seeding

This study constitutes another confirmation of the results of Israeli II, with regard to the effect of seeding in the LKC, as it is based on a totally independent dataset of rainfall measurements. Its main contributions are to provide additional evidence for the physical plausibility of the statistical findings of both Israeli I and II and to add strength to the theory of "static" cloud seeding for rainfall enhancement from continental clouds by corroborating the basic assumptions underlying this theory.

7.3. Evaluation of the basic tenets of "dynamic" seeding as inferred from an analysis of FACE-2

7.3.1. General remarks

The Florida Area Cumulus Experiment (FACE) was a two-stage program to investigate the potential of dynamic-mode seeding for enhancing convective rainfall in a fixed target area in South Florida. The first, or exploratory phase (FACE-1, 1970–76), produced indications of increased rainfall in the target area (Woodley et al., 1982). The second, or confirmatory phase (FACE-2, 1978–80), did not confirm the results of FACE-1, although it did produce indications of a possible seeding effect (Woodley et al., 1983).

As specified by Woodley et al. (1978), the FACE-2 program contained an ambitious exploratory component in order to obtain better physical insights as to the effects of seeding. A major component of the FACE-2 exploratory studies was the volume-scan radar program. The radar-volume-scan (RVS) technique, which was developed specifically for these studies, has been described in detail by Gagin et al. (1985). The major objective of this study was to provide a more detailed scientific basis for the theory of rainfall enhancement by seeding for dynamic effects. The basic tenet of dynamic seeding is that the production of a taller cloud results in more rainfall. While previous studies by Simpson et al. (1967), Dennis et al. (1975) and Gagin (1980), have all shown that taller convective cells produce more rain, the Gagin et al. (1985) study was aimed at finding out how the larger rain volumes were related to the larger areas, longer rain durations, and stronger rain rates of the taller clouds.

7.3.2. Methodology of radar-volume-scan observations

Details of the methodology of observing the evolution of convective rain cells by using radar-volume-scan techniques have been given in Gagin et al. (1986). The following is only a summary describing this technique in general terms.

The determination of the relationship between the vertical dimension of convective cells and their other basic properties, such as area, intensity and duration, requires the systematic collection of extensive radar observations of cell characteristics, along with analysis procedures capable of tracking these cells throughout their lifetime. Digitized radars, with adequate data-recording facilities, are required for data collection, and high-speed, large-memory computers are needed for the subsequent analysis.

These studies were carried out in South Florida within the framework of FACE-2. They made use of two different, but virtually collocated, radar systems. The first was a WSR-57 S-band radar operated by the National Hurricane Center, and the second was an MPS-4 C-Band radar operated by the University of Miami. The C-Band radar provided the volume scan data, while the S-band radar provided measurement of the surface rainfall characteristics from low-level scans.

A set of computer programs was developed to identify radar echo cells and to track them throughout their lifetimes. In general terms, an echo is isolated by determining a set of contiguous grid elements containing a radar return above the noise threshold. It is then inspected for radar reflectivity maxima. These maxima are identified and labeled as centers of cells. The total echo area is then divided into the different cells along the lines of minimum reflectivity between the peaks. Grid elements that fall on those lines are assigned to the cell whose peak is closest. On rare occasions, the peaks are embedded in a matrix of uniform but lower reflectivity. In these cases, the dividing line is drawn midway between the reflectivity maxima.

As defined by this procedure, a cell corresponds to a precipitation area with a distinct maximum. These precipitation areas very often correspond to distinct height features in the echo-top height maps. Echoes larger than about 100 km² in area tend to have two or more maxima in reflectivity and height. This characteristic has been observed since the days of the Thunderstorm Project (Byers and Braham, 1949) in Florida and elsewhere and seems to be a general property of convective systems (Lopez et al., 1984). These echo regions can be tracked and are seen to form, reach a maximum reflectivity, decay, and eventually dissipate, with life cycles generally independent of each other. They also retain their relative positions and their general configurations from one time period to the next. Lopez (1978) has associated these regions or cells with individual convective elements. Once a rain cell was identified and tracked throughout its lifetime, the volume-scan data were used to provide the maximum vertical dimension (H_{max}) as well as its time–height history. In addition, the low-level scan radar provided the maximum and lifetime properties of the rain cell, such as its instantaneous and maximum area (A and A_{max}), the corresponding values of reflectivity (Z and Z_{max}), and cell duration (DUR). The rainfall values, R, were calculated from the FACE Z–R relationship ($Z = 300R^{1.4}$) and used to compute the other integrative properties of the cells, such as the rainfall volume rate (RVR) and the total rain volume (RV).

7.3.3. Regression analyses of the data

A total of 349 "natural" unseeded convective cells that occurred during July and August 1979 were studied. The cells were all located at distances of between 60 and 100 km from the radar site. Regression analyses were applied to these data in an attempt to relate the various properties of these cells to the maximum vertical extent attained by them during their lifetimes.

A set of highly correlated power-law relationships was found. These relationships show that the taller convective rain cells produced the larger total rainfall volumes by virtue of larger precipitating areas, rain rates and rain durations. The power law constants, as well as the actual scattergrams, are given in Gagin et al. (1985). Figure 7.4 displays the relationship between the maximum height and the total rain volume of seeded and unseeded convective rain cells. One notes that a twofold change in maximum cell height corresponds roughly to a factor of 10 increase in total rain volume for both types of clouds. This result confirms one of the basic tenets of the theory of rainfall augmentation by cloud seeding through the production of "dynamic" effects, as hypothesized by Simpson and Woodley (1971).

Having described the natural relationship between cell depth and other cell properties in a set of power laws, it

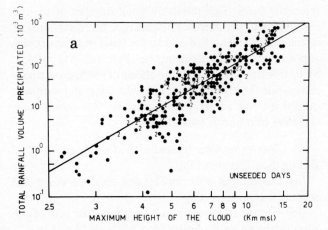

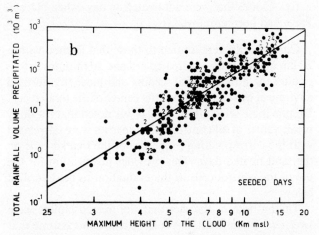

FIG. 7.4. The relationship between the lifetime total rainfall volume precipitated by the cells, RV, and the factor (H_{max}-h) defined by the lifetime maximum height, H_{max}, of the cells. (a) Relates to cells on sand-treated days (unseeded) and (b) to cells on AgI-treated days (seeded).

was then possible to investigate whether seeding affected the cell properties and whether any such changes were produced by changing the basic natural relationships between cell depth and these other cell properties.

7.3.4. The data and description of the analytical approach

The results and database for the present summary came from a study performed by Gagin et al. (1986) using data from 4 days from the 1979 field season and 17 days from the 1980 field season. Unfortunately, all other FACE-2 days could not qualify for this study because of technical difficulties either in recording the data or in the operation of the MPS-4 radar. Out of the 21 days, only 3 days were in the FACE "A" day category (<60 seeding flares expended); the rest fall in the "B" day category (>60 flares expended). Only category "B" days were used in this study.

A total of 2369 convective cells between 60 and 100 km from the radar were studied. Of these, 2069 were untreated, and 300 cells were treated with either AgI or sand (placebo) flares. This relatively small number of treated cells is due to the narrow annulus (i.e., 60 to 100 km) in which cells were studied and to the strict seeding criteria applied during the experiments (see Barnston et al., 1983). Since this study deals with the effects of seeding on single cells, while the experimental unit was a day, the numbers quoted provide an acceptable database. All clouds were classified into one of four groups:

(i) Clouds that were treated with at least one AgI flare during their lifetimes (T_{AgI}).

(ii) Clouds that were not treated on days when AgI seeding was applied (C_{AgI}).

(iii) Clouds that were treated with at least one of the sand (placebo) flares during their lifetimes (T_{sand}).

(iv) Clouds that were not treated on days when placebo flares had been employed (C_{sand}).

Control cells were limited to those that formed within 20 km of the center of a treated cell and 30 minutes before or after the time of initial seeding and those that formed anytime at least 20 km from the center of the treated cell. Despite these selection criteria, it was noted that the daily mean values of all control cell properties were correlated with the corresponding properties of the treated cells on the sand-treated days. This provides justification for the use of the control cells in the evaluation to be described below.

Taking note of the fact that the treated clouds were *selected* for treatment, it would be natural to assume that they are biased (whether treated with AgI or sand) with respect to all untreated clouds. Nevertheless, untreated cells can be used as internal controls, as a means of reducing the day-to-day variability or bias. Thus, if one could guarantee that on all days relatively equal proportions of treated and untreated control cells were selected, then the double-ratio estimate of seeding effect could be utilized to compensate for this bias. The double ratio is calculated from Eq. (7.3)

$$DR = \frac{(X)T_{AgI}}{(X)C_{AgI}} \times \frac{(X)C_{sand}}{(X)T_{sand}} \qquad (7.3)$$

where the values of X, in parentheses, can be the mean of any of the following lifetime properties of the cells: $Z_{max}, H_{max}, A_{max}, RV, RVR_{max}$ and DUR. The subscripts that relate the cells to the various four groups were defined earlier.

If DR exceeds the value of one, it suggests positive effects of seeding on the particular parameter under study. Again, it must be emphasized that the DR estimate can be used only if the right-hand ratio of Eq. (7.3) compensates fully for the bias introduced by the selection of the clouds and if the samples are representative of the four groups.

In addition to the use of the double ratio, use was made of the "single ratio" (SR), Eq. (7.4), for the estimation of AgI treatment. Obviously, SR lacks the compensation for day-to-day variations if they occur. Its use is therefore limited by the assumption that, in the long run, if seeding had no effect, SR should have attained the value of unity:

$$SR = (X)T_{AgI}/(X)T_{sand}. \qquad (7.4)$$

7.3.5. Results

A list of the dates used in the study and the treatment decision for each day are given in Gagin et al. (1986). The relationships between the maximum vertical dimensions of the cells and their other lifetime properties, for both AgI- and sand-treated days, are given in Table 7.7. As can be seen, these relationships are not only well correlated, but the AgI and sand relationships are practically the same.

The similarities of the sand and AgI regressions for all cell properties have two possible interpretations.

(i) There is no effect of AgI treatment.

(ii) There may be an effect of AgI treatment which results in smaller or taller clouds that behave as smaller or taller natural clouds. In other words, the effect of seeding may be to move the points along the regression lines. This possibility is investigated through the use of single and double ratios.

Table 7.8 provides mean cell properties for the four categories ($T_{sand}, C_{sand}, T_{AgI}, C_{AgI}$) for all treated cells that

TABLE 7.7. Correlation coefficients and the regression constants defining the relationship between the maximum "depth" of convective cells and their other properties for cells on sand-treated days and AgI-treated days. FACE-2 data.

| Regression between: | Number of cells | | Regression coefficients | | | | | | | |
| | | | Slope | | Intercept | | Correlation coefficient | | Standard error of estimate | |
	Sand	AgI	Sand	AgI	Sand	AgI	Sand	AgI	Sand	AgI
$\log A_{max}$ and $\log(H_{max} - h)$	349	385	0.97	0.92	1.02	1.01	0.79	0.77	0.19	0.18
$\log Z_{max}$ and $\log(H_{max} - h)$	349	385	2.20	2.33	2.68	2.45	0.78	0.78	0.46	0.44
$\log RVR_{max}$ and $\log(H_{max} - h)$	349	385	2.14	2.20	0.82	0.65	0.84	0.84	0.35	0.34
$\log DUR$ and $\log(H_{max} - h)$	349	385	0.60	0.70	1.00	0.84	0.59	0.66	0.14	0.15
$\log RV$ and $\log(H_{max} - h)$	349	385	2.44	2.63	−0.02	−0.28	0.84	0.85	0.40	0.39

TABLE 7.8. Means, single ratios and double ratios for the various cell properties on 18 "B" days of FACE-2. The significance levels, SL (%), were obtained by 3000 rerandomizations of the data.

Cell property	T_{AgI}	C_{AgI}	T_{sand}	C_{sand}	SR	DR	Rerandomization SL (%) SR	DR
Rv [10^3 m^3]	200.6	83.7	175.3	104.0	1.14	1.42	25.4	4.2
Z_{max} [dBZ]	44.9	38.5	44.4	39.9	1.01	1.05	25.5	4.0
H_{max} [km]	9.1	8.0	8.6	8.1	1.06	1.07	19.6	10.0
A_{max} [km^2]	75.8	59.6	74.5	60.3	1.02	1.03	40.0	38.0
RVR_{max} [10^3 m^3 h^{-1}]	616.2	320.1	612.6	393.1	1.01	1.24	47.8	12.1
DUR [min]	34.1	23.4	32.3	23.9	1.06	1.08	20.0	18.4
Cell sample size	161	1066	139	1003				

occurred within the annulus of 60 to 100 km from the radar on the 18 "B" days (9 AgI-treated and 9 sand-treated) of FACE-2. This table also provides the double ratios and single ratios as defined above. The significance levels were obtained by the Monte Carlo rerandomization technique (3000 iterations). The rerandomizations were done on a daily basis, in adherence with the original design of FACE-2.

There are several items of interest in Table 7.8. In every instance, the mean values for the control cells are smaller than the means for the treated (by either AgI or sand) cells. This difference is not surprising since the best clouds were selected for treatment. Nevertheless, the properties of the control cells are strongly correlated with the corresponding properties of the treated cells and can therefore be used validly as internal controls.

A second item of interest in Table 7.8 is that, on average, the properties of the control cells on AgI days are somewhat smaller than the corresponding properties of the control cells on sand days. (This disparity is greatest for cell rain volumes to which rain rate, area and duration contribute in a multiplicative fashion.) This could be due either to chance or to suppression of the control cells by compensating currents around the AgI-treated cells. Acceptance of the latter possibility, however, necessarily means acceptance of the premise that AgI seeding has enhanced the growth of the treated cells.

Evidence for an overall effect of AgI seeding, in terms of the double ratio, is strongest for cell rain volume and for maximum reflectivity. However, because of the imbalance between control cells on sand- and AgI-treated days, it is not clear what interpretation should be given to this result. In terms of the single ratio, the indicated effect is smaller, but still greater than one. This indicates a positive overall effect of seeding.

Greater physical insight is obtained by partitioning the data on the amount of material (AgI or sand) injected into the cloud and on the stage of the cell life cycle in which the treatment took place. Throughout FACE, an attempt was made to seed supercooled clouds when they were young and vigorous, with strong updrafts, and with much of the liquid water near cloud top (Woodley et al., 1982; Woodley et al., 1983). Typically, such cells may not have a radar echo or will have had one for only a short time. Thus, there is a solid physical basis for partitioning the results in terms of the time, within the life cycle, when cells were seeded. Table 7.9 summarizes the mean properties of the treated cells into two groups. The early treatment group contains clouds that were treated within 5 minutes of their first appearance as an echo on the S-band radar. The second group contains clouds that were treated later in their lifetimes. It can be noted that there is a distinct difference in the behavior of these two groups of cells. Early treatment seems to have produced

TABLE 7.9. Means, single ratios and significance levels (SL) for the various properties of cells from the 18 "B" days of FACE-2, stratified by time of treatment. Significance levels were obtained by 3000 rerandomizations of the data.

Cell property	Early treatment T_{AgI}	T_{sand}	SR	Randomization SL (%)	Late treatment T_{AgI}	T_{sand}	SR	Randomization SL (%)
RV (10^3 m^3)	248.9	157.2	1.58	4.4	165.4	186.2	0.89	65.0
Z_{max} (dBZ)	46.4	44.3	1.05	1.8	43.8	44.5	0.98	66.0
H_{max} (km)	9.7	8.6	1.13	8.8	8.6	8.6	1.00	45.6
A_{max} (km^2)	82.4	72.4	1.14	11.8	70.9	75.8	0.94	61.0
RVR_{max} (10^3 m^3 h^{-1})	759.6	598.2	1.27	9.1	511.5	621.3	0.82	75.8
DUR (min)	32.2	27.9	1.15	14.6	35.5	34.8	1.02	41.2
Cell sample size	68	52			93	87		

consistent positive effects on all AgI-treated cell properties, of which effects on the total rain volume and the maximum reflectivity are strongly significant. In contrast, clouds treated later in their lifetimes indicate virtually no seeding effects.

Table 7.10 summarizes the means, single ratios and significance levels for the various properties of cells on all 18 "B" days of FACE-2. These data are stratified into two categories of time of treatment and three categories of amount of treatment. As before, early treatment refers to cells treated within 5 minutes of their appearance as a radar echo. Late treatment includes cells that were treated at times later than 5 minutes following the showing of an echo. An examination of Table 7.10 shows that H_{max} of the sand-treated cells (T_{sand}) generally decreases with increasing flare expenditure. This trend, which is apparent in both the early and late treatment categories, is particularly evident in the case of the early treatment. Moreover, the decrease in H_{max} of the sand-treated cells seems to be larger in moving from the group receiving the lowest dosages to those of intermediate dosages. This trend is not so systematic for the other cell properties; nevertheless, the same general trend is evident.

If one accepts the logical assumption that the sand treatment had no inhibitive effect on the clouds, then the remaining alternative for explaining this trend is that flare expenditure, in addition to being a measure of treatment dosage, reveals the existence of some sort of selection bias. Treatment dosage may also stratify the cells meteorologically. In this case, one should compare cells only within

a given dosage category since comparisons across dosage categories will be confounded by the effect of the dosage partitioning. For this reason, we examine the combined effect of flare expenditure and the timing of treatment by comparing early treatment with late treatment within the same flare expenditure categories. Comparisons between AgI- and sand-treated cells receiving the same dosage should exclude the possibility of any selection bias by the scientists carrying out the seeding experiments.

From an examination of the differences between cell properties, their SRs and their corresponding statistical significances within dosage categories, the combined effect of quantity of seeding material and its injection time becomes very clear and unequivocal. In the case of early treatment and large flare expenditure, the rather large SRs and their significances, relating increases of 160% in rain volume to increases of 22% in maximum cell height, 51% in maximum cell area, and 43% in rain duration, are quite impressive.

The results for cells that were treated 5 or more minutes after their first detection by the S-band radar show virtually no seeding effects, even for the high flare expenditure category. This underscores the importance of timing of treatment when conducting dynamic-mode seeding experiments.

Further quantification of the strength of the derived relationships between cell height and other cell properties, and their general applicability to both treated and untreated cells, is provided in Table 7.11. In the left portion of the table are observed cell properties and on the right

TABLE 7.10. Means, single ratios and significance levels (SL) for the various properties of cells from the 18 "B" days of FACE-2, stratified by time of treatment and number of treatment flares. Significance levels were obtained by 3000 rerandomizations of the data.

	Early treatment				Late treatment				
Cell property	T_{AgI}	T_{sand}	SR	SL (%)	T_{AgI}	T_{sand}	SR	SL (%)	Number of treatment flares
RV (10^3 m^3)	142.5	154.1	0.93	59.2	145.6	193.8	0.75	71.5	
Z_{max} (dBZ)	45.7	44.8	1.02	21.6	43.7	45.7	0.96	76.0	
H_{max} (km)	9.6	9.5	1.01	44.5	8.7	9.3	0.93	77.2	
A_{max} (km^2)	65.4	75.7	0.86	84.8	65.2	96.1	0.68	92.1	$0 < NF \leqslant 4$
RVR (10^3 m^3 h^{-1})	516.8	621.3	0.83	76.5	460.1	796.1	0.58	89.8	
DUR (min)	29.6	31.1	0.95	59.0	32.8	28.4	1.16	19.0	
Cell sample size	21	15			20	13			
RV (10^3 m^3)	289.4	234.9	1.23	26.8	163.2	200.4	0.82	42.0	
Z_{max} (dBZ)	45.9	46.2	0.99	54.0	43.9	43.7	1.00	41.0	
H_{max} (km)	9.7	8.5	1.14	16.7	8.7	8.0	1.07	15.8	
A_{max} (km^2)	85.1	85.7	0.99	51.5	72.0	70.3	1.03	43.0	$5 \leqslant NF \leqslant 8$
RVR (10^3 m^3 h^{-1})	809.0	732.8	1.10	34.7	525.4	622.7	0.83	64.5	
DUR (min)	32.4	31.3	1.04	43.7	33.3	32.6	1.02	44.5	
Cell sample size	25	13			37	30			
RV (10^3 m^3)	304.4	117.1	2.60	0.6	178.5	174.3	1.02	46.0	
Z_{max} (dBZ)	47.6	42.9	1.11	0.3	43.9	44.8	0.98	68.1	
H_{max} (km)	9.9	8.1	1.22	2.1	8.6	8.8	0.97	59.0	
A_{max} (km^2)	95.6	63.2	1.51	1.6	72.9	73.6	0.99	51.6	$NF \geqslant 9$
RVR (10^3 m^3 h^{-1})	935.1	510.9	1.83	2.0	525.7	568.7	0.92	59.5	
DUR (min)	34.3	24.0	1.43	1.8	39.3	38.4	1.02	40.7	
Cell sample size	22	24			36	44			

TABLE 7.11. A typical example comparing the observed properties of convective cells with those predicted solely from the observed maximum cell heights, based on the regression coefficients (Gagin et al., 1985). This table corresponds to a case where cells were treated within 5 minutes of initial appearance on the radar. The SR values are calculated for both cases.

Cell property	Observed			Predicted		
	T_{AgI}	T_{sand}	SR	T_{AgI}	T_{sand}	SR
RV [10^3 m^3]	248.9	157.2	1.58	224.8	156.0	1.44
Z_{max} [dBZ]	46.4	44.3	1.05	46.5	45.1	1.03
H_{max} [km]	9.7	8.6	1.13	9.7	8.6	1.13
A_{max} [km^2]	82.4	72.4	1.14	85.7	74.1	1.16
RVR$_{max}$ [10^3 m^3 h^{-1}]	759.6	598.2	1.27	763.7	554.4	1.38
DUR [min]	32.2	27.9	1.15	39.0	35.6	1.09
Cell sample size	68	52		68	52	

are cell properties that were predicted solely from the observed maximum cell heights and the regression relationships from Gagin et al. (1985). Note that in most instances the predicted cell values agree fairly well with the observed values. Even the results of treatment on rainfall are predicted with considerable accuracy by knowing the effect of treatment on cell height.

7.4. Concluding remarks

This study shows that the combined evolution of cell height, area, rain rate and duration as a precipitating entity in naturally forming convective rain cells can be described by a power law relating maximum cell height to total rain volume precipitated by it. An increase in precipitation by a factor of ~10 accompanies an increase by a factor of 2 in rain cell depth.

This study also examined the effects of dynamic-mode seeding on convective cell height, area, rain rate and duration. It suggests that by seeding with large amounts of nucleant early in the life cycle of Florida-type rain cells, it was possible to increase their heights by an average of 22% and their rain volumes by an average of over 100%. This result confirms the early single-cloud studies reported by Simpson and Woodley (1971) in Florida. Strong evidence has been presented that seeding effects are limited to cells treated early in their lifetimes. This result can be explained by microphysical studies in Florida which indicate that individual cells remain suitable (i.e., high water contents and strong updrafts) for only a short time. In view of the criticality of the timing of seeding, treatment of a broad tower with a small number of flares is not likely to be effective, because the time necessary to diffuse the nucleant throughout the tower may well exceed the period of time during which the tower is suitable.

A last matter of importance is the question of whether the control clouds in the vicinity of the AgI-treated clouds in FACE-2 were suppressed by the invigorated seeded clouds. These data suggest that this might have been the case, since the control cells on the AgI days were weaker and less rain-productive than the comparable control clouds on the sand days. However, one cannot be certain that this control cloud disparity was not produced by natural causes. In other words, the "luck of the draw" may have dictated wetter days for sand treatment. Acceptance of the argument that AgI seeding suppressed the nearby control clouds means acceptance of a seeding effect. Acceptance of the argument that the draw favored the sand-treated days also indicates a seeding effect since its magnitude is increased when accounting for the natural bias.

These are encouraging results; they shed light on our understanding of the reasons for the inconclusive overall results of FACE-2. These findings also carry a fair degree of physical plausibility. However, they should be regarded only as suggestive, since they rest on a small sample of treated clouds. The validity of these results should be tested in a confirmative experiment.

REFERENCES

Barnston, A. G., W. L. Woodley, J. A. Flueck and M. H. Brown, 1983: The Florida Area Cumulus Experiment's second phase (FACE-2). Part I: The experimental design, implementation, and basic data. *J. Climate Appl. Meteor.,* **22**, 1504–1528.

Byers, H. R., and R. R. Braham, Jr., 1949: *The Thunderstorm.* Washington, DC, U.S. Govt. Printing Office, 287 pp.

Dennis, A. S., A. Koscielski, D. E. Cain, J. H. Hirsch and P. L. Smith, Jr., 1975: Analysis of radar observations of a randomized cloud seeding experiment. *J. Appl. Meteor.,* **14**, 897–908.

Gabriel, K. R., 1970: The Israeli rainmaking experiment 1961–67 final statistical tables and evaluation. (Tables prepared by M. Baras). Tech. Rep., Hebrew University, Jerusalem, 47 pp.

——, and P. Feder, 1969: On the distribution of statistics suitable for evaluating rainfall stimulation. *Technometrics,* **11**, 149–150.

Gagin, A., 1975: The ice phase in winter continental cumulus clouds. *J. Atmos. Sci.,* **32**, 1604–1614.

——, 1980: The relationship between depth of cumuliform clouds and their raindrop characteristics. *J. Rech. Atmos.,* **14**, 409–422.

——, 1981: The Israeli rainfall enhancement experiments—a physical overview. *J. Wea. Mod.,* **13**, 108–120.

——, and J. Neumann, 1974: Rain stimulation and cloud physics in Israel. *Weather and Climate Modification,* W. N. Hess, Ed., Wiley-Interscience, 454–494.

——, and I. Steinhorn, 1974: The role of solid precipitation elements in natural and artificial production of rain in Israel. *J. Wea. Mod.,* **6**, 216–228.

——, and J. Neumann, 1981: The second Israeli randomized cloud seeding experiment: Evaluation of the results. *J. Appl. Meteor.,* **20**, 1301–1311.

——, and M. Aroyo, 1985: Quantitative diffusion estimates of cloud seeding nuclei released from airborne generators. *J. Wea. Mod.,* **17,** 59–70.

——, D. Rosenfeld and R. E. Lopez, 1985: The relationship between height and precipitation characteristics of summertime convective cells in South Florida. *J. Atmos. Sci.,* **42,** 84–94.

——, ——, W. L. Woodley and R. E. Lopez, 1986: Results of seeding for dynamic effects on rain cell properties in FACE-II. *J. Climate Appl. Meteor.,* **25,** 3–13.

Lopez, R. E., 1978: Internal structure and development processes of C-scale aggregates of cumulus clouds. *Mon. Wea. Rev.,* **106,** 1488–1494.

——, D. O. Blanchard, D. Rosenfeld, W. L. Hiscox and M. J. Casey, 1984: Population characteristics, development processes and structure of radar echoes in South Florida. *Mon. Wea. Rev.,* **112,** 56–75.

Neumann, J., K. R. Gabriel and A. Gagin, 1967: Cloud seeding and cloud physics in Israel: Results and problems. *Proc. Int. Conf. on "Water Peace,"* Washington, DC, **2,** 375–388.

Simpson, J., and W. L. Woodley, 1971: Seeding cumulus in Florida; new 1970 results. *Science,* **172,** 117–126.

——, G. W. Brier and R. H. Simpson, 1967: Stormfury cumulus seeding experiments 1965: Statistical analysis and main results. *J. Atmos. Sci.,* **24,** 508–521.

Smith, P. L., A. S. Dennis, B. A. Silverman, A. B. Super, E. W. Holroyd, W. A. Cooper, P. W. Mielke, K. J. Berry, H. D. Orville and J. R. Miller, 1984: HIPLEX 1: Experimental design and response variables. *J. Climate Appl. Meteor.,* **23,** 497–512.

Woodley, W. L., R. I. Sax, J. Simpson, R. Biondini, J. A. Flueck and A. Gagin, 1978: The FACE confirmatory program (FACE-2): Design and evaluation specifications. NOAA Tech. Memo., ERL, NHEML-2, 51 pp.

——, J. Jordan, J. Simpson, R. Biondini, J. A. Flueck and A. Barnston, 1982: Rainfall results of the Florida Area Cumulus Experiment, 1970–1976. *J. Appl. Meteor.,* **21,** 139–164.

——, A. Barnston, J. A. Flueck and R. Biondini, 1983: The Florida Area Cumulus Experiment's second phase (FACE-2). Part II: Replicated and confirmatory analyses. *J. Climate Appl. Meteor.,* **22,** 1529–1540.

CHAPTER 8

Modification of Mesoscale Convective Weather Systems

J. MICHAEL FRITSCH

Pennsylvania State University, University Park, Pennsylvania

ABSTRACT

Modification of mesoscale convective weather systems through ice-phase seeding is briefly reviewed. A simple mathematical framework for estimating the likely mesoscale response to convective cloud modification is presented, and previous mesoscale modification hypotheses are discussed in the context of this mathematical framework. Some basic differences between cloud-scale and mesoscale modification hypotheses are also discussed. Numerical model experiments to test the mesoscale sensitivity of convective weather systems are reviewed, and several focal points for identifying mesoscale modification potential are presented.

8.1. Introduction

When considered together, the numerous observational and modeling studies discussed in Orville's review of dynamic-mode seeding of summer cumuli make a strong case for the existence and experimental realization of dynamic-seeding potential of *individual* convective clouds. Yet, Orville points out the tremendous complexity of treating individual clouds and developing comprehensive physical hypotheses on which to base experiments. This complexity is reflected by the fact that after 37 years of research, the most recently completed weather modification field experiment, HIPLEX-1, still focused on understanding the most rudimentary of precipitating convective clouds, the small, semi-isolated, cumulus congestus (Smith, et al., 1984). Given, then, the additional complexity of the mesoscale, the overwhelming costs of mesoscale field experiments, and the absence of any clear ratonale on which to base modification attempts, why even consider *mesoscale* modification?

The answer is basically twofold.

(i) *Cloud-scale and mesoscale are inseparable.* It is becoming increasingly evident that the mesoscale environment and the characteristics of individual clouds are inseparably and nonlinearly linked together as a single functioning unit. The mesoscale environment responds quickly and strongly to heating from moist convection (see Kreitzberg and Perkey, 1977; Brown, 1979; Fritsch and Chappell, 1980b; Maddox et al., 1981; Xu and Clark, 1984) and has a correspondingly large impact in determining the location and amount of convection and type of convective cloud (see Moncrieff and Green, 1972; Cotton and Boulanger, 1975; Cotton et al., 1976; Klemp and Wilhelmson, 1978; Maddox et al., 1979; Emanuel, 1979; Chen and Orville, 1980; Tripoli and Cotton, 1980; Matthews, 1981; Cunning et al., 1982; Foote and Wade, 1982; Weisman and Klemp, 1982; Parsons and Hobbs, 1983).

(ii) *The payoff is much greater.* Recent studies (summarized in Houze and Hobbs, 1982) indicate that mesoscale convective weather systems play a very important, if not dominant role in production of warm-season precipitation. The impact of mesoscale *organization* of convective clouds on precipitation efficiency and production is most clearly demonstrated by Simpson et al. (1980). They calculated that merged convective echoes result in order-of-magnitude increases in precipitation over that which occurs from similar, but unmerged echoes. Thus, a "small" enhancement of a mesoscale system may result in a much greater increase in rainfall than that from larger changes in an entire population of separate, individually precipitating, convective clouds.

In principle, studies of modification of mesoscale systems are no different from countless other studies of large-scale weather systems, global circulations and climate wherein small-scale physical processes are approximated, parameterized, or introduced through reasonable assumptions based upon physical or empirical formulations. To simply introduce an adjustment in the formulations (based upon knowledge gained from observations and/or modeling of seeded clouds or otherwise man-altered processes) does not change the approach, nor does it make the results any less valid than those from studies of "unmodified" large-scale systems. Moreover, it is certainly valid for scientific inquiry to proceed on the basis that understanding and operational technology for treatment of individual clouds will eventually be forthcoming and that knowledge of modification sensitivity of mesoscale systems can only expedite the development of viable modificaton hypotheses.

It should also be pointed out that modification of mesoscale systems may sometimes be easier than modification of a single cloud. The development of precipitation in an individual cloud involves many complex microphysical and dynamical processes that occur in very short time periods, i.e., on the order of a few minutes. Successful cloud modification, then, requires intervention with the proper amount of seeding material, at the right place, and within a small time window. On the other hand, for mesoscale systems, it is likely that significantly longer time periods are available for intervention. Moreover, cumulative responses to seeding can be considered. Therefore, it is possible that what may be deemed a negative effect on rainfall from an *individual cloud* could be a positive contribution toward increasing rainfall from a *mesoscale system*. For example, consider the situation where individual clouds within a mesoscale cloud system are selected for modification treatment. If in some cases the application of the seeding agent fails to arrive within the proper time and space window, overseeding and a corresponding increase in evaporation (sublimation) of condensate may result. Thus, there would be an initial reduction in "natural" rainfall accompanied by an increase in environmental humidity. Lopez (1973a,b) has shown that an extremely important parameter for increasing rainfall from a population of convective clouds is to increase cloud-layer environmental humidity. The work of Simpson et al. (1980) supports this result. Therefore, an initial reduction in precipitation and increase in environmental humidity through the evaporation of cloud condensate may lead to significantly enhanced rainfall from subsequent clouds. In support of this argument, it is well known that tropical mesoscale convective systems typically require 3–6 h in which small convective clouds gradually moisten the environment and "pave the way" for development of the deep convective clouds that produce the bulk of the rain (see Frank, 1978). Moreover, Simpson (1980) pointed out that in FACE-1 (Woodley et al., 1982) it was often necessary for midlevel dry layers to be gradually moistened by successive small cloud towers before explosive growth and deep convection could occur later in the afternoon. One could also understand, then, how in some cases "successful" modification of individual clouds could lead to an areawide decrease in precipitation for several hours following cloud treatment (such as observed in Project Whitetop; see Flueck, 1971). The main point here is that mesoscale modification has the luxury of additional time to allow for compensating processes that change the environment in ways that may enhance or offset cloud-scale treatments. Thus, there may be certain treatments that are generally favorable (or detrimental) for enhancing rainfall whether or not a particular response in individual clouds is consistently achieved.

The preceding discussion poses a mesoscale "static-mode" seeding hypothesis, i.e., the environment is moistened, rainfall increases, and no dynamic response is required. Mesoscale dynamic-seeding hypotheses, e.g., STORMFURY and FACE, are considered following a discussion in section 8.2 of how convection interacts with its environment. The dynamic-mode seeding hypothesis for FACE (see Woodley et al., 1982; or Orville's review) is discussed in the context of the role of the mesoscale in formulating viable modification hypotheses. Finally, the use of mesoscale-model modification sensitivity studies as a tool to help establish meaningful modification hypotheses for both cloud-scale and mesoscale modification treatments is considered.

8.2. Interaction of moist convection with its environment

The problem of understanding how deep convection interacts with its environment has been widely explored in the context of convective parameterization for numerical models. Consequently, there is a great body of information available for consideration (for a review, see Frank, 1983). It is neither possible nor desirable in this brief review of mesoscale modification to present an in-depth discussion of this extremely complex problem; however, it will be helpful to consider some simple relationships that describe how convective clouds change the mesoscale environment. With these relationships, it is possible, in principle, to estimate the impact of modification of individual convective clouds on the mesoscale environment. Conversely, it is also important to consider how changes in the mesoscale environment will affect the characteristics of subsequent convective clouds. Both of these points are discussed below.

8.2.1. Cloud-forced changes in the mesoscale environment

In order to estimate how the mesoscale environment may change in response to modification of convective clouds, it is necessary to develop a relationship between cloud properties and the vertical distribution of local changes of the mesoscale environment. Temperature is the most common parameter considered in such relationships and will be the parameter considered here. Once convectively forced local temperature changes are determined, the thermodynamic equation can be used to compute vertical motion and, with the help of the continuity equation, the mesoscale convergence. Thus, changes in cloud properties can be linked to mesoscale changes that impact subsequent convective clouds.

Following Ogura (1975) and Anthes (1977), the convective-cloud contribution to the local rate of change of the mesoscale temperature can be expressed as

$$\frac{\partial \bar{T}}{\partial t}(z)_c = \frac{L}{c_p}\bar{C}(z) - \frac{\partial}{\partial z}\overline{w'(z)T'(z)},   (8.1)$$

where $\bar{C}$ is the local condensation or evaporation rate, and the overbar and prime indicate the mean and eddy components, respectively, for a grid element with area A. Other variables have their usual interpretation, i.e., T is temperature, p is pressure, etc. The overbar quantities can be considered as representative of the mesoscale, while primed quantities can be thought of as the cloud-scale departures from the mesoscale. The first term on the rhs of (8.1) is the temperature change from latent heating. The fraction of latent heat release corresponding to the amount of condensate or sublimate reaching the ground represents the *net* heating by convection. The second term is the eddy-flux divergence, and, although it does not contribute to any *net* warming or cooling, it sometimes produces large vertical redistributions of heating. Equation (8.1) can be written in an expanded form to reflect the various components that contribute to changes in the mesoscale temperature. Specifically,

$$A\frac{\partial \bar{T}}{\partial t}\Big|_c = \frac{L}{c_p}(A_u C_u + A_d C_d + A_e C_c) - \frac{\partial}{\partial z}[A_u \overline{(w_u - \bar{w})(T_u - \bar{T})}$$
$$+ A_d \overline{(w_d - \bar{w})(T_d - \bar{T})} + A_e \overline{(w_e - \bar{w})(T_e - \bar{T})}], \quad (8.2)$$

where $A = A_u + A_d + A_e$ and the subscripts u, d and e indicate properties of cloud updrafts, moist downdrafts and the region of compensating vertical motions in the near-cloud environment. In practice, the cloud and near-environment properties are usually determined from simple cloud models that either are initialized with observed mesoscale conditions or are predicted by a mesoscale model. An estimate of the mesoscale vertical motion forced by the convective heating or cooling ($\dot{Q}_c$) can then be obtained from the thermodynamic equation in the form

$$\dot{Q}_c \equiv \frac{\partial \bar{T}}{\partial t}\Big|_c = \bar{w}(\Gamma - \gamma), \quad (8.3)$$

where Γ is the dry adiabatic lapse rate and $\bar{\gamma}$ is defined by $-\partial \bar{T}/\partial z$. This expression arises from the Boussinesq approximation and assumes that the mesoscale is hydrostatic and steady and that horizontal temperature advection is insignificant. It also assumes that the response of the mesoscale to the convective heating will be instantaneous. Clearly, great care must be taken not to apply (8.3) in situations where the mesoscale lapse rate is dry adiabatic (more general versions of this formulation to determine $\bar{w}$ avoid this difficulty). Finally, the mesoscale convergence field can be derived through the continuity equation expressed as

$$\nabla_2 \cdot \mathbf{V} = -\frac{\partial \bar{w}}{\partial z} - \frac{\bar{w}}{\bar{\rho}}\frac{\partial \bar{\rho}}{\partial z}. \quad (8.4)$$

Equations (8.2)–(8.4) constitute a closed system wherein cloud properties from seeded and unseeded clouds can be used to estimate the response of the mesoscale to convective heating and cooling. Similar expressions can be derived for moisture, momentum, etc.

Inspection of this simple system provides much-needed insight into what the likely mesoscale response will be to convective heating and cooling. For example, examination of (8.3) indicates that if the convectively produced local temperature change is positive (negative), the mesoscale will respond with an upward (downward) vertical circulation. Moreover, when Eq. (8.3) is considered with (8.4), it is apparent that the vertical variation of convectively produced local temperature change determines whether mesoscale convergence or divergence will occur in response to deep convection (provided that the local lapse rate does not vary substantially with height—a reasonable assumption in many convection situations). It is also apparent from (8.3) and (8.4) that convectively-driven mesoscale convergence (divergence) is likely when $\partial \bar{T}/\partial t|_c$ is increasing (decreasing) with height. Therefore, in order for modification treatments to produce enhanced *low-level* mesoscale convergence that will result in additional convective clouds or clouds with broader bases, the treatments must alter the clouds in such a way as to increase the vertical gradient of the local convective heating rate in the vicinity of cloud base and immediately above. Increased convergence would also occur if the vertical gradient of local convective cooling were increased (cooling decreasing with height), but this would primarily support the mesoscale downdrafts that develop in response to divergence of moist downdraft air in the boundary layer. For example, consider the vertical profiles of convectively produced local temperature change shown in Fig. 8.1. Ac-

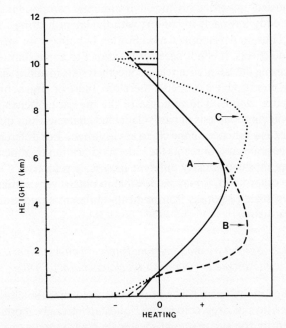

FIG. 8.1. Vertical profiles of convectively produced local temperature change. Curves A, B and C indicate profiles with middle-, lower- and upper-level heating maxima, respectively.

cording to (8.3), $\bar{w}$ for profile A would be directed upward in the layer from about 1 to 9 km and downward above and below this layer. In order to enhance mesoscale ascent in the vicinity of cloud base and possibly produce more or larger convective clouds, it would be most beneficial if seeding would increase the vertical gradient of $\partial\bar{T}/\partial t|_c$ in the lower portion of the cloud layer; e.g., see curve B. If the seeding increased $\partial\bar{T}/\partial t|_c$ at high levels, such as shown by C, then *midlevel* convergence would be enhanced.

If (8.2) is now considered, it is possible to examine how specific changes in cloud-scale characteristics impact $\partial\bar{T}/\partial t|_c$ and therefore the mesoscale environment. For example, it is clear from the first term on the rhs of (8.2) that an increase in the rate of release of latent heat will contribute toward increasing $\partial\bar{T}/\partial t|_c$. But, as indicated in the discussion above, it is not sufficient just to increase $\partial\bar{T}/\partial t|_c$ in order to enhance *low-level* convergence. The vertical distribution of the convective heating determines the levels where convectively driven mesoscale convergence and divergence occur. Thus, the height at which latent heat release is increased is also an important determining factor. Furthermore, even if latent heat release is increased in a layer favorable for enhancing low-level convergence, the second term in (8.2), the eddy flux, can adjust the heating profile so that any additional convergence will not necessarily occur below the layer of increased latent heating. Specifically, if seeding results in an updraft temperature increase, then, through the resulting additional buoyancy, w_u increases and the updraft component of the eddy-flux term can become large and shift the convective heating upward (see Anthes, 1977). Similarly, if downdrafts are enhanced, convective heating is adjusted upward. Only the environmental component of the eddy term acts in concert with the latent heating.

Thus, in developing a modification hypothesis for area-wide effects, it is not necessarily correct to argue that increasing the latent heat release directly results in additional low-level convergence and, therefore, more or larger convective clouds. The response of the mesoscale depends upon combinations of particular cloud characteristics that vary as a function of the cloud environment. For different large-scale environments the same modification treatment may produce entirely different mesoscale responses. On the other hand, it may be possible to obtain very similar mesoscale responses from radically different cloud-scale treatments.

8.2.2. Secondary modification: Impact of alteration of the mesoscale environment on cloud characteristics

For many years it has been recognized that the characteristics of deep convective clouds are strongly modulated by the synoptic and mesoscale environment (e.g., see Fawbush and Miller, 1954; Browning, 1964; Marwitz, 1972a,b,c; Moncrieff and Green, 1972; Cotton et al.,

1976). More explicitly, diagnostic and modeling studies have related the precipitation efficiency of convective clouds to, among other things, vertical wind shear (Marwitz, 1972d; Foote and Fankhauser, 1973; Cotton and Tripoli, 1978; Weisman and Klemp, 1982), low-level convergence (Lopez, 1973a; Anthes, 1977; Tripoli and Cotton, 1980), and environmental humidity (Lopez, 1973a,b). Therefore, if modification of individual clouds results in mesoscale changes in low-level forcing, cloud-layer humidity, vertical wind shear, etc., a second phase of modification of individual clouds occurs as new cloud growth adjusts to the modified environment. At this time, little is known about the magnitude of such "second phase" effects. However, the extreme sensitivity apparent in the modeling studies mentioned above suggests it may be significant.

8.3. Mesoscale modification experiments

8.3.1. STORMFURY

Modification hypothesis. Probably the earliest and most notable studies of mesoscale modification through seeding were those that focused on hurricanes. As far back as 1947, scientists considered the possibility of modifying hurricanes with the intent of reducing the storm's intensity or changing its path (see Braham and Neil, 1958; Simpson and Malkus, 1964). However, so little was known about the structure and energetics of hurricanes at that time that it was not until 1964, following nearly two decades of intensive observations, that a mesoscale modification hypothesis was conceived. Simpson and Malkus (1964) proposed that additional latent heat released by seeding the supercooled water present in the wall cloud may produce a hydrostatic pressure fall of sufficient magnitude to modestly reduce the surface pressure gradient and, therefore, the wind speed. Calculations suggested that reductions of 15% to 20% were possible. This hypothesis was modified somewhat when numerical experiments indicated that the modification treatment may have greater impact if latent heat release is enhanced at radii greater than that of the maximum surface wind (see Gentry, 1969). Subsequent numerical experiments by Rosenthal (1971) tended to support this latter hypothesis and indicated wind speeds would indeed decrease at radii interior to the "seeded" radii. As hypothesized, the reductions in wind speed in the numerical simulations were modest ($\approx$3–6 m s^{-1}) when appropriate heating rates were applied and, interestingly, did not decrease further even when extreme heating rates were introduced.

Since the characteristics of the boundary layer air and its rate of supply to the central portion of a hurricane are crucial to the strength and structure of the storm (Frank, 1977a,b; McBride, 1981), it is worthwhile to consider how seeding effects could significantly impact the boundary layer. Two likely possibilities are

1) hydrostatically, i.e., no significant vertical motions develop in response to the additional latent heat release and the warming is hydrostatically transmitted downward to the boundary layer as a pressure perturbation;

2) dynamically, i.e., enhanced vertical circulations in response to the seeding extend into the boundary layer and negatively alter its characteristics or circulation.

Rosenthal (1971) pointed out that the numerical model simulation responded to additional heat with a sufficient increase in vertical circulations so that no local "accumulation" of heat occurred, i.e., there was a predominantly *dynamic* response where adiabatic cooling by vertical motion offset the local warming. Moreover, since the heating was introduced at 500 mb and above and there was no attempt to include the effects of moist downdrafts, the mesoscale circulation response largely appeared in the middle and upper levels (see section 8.2). Thus, for these simulations, it is difficult to see how the seeding effects could significantly interfere with boundary-layer processes and substantially alter the storm. Of course, as pointed out in the original hypotheses (Simpson and Malkus, 1964; Gentry, 1969), only a slight reduction in wind speed would very likely produce a great reduction in damage.

Anthes (1971) also performed sensitivity experiments on hurricanes using a three-layer, two-dimensional (axially symmetric) numerical model that was an analog of the Anthes et al. (1971) three-dimensional model. In these experiments, seeding effects were not explicitly considered. Rather, it was hypothesized that deep convection, and therefore convective heating, could somehow be increased at prescribed radii and levels. The amount of heating was approximately the same as Gentry (1969) estimated could be obtained from the release of the latent heat of fusion. Anthes argued that if additional *condensation* heating is introduced, then a corresponding amount of moisture must be removed (net drying) and the bulk of this moisture would very likely come from the boundary layer and/or middle levels. The results of his experiments clearly indicated that the largest changes in central pressure and surface wind speed occurred when the moisture was removed from the boundary layer. Maximum wind speeds decreased by approximately 40% and central pressures rose as much as 26 mb. On the other hand, when the moisture was primarily removed from the middle layer, the results were similar to Rosenthal's (1971), i.e., only a 3 to 4 m s^{-1} reduction in maximum wind speed occurred.

Admittedly, it cannot easily be argued how additional deep convection can be produced at arbitrary radii. However, the reason Anthes' experiments are important from the standpoint of cloud seeding is that a comparable effect (i.e., drying of the boundary layer) may occur as a result of seeding if the seeding produces or enhances moist downdrafts. This is especially true for downdrafts originating at midlevels, such as suggested by Simpson (1980)

and simulated by Fritsch and Chappell (1981). Moreover, moist downdrafts would introduce the additional negative impact of *cooling* the boundary layer. While the potential for moist downdrafts in the eyewall region of a hurricane is very small, the same is not true for the bands of deep convection propagating toward the storm center. In fact, in a recent analysis (Barnes et al., 1983) of data taken from low-level (150–6400 m) aircraft passes through Hurricane Floyd (7 September 1981), it was found that cloud-scale moist downdrafts from convection in a rainband decreased the equivalent potential temperature in the lowest kilometer by as much as 12 K. This cooler/drier air then *spiraled inward toward the eyewall*. Thus, it may be possible that modification treatments which enhance moist downdrafts in the rainbands could result in a slight cooling or drying of the eyewall region and therefore a reduction in the storm's intensity. On the other hand, one could also hypothesize that the stabilizing action of the rainband downdrafts enhances the likelihood for the storm to deepen. Specifically, the downdrafts may be at least partially responsible for the fact that the eyewall region is primarily nonconvective (i.e., mean moist mesoscale ascent). Nonconvective saturated ascent tends to produce a vertical distribution of latent heating with a low-level maximum, and theoretical studies of mesoscale vortices (e.g., Koss, 1976) as well as numerical studies (Anthes and Keyser, 1979; Sardie, 1984; Zhang, 1985) have shown that the most rapid deepening occurs when the latent heat release maximum is in the lower troposphere.

Field experiments. In addition to the numerical-simulation sensitivity studies, Project STORMFURY (Gentry, 1969) conducted several field experiments to test the hurricane-modification hypotheses. The results of these experiments (see Simpson and Malkus, 1964; Gentry, 1970; Hawkins, 1971) were generally encouraging and in fair agreement with the hypotheses and numerical simulations. If anything, the impact of the seeding appeared to be stronger than hypothesized, possibly due to the inadvertent strengthening of moist downdrafts or to chance since the sample was small. However, as can be expected with weather-modification field experiments of this magnitude, there are virtually always large natural variabilities that are difficult to evaluate and factors external to the science that preclude continuing long-term investigation. Consequently, the quantitative impact of the seeding of hurricanes remains a potentially exciting, unresolved problem.

8.3.2. *The Florida Area Cumulus Experiment (FACE)*

Modification hypothesis. The FACE hypothesis, as outlined by Woodley et al. (1982), is composed of a complicated, multiscale chain of events. It begins with the rapid glaciation of the updraft regions of supercooled water by silver iodide seeding and culminates with an enhanced mesoscale circulation providing additional mass and moisture to additional large clouds precipitating more efficiently in a more moist environment. The hypothesis concludes with the statement that the seeding increases rainfall over the experimental area by

(i) obtaining more rain from the available moisture than would have occurred naturally, and/or

(ii) enhancing the moisture supply to the target area.

Item (i) requires either the development of more clouds than would have happened naturally or greater precipitation efficiency from individual clouds or both. It does not necessarily require an enhanced mesoscale circulation if cloud-scale circulations can initiate new growth. On the other hand, item (ii) clearly does require an enhanced mesoscale circulation. Using the simple set of equations from section 8.2.1, it is possible to estimate the mesoscale response to changes in cloud characteristics for seeded versus nonseeded convection in the Florida area. To do this, the cloud model results of Simpson and Wiggert (1971) are most suitable. Specifically, Figs. 8.2 and 8.3 show model-predicted seeded and unseeded vertical profiles of $T_u - \bar{T}$ and $w_u - \bar{w}$ for two experimental days in the 1968 Florida Cumulus Seeding Experiment. The profiles in Fig. 8.2 are for a seeded case (cloud 3, 16 May 1968) in which large increases in cloud height and precipitation were predicted by the model and observed by aircraft and radar. The profiles in Fig. 8.3 are also for a seeded case (cloud 17, tower B, 30 May 1968), but with

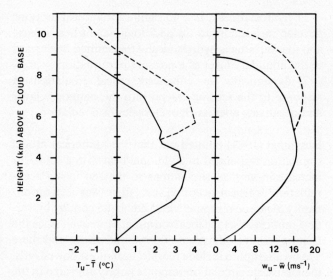

FIG. 8.3. As in Fig. 8.2 except for cloud 17, tower B, 30 May 1968.

naturally occurring, much deeper clouds and considerably smaller height and precipitation increases from seeding. Figure 8.4 shows the corresponding profiles of the product $(T_u - \bar{T})(w_u - \bar{w})$, where it is assumed that $\bar{w}$ is small relative to w_u. If these profiles (Figs. 8.2–8.4) are considered in the context of Eqs. (8.2)–(8.4), it is apparent that in both cases the naturally occurring convection will enhance low-level convergence, but that neither the additional condensation heating from seeding nor the eddy-flux adjustments will result in additional *low-level* me-

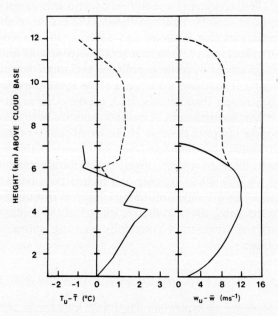

FIG. 8.2. Model-predicted seeded and unseeded vertical profiles of $T_u - \bar{T}$ (°C) and $w_u - \bar{w}$ (m s⁻¹) for cloud 3, 16 May 1968 (from Simpson and Wiggert, 1971). Solid lines indicate unseeded profiles; dashed lines show profiles for the additional growth from seeding.

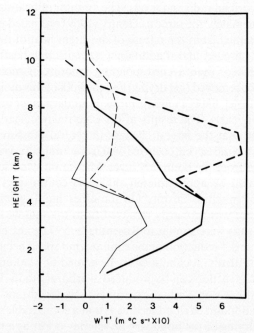

FIG. 8.4. Vertical profiles of $(T_u - \bar{T})(w_u - \bar{w})$. Solid lines indicate unseeded profiles; dashed lines show profiles for the additional growth from seeding. Thin lines are for cloud 3, 16 May 1968; heavy lines are for cloud 17, tower B, 30 May 1968.

soscale convergence—only middle- and upper-level convergence will be increased (see Levy and Cotton, 1984, for a "cloud scale" discussion of this same point). On the other hand, if *moist downdrafts* are enhanced by seeding and *additional clouds* develop as a result of the lifting along the outflow boundaries (such as simulated by Orville and Chen, 1982), then additional heating would occur in virtually the entire cloud layer and enhanced subcloud-layer convergence would result. Moreover, the additional low-level inflow would be forced to ascend over the shallow mesoscale dome of cool downdraft air, so lifting would be further enhanced. This argument not only supports the concepts proposed by Simpson (1980) and Tripoli and Cotton (1980), but it also suggests how seeding-induced cloud changes can enhance the precipitation efficiency of subsequent towers and clouds; i.e., increased midlevel convergence and ascent would tend to moisten the convective cloud environment and thereby reduce the negative effect of dry-air entrainment. Furthermore, in some instances, the midlevel convergence and lifting is sufficiently strong to saturate the midlevel environment and directly contribute to the overall production of precipitation (see Brown, 1979; Houze, 1977; Leary and Houze, 1979; Maddox, 1980; Gamache and Houze, 1983). As such a layer of midlevel saturated ascent develops, it adjusts the convectively dominated level of maximum latent heat release and the layer of maximum vertical gradient in latent heat release downward, which tends to enhance convergence in progressively lower levels of the convective cloud system.

From the preceding brief analysis of the mesoscale response to convective heating, and in view of the order-of-magnitude increases in precipitation that occur as a result of merger (Simpson et al., 1980), it would seem that the first modification treatments on an experimental day in the Florida area should not necessarily concentrate on increasing rainfall from individual clouds or towers. It may be that seeding to enhance moist downdrafts and thereby increase the number (area) of new clouds and towers will, when considered over mesoscale time and space, prove much more fruitful. If maximum downdraft enhancement happens to correspond with maximum cloud-scale rainfall enhancement, so much the better.

In the context of the discussion immediately above, three focal points become important in identifying mesoscale modification potential:

1) the enhancement of downdrafts,
2) the extent of *mid*level moistening by mesoscale ascent, and
3) the ease with which new clouds or towers can be initiated.

This latter point is very important since previous studies have shown how the timing, location and even the type of convective cloud is strongly influenced by the amount of lift required to overcome stable layers and force boundary-layer air to its level of free convection (e.g., see Foote and Fankhauser, 1973; Browning and Foote, 1976; Cotton et al., 1976; Matthews and Silverman, 1980; Caracena and Fritsch, 1983). In this regard, it is not at all difficult to imagine situations in which the characteristics of the clouds are very much the same on two different days but the required lift to initiate new growth may be drastically different. Consequently, there may be days when the modification potential for individual clouds and towers is great, but the potential for new growth, merger and mesoscale organization is poor. Obviously, the reverse is also true. Clearly then, field experiments must distinguish between cloud-scale and mesoscale modification potential and, most importantly, when the two scales work synergistically.

Field results. The early results from FACE (1970–76) were exciting and highly encouraging. Woodley et al. (1977, 1982) presented strong evidence that dynamic seeding significantly augmented areawide rainfall over portions of South Florida on selected experimental days (see Orville's review). However, the second phase of FACE, conducted during the summers of 1978–80, failed to confirm the results of the first experimental period (see Woodley et al., 1983; Barnston et al., 1983). Woodley et al. (1983) attributed the failure to

1) an unknown and/or possibly intermittent seeding effect,
2) a limited sample size, and
3) inadequate predictor equations.

The latter two reasons reflect the difficulty of eliminating natural variability in attempting to isolate a seeding effect. Estoque and Partagas (1974) and Nickerson (1979, 1981) pointed out the overwhelming impact of synoptic and mesoscale forcing (or its absence) in determining the characteristics and evolution of a convective event. In this context, perhaps a most important result of the FACE field projects would be the recognition that the "seedability" of a given day should not be defined in terms of just the individual cloud seedability (see Tripoli and Cotton, 1980). Until a reliable, quantitative assessment of the natural evolution of mesoscale and cloud-scale interactions is possible, i.e., until better multiscale predictability is developed, it is not likely that enough of the natural variability of these extremely complex convective systems will be removed to confidently establish accepted modification treatments for convective clouds and cloud systems.

8.3.3. Other studies

After the apparent successes of the early cloud-seeding experiments in the Florida area, Fritsch and Chappell (1981) conducted numerical tests of the modification sensitivity of mesoscale convective weather systems. Their studies addressed two modification possibilities: 1) dynamic seeding and 2) alteration of the timing and location of initial convection. Only the dynamic-seeding results will be discussed here.

In order to perform the sensitivity experiments, the effects of deep convection were parametrically included (Fritsch and Chappell, 1980a) in a fine mesh ($\Delta x = 20$ km), 20-level, primitive equation model (Fritsch and Chappell, 1980b). The convective parameterization contained a cloud model in which condensate-freezing in updrafts could be specified as a function of temperature. Dynamic seeding was simulated by specifying freezing at temperatures above $-25°C$. This resulted in additional buoyant energy being available to convective updraft parcels. Moreover, because downdrafts are linked to updrafts in the Fritsch–Chappell cloud model, additional updraft mass flux increases moist downdraft mass flux when "dynamic seeding" is introduced. Consequently, in the model, the "seeding" automatically has a link to the boundary layer.

Starting from identical initial conditions, two simulations were performed: a control run with updraft freezing at $-25°C$ and a dynamic-seeding run with freezing at $-10°C$. The impact of the seeding produced several obvious differences in the mesoscale structure and circulation. The most significant of these are

- stronger pressure systems (i.e., larger pressure gradients),
- stronger convergence, divergence and vertical motion,
- more rapid propagation,
- increased area of precipitation, and
- increased precipitation.

Fritsch and Chappell pointed out that although mesoscale convergence was increased by the additional latent heat release, the most promising link to additional precipitation seemed to be the enhancement of new growth by strengthening mesohigh outflow. This was based upon the observation that the additional latent heat release in the updraft primarily increased *middle-* and *upper*-level convergence. Enhanced *low-level* forcing seemed to be a result of the stronger downdrafts.

8.4. Concluding remarks

The significance of the mesoscale environment in determining the characteristics of individual convective clouds has long been recognized. Moreover, there is now strong evidence that in many instances convective clouds have a comparable impact on the mesoscale environment. Consequently, convective-cloud modification hypotheses for precipitation enhancement should consider not only cloud-scale microphysical changes, but also whether or not these changes interact constructively with their environment to enhance subsequent clouds, cloud towers and the mesoscale circulations that support them. Before attempting to produce areawide precipitation increases, great care must be taken to ascertain if the modification treatment works in cooperation with natural mesoscale processes or instabilities (e.g., cloud streets, symmetric instability, gravity waves, sea breeze circulations, etc.) and if the cumulative effect of seeding enhances or impairs the environment for precipitation production from subsequent clouds.

It is also becoming increasingly evident that the role of moist downdrafts in initiating, organizing and maintaining deep convection needs much more clarification. Diagnostic studies and cloud-scale and mesoscale numerical simulations indicate that moist downdrafts significantly impact the evolution of convective weather systems. Yet little is known about the modification potential of moist downdrafts and its inherent relationship to updraft modification. Moreover, in this same vein, little is known about the modification potential (microphysical characteristics) of the midlevel, quasi-stratiform, cloud layer that develops in response to the convective cloud heating. Since this cloud layer contributes up to half of the rainfall from mesoscale convective weather systems, it also deserves additional attention in future studies.

In summary, it is time for a broader view of modification of convective clouds where it is recognized that the updraft microphysics are only one portion (albeit an important one) of a complex, multiscale process. Progress in understanding a single component of this process is intrinsically limited if understanding of all major components is not developed to a comparable level. The technological tools (both observational and numerical) to comprehensively address the multifaceted problem of convective cloud modification are now at hand. The coming decade is likely to see great strides in developing operational modification technology as new understanding of the multiscale nature of convection is brought to bear on the problem.

Acknowledgments. I am grateful to Bill Frank, Nels Shirer, Paul Hirschberg, Peter Webster, Da-lin Zhang and Harry Orville for their helpful discussions and suggestions, and to Sue Frandsen for her expert assistance in preparing the manuscript and figures. This work was supported by USAF AFOSR-83-0064 and NSF Grants ATM-8218208 and 8418995.

REFERENCES

Anthes, R. A., 1971: The response of a 3-level axisymmetric hurricane model to artificial redistribution of convective heat release. NOAA Tech. Memo. ERL NHRL-92, 14 pp. [NTIS COM-71-00711.]
——, 1977: A cumulus parameterization scheme utilizing a one-dimensional cloud model. *Mon. Wea. Rev.,* **105,** 270–286.

——, and D. Keyser, 1979: Tests of a fine-mesh model over Europe and the United States. *Mon. Wea. Rev.,* **107,** 963–984.

——, J. W. Trout and S. L. Rosenthal, 1971: Comparisons of tropical cyclone simulations with and without the assumption of circular symmetry. *Mon. Wea. Rev.,* **99,** 759–766.

Barnes, G. M., E. J. Zipser, D. Jorgensen and F. Marks, Jr., 1983: Mesoscale and convective structure of a hurricane rainband. *J. Atmos. Sci.,* **40,** 2125–2137.

Barnston, A. G., W. L. Woodley, J. A. Flueck and M. H. Brown, 1983: The Florida Area Cumulus Experiment's second phase (FACE-2). Part I: The experimental design, implementation and basic data. *J. Appl. Meteor.,* **22,** 1504–1528.

Braham, R. R., and E. A. Neil, 1958: Modification of hurricanes through cloud seeding. U.S. Dept. of Commerce, Nat. Hurr. Res. Proj. Rep. No. 16, 12 pp.

Brown, J. M., 1979: Mesoscale unsaturated downdrafts driven by rainfall evaporation: A numerical study. *J. Atmos. Sci.,* **36,** 313–338.

Browning, K. A., 1964: Airflow and precipitation trajectories with severe storms which travel to the right of the winds. *J. Atmos. Sci.,* **21,** 634–639.

——, and G. B. Foote, 1976: Airflow and hail growth in supercell storms and some implications for hail suppression. *Quart. J. Roy. Meteor. Soc.,* **102,** 499–533.

Caracena, F., and J. M. Fritsch, 1983: Focusing mechanisms in the Texas Hill Country flash floods of 1978. *Mon. Wea. Rev.,* **111,** 2319–2332.

Chen, C. H., and H. D. Orville, 1980: Effects of mesoscale convergence on cloud convection. *J. Appl. Meteor.,* **19,** 256–274.

Cotton, W. R., and A. Boulanger, 1975: On the variability of "dynamic seedability" as a function of time and location over south Florida: Part I. Spatial variability. *J. Appl. Meteor.,* **14,** 710–717.

——, and G. J. Tripoli, 1978: Cumulus convection in shear-flow three-dimensional numerical experiments. *J. Atmos. Sci.,* **35,** 1503–1521.

——, R. A. Pielke and P. T. Gannon, 1976: Numerical experiments on the influence of the mesoscale circulation on the cumulus scale. *J. Atmos. Sci.,* **33,** 252–261.

Cunning, J. B., R. L. Holle, P. T. Gannon and A. I. Watson, 1982: Convective evolution and merger in the FACE experimental area: Mesoscale convection and boundary layer interactions. *J. Appl. Meteor.,* **21,** 953–977.

Emanuel, K. A., 1979: Inertial instability and mesoscale convective systems, Part I: Linear theory of inertial instability in rotating viscous fluids. *J. Atmos. Sci.,* **36,** 2425–2449.

Estoque, M. A., and J. J. Fernandez-Partagas, 1974: Precipitation dependence on synoptic-scale conditions and cloud seeding. *Geofis. Int.,* **14,** 181–206.

Fawbush, E. J., and R. C. Miller, 1954: The types of airmasses in which North American tornadoes form. *Bull. Amer. Meteor. Soc.,* **35,** 154–165.

Flueck, J. A., 1971: Statistical analyses of the ground level precipitation data. Project Whitetop, Part V. Dept. of Geophysical Sciences, University of Chicago.

Foote, G. B., and J. C. Fankhauser, 1973: Airflow and moisture budget beneath a northeast Colorado hailstorm. *J. Appl. Meteor.,* **12,** 1330–1353.

——, and C. G. Wade, 1982: Case study of a hailstorm in Colorado. Part I: Radar echo structure and evolution. *J. Atmos. Sci.,* **39,** 2828–2846.

Frank, W. M., 1977a: The structure and energetics of the tropical cyclone, Paper I: Storm structure. *Mon. Wea. Rev.,* **105,** 1119–1135.

——, 1977b: The structure and energetics of the tropical cyclone, Paper II: Dynamics and energetics. *Mon. Wea. Rev.,* **105,** 1136–1150.

——, 1978: The life cycles of GATE convective systems. *J. Atmos. Sci.,* **35,** 1256–1264.

——, 1983: The cumulus parameterization problem. *Mon. Wea. Rev.,* **111,** 1859–1871.

Fritsch, J. M., and C. F. Chappell, 1980a: Numerical prediction of convectively driven mesoscale pressure systems. Part I: Convective parameterization. *J. Atmos. Sci.,* **37,** 1722–1733.

——, and ——, 1980b: Numerical prediction of convectively driven mesoscale pressure systems. Part II: Mesoscale model. *J. Atmos. Sci.,* **37,** 1734–1762.

——, and ——, 1981: Preliminary numerical tests of the modification of mesoscale convective systems. *J. Appl. Meteor.,* **20,** 910–921.

Gamache, J. F., and R. A. Houze, Jr., 1983: Water budget of a mesoscale convective system in the tropics. *J. Atmos. Sci.,* **40,** 1835–1850.

Gentry, R. C., 1969: Project STORMFURY. *Bull. Amer. Meteor. Soc.,* **50,** 404–409.

——, 1970: Hurricane Debbie modification experiments, August 1969. *Science,* **168,** 473–475.

Hawkins, H. F., 1981: Comparison of results of the hurricane Debbie (1969) modification experiments with those from Rosenthal's numerical model simulation experiments. *Mon. Wea. Rev.,* **99,** 427–434.

Houze, R. A., Jr., 1977: Structure and dynamics of a tropical squall-line system. *Mon. Wea. Rev.,* **105,** 1540–1567.

——, and P. V. Hobbs, 1982: Organization and structure of precipitating cloud systems. *Advances in Geophysics,* Vol. 24, Academic Press, 225–315.

Klemp, J. B., and R. B. Wilhelmson, 1978: Simulations of right- and left-moving storms produced through storm splitting. *J. Atmos. Sci.,* **35,** 1097–1110.

Koss, W. J., 1976: Linear stability of CISK-induced disturbances: Fourier component eigenvalue analysis. *J. Atmos. Sci.,* **33,** 1195–1222.

Kreitzberg, C. W., and D. J. Perkey, 1977: Release of potential instability: Part II. The mechanism of convective–mesoscale interaction. *J. Atmos. Sci.,* **34,** 1569–1595.

Leary, C. A., and R. A. Houze, Jr., 1979: Melting and evaporation of hydrometeors in precipitation from the anvil clouds of deep tropical convection. *J. Atmos. Sci.,* **36,** 669–679.

Levy, G., and W. R. Cotton, 1984: A numerical investigation of mechanisms linking glaciation of the ice-phase to the boundary layer. *J. Climate Appl. Meteor.,* **23,** 1505–1519.

Lopez, R. E., 1973a: A parametric model of cumulus convection. *J. Atmos. Sci.,* **30,** 1354–1373.

——, 1973b: Cumulus convection and larger scale circulations. II: Cumulus and mesoscale interactions. *Mon. Wea. Rev.,* **101,** 856–870.

McBride, J. L., 1981: Observational analysis of tropical cyclone formation. Part III: Budget analysis. *J. Atmos. Sci.,* **38,** 1152–1166.

Maddox, R. A., 1980: Mesoscale convective complexes. *Bull. Amer. Meteor. Soc.,* **61,** 1374–1387.

——, C. F. Chappell and L. R. Hoxit, 1979: Synoptic and mesoscale aspects of flash flood events. *Bull. Amer. Meteor. Soc.,* **60,** 115–123.

——, D. J. Perkey and J. M. Fritsch, 1981: Evolution of upper tropospheric features during the development of a mesoscale convective complex. *J. Atmos. Sci.,* **38,** 1664–1674.

Marwitz, J. D., 1972a: The structure and motion of severe hailstorms. Part I: Supercell storms. *J. Appl. Meteor.,* **11,** 166–179.

——, 1972b: The structure and motion of severe hailstorms. Part II: Multi-cell storms. *J. Appl. Meteor.,* **11,** 180–188.

——, 1972c: The structure and motion of severe hailstorms. Part III: Severely sheared storms. *J. Appl. Meteor.,* **11,** 189–201.

——, 1972d: Precipitation efficiency of thunderstorms on the High Plains. *J. Rech. Atmos.,* **6,** 367–370.

Matthews, D. A., 1981: Natural variability of thermodynamic features affecting convective cloud growth and dynamic seeding: A comparative summary of three High Plains sites from 1975 to 1977. *J. Appl. Meteor.,* **20,** 971–996.

——, and B. A. Silverman, 1980: Sensitivity of convective cloud growth to mesoscale lifting: A numerical analysis of mesoscale convective triggering. *Mon. Wea. Rev.,* **108,** 1056–1063.

Moncrieff, M. W., and J. S. A. Green, 1972: The propagation and transfer properties of steady convective overturning in shear. *Quart. J. Roy. Meteor. Soc.,* **98,** 336–352.

Nickerson, E. C., 1979: FACE rainfall results: Seeding effect or natural variability? *J. Appl. Meteor.,* **18,** 1097–1105.

——, 1981: The FACE-1 seeding effect revisited. *J. Appl. Meteor.,* **20,** 108–114.

Ogura, Y., 1975: On the interaction between cumulus clouds and the large scale environment. *Pure Appl. Geophys.* **113,** 869–890.

Orville, H. D., and J. M. Chen, 1982: Effects of cloud seeding, latent heat of fusion, and condensate loading on cloud dynamics and precipitation evolution: A numerical study. *J. Atmos. Sci.,* **39,** 2807–2827.

Parsons, D. B., and P. V. Hobbs, 1983: The mesoscale and microscale structure and organization of clouds and precipitation in mid-latitude cyclones. XI: Comparison between observational and theoretical aspects of rainbands. *J. Atmos. Sci.,* **40,** 2377–2397.

Rosenthal, S. L., 1971: A circularly symmetric primitive equation model of tropical cyclones and its response to artificial enhancement of the convective heating functions. *Mon. Wea. Rev.,* **99,** 414–426.

Sardie, J. M., 1984: On development mechanisms for polar lows. Ph.D. dissertation, The Pennsylvania State University, 220 pp.

Simpson, J., 1980: Downdrafts as linkages in dynamic cumulus seeding effects. *J. Appl. Meteor.,* **19,** 477–487.

——, and V. Wiggert, 1971: 1968 Florida cumulus seeding experiment: Numerical model results. *Mon. Wea. Rev.,* **99,** 87–118.

——, N. E. Westcott, R. J. Clerman and R. A. Pielke, 1980: On cumulus mergers. *Arch. Meteor. Geophys. Bioklim.,* **A29,** 1–40.

Simpson, R. H., and J. S. Malkus, 1964: Experiments in hurricane modification. *Sci. Amer.,* **211,** 27–37.

Smith, P. L., A. S. Dennis, B. A. Silverman, A. B. Super, E. W. Holroyd III, W. A. Cooper, P. W. Mielke, Jr., K. J. Berry, H. D. Orville and J. R. Miller, Jr., 1984: HIPLEX-1: Experimental design and response variables. *J. Appl. Meteor.,* **23,** 497–512.

Tripoli, G. J., and W. R. Cotton, 1980: A numerical investigation of several factors contributing to the observed variable intensity of deep convection over south Florida. *J. Appl. Meteor.,* **19,** 1037–1063.

Weisman, M. L., and J. B. Klemp, 1982: The dependence of numerically simulated convective storms on vertical wind shear and buoyancy. *Mon. Wea. Rev.,* **110,** 504–520.

Woodley, W. L., J. Simpson, R. Biondini and J. Jordan, 1977: Rainfall results, 1970–1975: Florida Area Cumulus Experiment. *Science,* **195,** 735–742.

——, J. Jordan, A. Barnston, J. Simpson, R. Biondini and J. Flueck, 1982: Rainfall results of the Florida Area Cumulus Experiment, 1970–76. *J. Appl. Meteor.,* **21,** 139–164.

——, A. Barnston, J. Flueck and R. Biondini, 1983: The Florida Area Cumulus Experiment's second phase (FACE-2). Part II: Replicated and confirmatory analyses. *J. Appl. Meteor.,* **22,** 1529–1540.

Xu, Q., and J. H. E. Clark, 1984: Wave CISK and mesoscale convective systems. *J. Atmos. Sci.,* **41,** 2089–2107.

Zhang, D.-L., 1985: Nested-grid simulation of the meso-β scale structure and evolution of the Johnstown flood of July 1977. Ph.D. dissertation, The Pennsylvania State University, 251 pp.

CHAPTER 9

Review of Wintertime Orographic Cloud Seeding

ROBERT D. ELLIOTT

North American Weather Consultants, Salt Lake City, Utah

ABSTRACT

This review provides a sketchy background of orographic weather modification activities prior to the 1960s, followed by a more critical review of major orographic projects carried out and reported in the scientific literature during the past 25 years. In the earlier of these major projects, evaluation of results had been based largely upon comparisons of seeded and nonseeded precipitation experimental units stratified by various sounding-derived parameters in an attempt to amplify the physical significance of the seeding effects within various sub-types of orographic clouds.

The later major projects are still underway with no final evaluations having been presented. However, a wealth of significant data analyses have been reported that provide important insights into the various natural and seeding precipitation mechanisms. Much of this is attributable to the new observational tools in use, which include airborne and ground microphysical sensors, doppler radar, and microwave radiometers.

9.1. Early seeding projects

9.1.1. Initial projects

The scientific basis for precipitation enhancement occurred with the advent of the Bergeron–Findeisen theory, first presented by Bergeron in 1935 (Bergeron, 1935), and expanded by Findeisen in 1938 (Findeisen, 1938). A deficit of atmospheric ice-forming nuclei was recognized, there being examples of supercooled water appearing in clouds at temperatures as cold as $-20°C$ or even $-30°C$. Bergeron estimated that a single ice crystal in a cubic centimeter of cloud having a liquid water content of 4.2 g m^{-3} dispersed in 1000 droplets of 20 m diameter would freeze all of the water (by diffusion) in 10–20 minutes.

A thorough review of this and other background by Byers appears in the first chapter of *Weather and Climate Modification* (Hess, 1974). In this review the focus will be on developments in the general area of wintertime orographic weather modification. Even with this sharp a focus, restrictions prevent complete coverage of this special area.

The first tests of both dry ice and AgI seeding were carried out during 1946 and 1947 by a General Electric Laboratory group that included Langmuir, Schaefer, and Vonnegut. Schaefer's famous dry ice seeding in a stratocumulus cloud occurred near Schenectady, New York, on 13 November 1946. It followed on his discovery during the summer of 1946 that dry ice could be used to "seed" a cloud of small supercooled cloud droplets produced in a deep freezer. Dry ice particles chilled the air near them to the critical temperature level of $-40°C$, below which spontaneous formation of small ice embryos occurred (Schaefer, 1946). Shortly after, Vonnegut discovered that

silver iodide smoke was effective as an ice-forming nucleus at temperatures as high as $-4°C$ because its crystalographic properties resemble those of ice (Vonnegut, 1949). In February 1947, Project Cirrus began, with intensive field and laboratory experiments of seeding effects being carried out by an expanded General Electric research staff over the next five years. Many interesting details of this famous project have been reviewed by several of the original participating scientists (Schaefer, 1981; Blanchard, 1981; Falconer, 1981; and especially Havens et al., 1981).

The earliest weather modification projects included some carried out on a strictly operational basis, as well as those designed for randomized testing. The former were the only ones available to the Advisory Committee on Weather Control (1954–57) (Advisory Committee, 1957). The National Academy of Sciences (NAS, 1966) reevaluated some of the same projects and several new ones. The conclusions were quite favorable with respect to orographic seeding for enhancing winter snowpack in western mountain areas. As a result of their reports, support for scientifically designed seeding experiments was stimulated.

A series of Australian randomized experiments (mostly 1950s and early 1960s) included two within a strongly orographic setting. These were the Snowy Mountain project (Smith et al., 1963) and the Tasmanian project (Smith et al., 1971 and 1979). The seeding in these and other experiments was carried out by aircraft equipped with high-output silver iodide generators. Randomization used a crossover design, involving two alternating target areas. The response variate was 24-hour precipitation blocks, but since the seeding activity was limited to about a quarter of the 24-hour block, there was considerable dilution of the effects of seeding. More details on the Australian experiments will be found in Hess (1974, Chap. 12).

Evaluations based upon target/control area seasonal streamflow comparisons, where historical correlation coefficients far exceeded 0.9, were published for two long-term orographic projects in the southern Sierra Nevada (upper San Joaquin from 1950 and Kings River from 1954) by the operators and sponsors (Henderson, 1966; Elliott and Lang, 1967). A significant early evaluation of a randomized project in the Lake Almanor Basin, Sierra Nevada, stratified cases by air mass temperature and found positive seeding response with the colder air masses (Mooney and Lunn, 1968).

In what follows, the term "project" will refer to long-term field investigations of seeding effects and of background natural precipitation, which may and may not contain designed experiments. The term "experiment" is reserved exclusively for fully designed randomized experiments. The term "operational" pertains to those projects designed to use state-of-the-art methods to enhance precipitation. Some of the latter contain knowledge-enhancement features such as special data collections.

In the interests of brevity, the acronyms shown in Table 9.1 will be employed. These acronyms pertain to various scientific projects, most of which involved cloud seeding experiments. The locations of the ones being given major attention in this review are shown in Fig. 9.1.

9.1.2. Climax

Climax I and II, an exploratory and a following confirmatory experiment, were carried out in the Colorado Rockies near the town of Climax over a target area at altitudes ranging from 3 to 4 km. The terrain was complex

FIG. 9.1. Location of key orographic seeding experiments.

TABLE 9.1. Acronyms employed in text.

Climax	The randomized seeding experiment carried out in the vicinity of Climax, Colorado (1960 to 1970).
CRBPP	The Colorado River Basin Pilot Project, a Bureau of Reclamation experiment conducted in the San Juan Mountains of Colorado (1965–71).
SCPP	The Sierra Cooperative Pilot Project, a Bureau of Reclamation experiment being carried out over the central Sierra Nevada range of California (1976 to date).
COSE	The Colorado Seeding Experiment carried out in the northern Colorado Rockies by a Colorado State University group under a NSF grant.
SBA II	The experiment carried out over Santa Barbara and adjoining counties in California from 1967 to 1973, supported by the Earth and Planetary Sciences Division, Naval Ordinance Test Station, China Lake, California, and by the Bureau of Reclamation.
Utah Federal–State	The Utah "piggy back" research project carried out under NOAA support.

but generally featured north–south ridgelines. Both of these experiments used 24-hour experimental units. A network of strategically placed ground-based AgI–NaI generators provided the artificial nuclei and were operated whenever the official Weather Bureau forecast indicated snowfall was expected during the 24-hour period and the randomization called for treatment.

Mielke and Grant (1967) presented the results for Climax I (1960–65), finding that one of the more important positive responses to the seeding lay within a 500 mb temperature stratification from −10° to −25°C. Other stratifications included wind direction and equivalent potential temperature. Mielke et al. (1970) presented evidence of the spatial distribution of effects over a number of passes in the general vicinity of Climax. Chappell et al. (1971) presented strong evidence that an important seeding effect is to increase the duration of precipitation. Precipitation was initiated many hours earlier in the warmer clouds in the seeded than in the nonseeded cases. For the following similar confirmatory project, Climax II (1965–70), Mielke et al. (1971) presented results that essentially confirmed the findings for Climax I.

The 500 mb temperature was interpolated, as were other sounding parameters, from surrounding National Weather Service rawinsonde stations. This scheme was considered to represent approximately the cloud-top temperature in the Climax area, which in turn was believed to be an index of the supply of natural nuclei to the cloud. This nuclei supply was postulated to have an important bearing on the natural rate of removal of water by ice particles, and seeding was intended to supplement this supply in warmer top clouds, where it was low. Chappell and Johnson (1974) presented results from a numerical model for a mixed-phase, cold orographic cloud that showed a threshold cloud-top temperature of $-25°C$, above which precipitation augmentation is not possible due to an adequate supply of natural nuclei.

The statistical task force of the Weather Modification Advisory Board (Tukey et al., 1978) reported that, along with most other experiments, Climax did not meet the strict confirmatory criteria that they had outlined. At the time of their report the original Climax investigators were reanalyzing the data to review a possible source of error (Meltesen et al., 1977)—a regionwide precipitation pattern that could have, by chance, favored randomly selected seeding days in some data partitions. In addition, Hobbs and Rangno (1979) disputed the physical hypothesis, which they believed did not adequately consider secondary ice production. They also challenged the use of the 500 mb temperature to represent cloud-top temperature.

Mielke et al. (1981) used covariate (control area) relationships developed before Climax II in a reanalysis of the replicated Climax I and II experiments. The introduction of such covariates actually strengthened the statistical evidence of effects. But the reanalysis also showed that previous estimates of increases based upon seeded to nonseeded precipitation ratios were apparently too large. The combined Climax I and II increase for the warm 500 mb temperature category was reduced from 41% to 25%.

Mielke et al. (1982) developed a distortion-resistant residual analysis and applied it to Climax I and II. The shortcomings of the use of least-squares fits and linear-rank test statistics used in the previous reanalysis led to the development of a new inference technique that resolved the shortcomings. When these multiresponse permutation procedures were applied, the evidence for seeding effect was a little stronger, i.e., P values were smaller, than in the previous analysis. Mielke and Medina (1983) applied new covariate ratio procedures to Climax I and II experiments. This adjusted for disproportionate allocation on treated and nontreated experimental units.

Rhea's (1983) comments on the above reanalyses pointed out differences between them and an earlier reanalysis (Grant et al., 1979) that found seed–no seed differences had been due, at least in part, to regionwide natural differences in precipitation patterns. He showed that by using more nearly synchronized control station precipitation data, double ratios were smaller and P values higher compared to results in Mielke et al. (1981). In a reply, Grant et al. (1983) asserted that they believe the differences Rhea observed were due to his use of an inferior

dataset for control stations and the multiplicity introduced.

9.1.3. Wolf Creek Pass Experiment

An experiment carried out over Wolf Creek Pass targeted a small part of the later CRBPP area. Instead of being randomized by daily experimental units, it was randomized by years, three of which were seeded at every opportunity and three nonseeded. The seeded years were 1965, 1967, and 1969. The purpose was to test the use of seasonal streamflow data in evaluating snowpack enhancement projects. An evaluation based on the 24-hour precipitation values, treated as independent events, indicated that statistically significant increases were present in precipitation when the 500 mb temperatures were $-23°C$ or warmer, and especially when they were $-20°C$ or warmer. Morel-Seytoux and Saheli (1973) found that runoff in the three seeded seasons was significantly greater than that predicted from historical regressions.

Rangno (1979) reanalyzed the data from this project and concluded that the seeding effects claimed were apparently due to natural causes. He made a detailed analysis of 24-hour precipitation over an area extending from about 100 km south of the target to nearly 200 km to the north and over 100 km to the west. The seeded–nonseeded ratio map, stratified by 500 mb temperature, suggested a substantial (large-scale) difference between seeded and nonseeded seasons that was related to seasonal wind flow differences. He demonstrated the importance of control precipitation sensors being highly correlated with target sensors, which would require them being located in similar terrain, with similar exposure to moisture-bearing winds.

With a variety of control areas available for this after-the-fact study, it is possible to fall into the multiplicity trap, and Rangno's apparent invalidation of the estimates of seeding effects is, of course, subject to this concern. However, he does appear to have demonstrated that a "lucky draw" may have occurred in this experiment's randomization because the 24-hour periods were not independent due to strong seasonal characteristics.

9.1.4. Colorado River Basin Pilot Project

The CRBPP was essentially a follow-on to Climax, being designed by the Colorado State University group (CSU) (Grant et al., 1969). The San Juan Mountain barrier had different orientations within the project area, with an east–west crestline extending to the west of Wolf Creek Pass, and a generally north–south crestline to the south of the pass. Thus, the Wolf Creek Pass lay at the point of a funnel. A ground-based silver iodide/sodium iodide generator network was again employed, as was a 24-hour experimental unit. The primary response variate was still the 24-hour precipitation amount garnered from a gage network that was unprecedentedly dense, considering the rugged terrain and adverse environment. The results showed that more precipitation (but not significantly so) fell on nonseeded than on seeded days. Only when the 24-hour experimental unit was broken down in an a posteriori analysis into six-hourly and three-hourly blocks did the reason for such overall results emerge (Elliott et al., 1978). The seeding protocol had excluded seeding of deep convective clouds and of clouds whose tops (at Durango) were colder than $-25°C$. It had proved operationally impossible to conform to this protocol with a 24-hour experimental unit, due to the difficulties of forecasting.

It was possible through the use of the shorter time blocks to categorize the types of air flow controlling transport of the ground-based generator plumes using the three-hourly soundings at Durango and the abundant anemometer records from sites covering a range of elevations. The plume tracking for seeded cases was then verified by reference to the records from eight nucleus counters (Bigg–Warner cold box type) scattered around the target area and to the west of it, at elevations ranging from 2 to 3 km above sea level. This analysis indicated that about 20% of the time the generators had been operated under conditions in which the effluent moved in the blocking flow from southeast to northwest around the end of the east–west oriented San Juan Mountain barrier. Thus, no seeding had actually occurred on many experimental days, and in some cases, the effluent severely contaminated a following nonseeded experimental day as the air mass destabilized and began to flow back over the target area. Aircraft measurements in the last year of the project confirmed the blocking flow effect and, in one case, the lingering of ice nuclei for up to 18 hours after its release.

Various sounding parameters were derived from the three-hourly upwind soundings at Durango. The most important was the cloud-top temperature, which was projected up over the barrier crest from the Durango sounding site by means of a simple flow model and designated the "lifted cloud-top temperature." The corresponding three-hour precipitation rates at each recording gage, properly time lagged from Durango, were then combined into six-hour blocks to reduce serial correlation (from 0.6 to 0.3). The six-hour blocks were used to perform various stratifications of the seeded and nonseeded precipitation on the sounding parameters.

Two main categories of cloud were separated. In the one called stable, the air mass was either stable (but not having blocking flow that prevented seeding) or the depth of the positive thermodynamic area (parcel-lift method) was less than 75 mb. The unstable category had a positive area 75 mb or more deep. The unstable category showed less precipitation, and the stable category more, during seeded than nonseeded blocks. In the stable category the difference was statistically significant when the cases with a high wind and a lifted cloud-top temperature below

$-29°C$ were excluded. The effect of seeding under favorable conditions was found to consist in part of an overall enhancement and in part of a redistribution toward higher elevations and to the lee side.

In the first of a three-part series covering the aircraft-based physical studies of the last year of the CRBPP, Marwitz (1980) analyzed the dynamical processes using extensive aircraft data and some ground-based data. Out of a total of 12 storms flown, 11 had an upper-level baroclinic zone, and in the 12th case the front extended to ground level. Four stages typified the storm: stable, with low-level diversion of flow; neutral, with deepest development; unstable, with a zone of horizontal convergence and a convective cloud line; and dissipation, with subsidence at mountaintop height. Stability was based upon sounding Θ_E lapse rate to point of minimum, whereas Elliott et al. used the parcel definition.

In the second article of this series, Cooper and Saunders (1980) described the microphysical structure. In the earlier stages growth by diffusion predominates, but during the later stages accretion is as important as diffusion. Liquid water was observed most frequently in later stages. Cooper and Marwitz (1980) concluded that the seeding potential is greatest during the latter portion of the typical storm, but that some potential exists during the neutral stage.

The null (actually somewhat negative) effect of seeding found by Elliott et al. in the convective stage, where Cooper and Saunders found greatest potential, was investigated more completely by Elliott (1984). A strong negative effect was confined to deep unstable conditions that occurred 3 to 6 hours after 700 mb trough passage. Elliott suggested that the seeding mode employed resulted in dynamic overseeding of the deep convection.

A reanalysis by Shaffer (1983) sharpened the results for the stable cloud stratification by excluding a number of cases where the ascending plumes would not have risen to the $-5°C$ level before the crest was reached. The importance of this requirement was stressed by Super and Heimbach (1983) in connection with their evaluation of the Bridger Range project (Montana) results.

Comments on Elliott et al. (1978) by Rangno and Hobbs (1980a) showed concern that too much diffusion had been allowed under stable conditions and that some of the observed high ice-nuclei counts could have come from dust storms or other sources. The reply (Elliott et al., 1980) pointed out the numerous cases where no such questions could arise. The use of the lifted cloud-top temperature rather than that observed at Durango was criticized. The reply indicated that the difference averaged about 3°C. The choice of which temperature to use depended upon the origin of particles reaching ground level on the barrier.

In general, the $-29°C$ threshold found in the CRBPP has been criticized as being too cold. But Hill (1980) suggests that due to rawinsonde errors it could be as much as 20°C colder. The cloud-top temperature window has received prominence in the weather modification literature because it is a measure of natural nucleation, i.e., primary ice particle production, the factor which seeding seeks to influence. Also important is the effect of secondary ice production (ice multiplication). This occurs further along the chain of events but can also play an important role in the precipitation process, especially in convection. Weickmann (Hess, 1974, Chap. 89) has categorized the Great Lakes cloud into three main categories as determined, in part, by the cloud temperature range. A variety of cases are discussed by Grant and Elliott (1974). Smith (see Hess, 1974, Chap. 12) relates success and failure of dry ice seeding of cumulus clouds to cloud-top temperature.

9.1.5. Vardiman–Moore composite analysis

During and after some of the CRBPP years, four other Bureau of Reclamation randomized projects were conducted in a variety of western mountain settings. Vardiman and Moore (1978) consolidated data from these and from Climax and the CRBPP to arrive at a composite seeding "window" for the crest zone based upon upwind sounding parameters representing time available for growth and fallout, water available, ice nuclei available, and degree of instability.

The results were marred by the inclusion of many extra zero precipitation cases, as Rangno and Hobbs (1980b) pointed out in their comments. Concerns by the original authors and Rottner about the use of Climax nonseeded cases that had not been subjected to randomization led to downward revisions of the significance of the categorizations (Rottner et al., 1980). The only significant window remaining was for a negative effect with cold top ($< -30°$C) unstable clouds. Rangno and Hobbs (1981) commented on this reanalysis also.

9.1.6. The Santa Barbara Experiment

The SBA II experiment, conducted during this same era (1967–73), was unique in that the object of study was seeding effects on convection bands, which provide most of the precipitation in that portion of California. The project area lay over the Santa Ynez Mountains, which are oriented east–west, parallel to the southern seacoast of Santa Barbara. A previous study had shown that these storm bands contained regions of liquid water, especially after they encountered the abruptly rising coastal mountains that triggered an enhancement of convection (Elliott and Hovind, 1964). The experiment was also unique in that the experimental unit was the duration of band precipitation at a precipitation station, generally around an hour. The experimental unit's timing therefore varied with location for each particular band. The earlier SBA I seeding experiment (1957–59) (Neyman et al., 1960) had used a 12-hour unit, and an a posteriori study (Elliott, 1961) suggested that a shorter time unit associated with convection bands might provide a more effective basis for evaluation of seeding effects. Both the project data collection and the evaluation covered the target area over the Santa Ynez Mountains and upwind area and over a downwind area to test for extended area effects.

This project was divided into two phases. In the first phase (1967–71) the seeding material was produced by a high-output pyrotechnic device located on a 1 km east–west-oriented mountain ridge paralleling the southern coastline of Santa Barbara County. If a radar and an upwind line of telemetered gages indicated that a band was approaching the site, this device was triggered (when the random draw indicated seed). The project director and those who analyzed the data were blind with respect to the seeding. During the first four years, 56 bands were seeded and 51 were left unseeded. Sixty recording gages spread over about 1500 square miles were used in the evaluation of results. Rawinsondes were launched at the Santa Barbara airport just prior to each band passage.

The results for the first phase showed an excess of seeded over nonseeded precipitation of around 50% for the bands, which accounted for about 60% of all precipitation (Elliott et al., 1971). This was determined to have an overall significance at the 0.06 level in a rerandomization test (Elliott and Brown, 1971). The bands were found to be most active on their backside, with seeding amplifying precipitation there, with resultant slowdown and widening of the band (Brown and Elliott, 1972). The seeding effect progressed eastward over 100 km, qualifying it for an extended-area effect. The progression was in the direction of band movement (normal to the band) at an angle to the right of the basic flow aloft. A dynamic effect appeared to be present, but not necessarily of the type associated with vertical breakthrough of a stable layer. Support for a dynamic effect was shown in a comparison of seeded and nonseeded band surface pressure patterns over the experimental area (Brown et al., 1974).

The second phase (1971–73) involved aerial seeding at cloud base along a 20 nautical mile north–south line off the western coast of the county, using a high-output continuous generator. The site of the seeding lay about 60 km upwind from that of the first phase. The entire seeding zone was under surveillance of the Vandenberg AFB radar and another radar located in the mountains to the east. Due to early termination, the second phase contained only 14 seeded and 17 nonseeded cases. However, these were enough to show seed–noseed precipitation ratio patterns over the entire area (target and extended) similar to those of Phase 1, but shifted 60 km to the west in accordance with the shift in the seeding location.

9.1.7. Extended area effects

Effects of seeding that may extend beyond the bounds of what had been thought of as the expected area of effect, or in the case of a purely operational project, the target area, have been the subject of concern during the past decade. A workshop on the total area effects of weather modification was held at Colorado State University in 1977 and resulted in a published report (Brown et al., 1978). It followed a more limited meeting covering the same topic held at Santa Barbara, California, in 1971. Several papers on the topic had already been presented at the first conference on weather modification in Albany, New York. The workshop concerned itself primarily with evidence supplied by six major projects, namely, Grossversuch; Israel, 1971–76; Santa Barbara, 1967–74; Climax, 1961–70; Whitetop, 1960–64; and Arizona, 1957–64. Hypotheses as to effects included dynamic effects, such as intensification of convection bands; downwind transport of ice nuclei; dynamic overseeding pumping water out of top; and dynamic downdraft stabilizing surrounding areas. The evidence was the best available and was entirely a posteriori in nature as the original experimental designs did not anticipate such effects. There appeared to be little evidence for the popular idea that seeding one area would rob a downwind area of precipitation. Other evidence showed that artificial nuclei are at times transported essentially unused over 100 miles downwind. Inadvertent large-area effects and social implications were considered. Recommendations of concern in orographic cloud seeding included model development for determining transbarrier water budget, model development of evaporation and re-precipitation processes, field measurements of transport of nuclei, and cloud physics (and dynamics) studies. Most recent designs of projects reflect concern about observing these suggested effects. For example, the downwind area under observation in SCPP extends well into Nevada.

9.2. Observational and seeding systems

9.2.1. Introduction

The evaluation of orographic seeding experiments is heavily dependent upon the flow of information from observational systems that are under heavy environmental stress. Because most of the observational systems and their attendent problems are common to all orographic projects, unnecessary repetition is being avoided by presenting within this section discussions of these systems.

9.2.2. Precipitation gages

The raingage is a source of key evaluation data for orographic seeding experiments. It normally collects water in frozen form and this greatly complicates the collection process. Special procedures are used to insure melt of the catch, and gages are designed to provide safeguards against bridging or capping of the orifice by snow. However, such safeguards are not completely effective, and a relatively large fraction of missing data results. Servicing of remote gages is infrequent and recording gages are a necessity.

An important term in the orographic water balance is the interception rate, i.e., the rate of accumulation of water (mostly in the form of snow) on a sloping surface. The conventional gage is designed to catch precipitation on a horizontal surface. It is possible under light snowfall and strong winds for such a gage to catch nothing, although the snowpack deepens. In practice, the gage is sited on a level portion of the terrain so that local flow tends to parallel the gage orifice. In addition, the gage catch is enhanced by shielding. Such strategies tend to bring gage catch up to interception, as has been shown by comparisons between gages and snowboards that slope with the ground. Super and Heimbach (1983) discussed problems encountered in the Bridger Range project in Montana and how siting gages in sheltered zones ameliorated the problems. In a randomized project it is believed there is no important gage bias favoring seeded over nonseeded precipitation, or vice versa. However, a systematic gage catch error could adversely affect computations of precipitation efficiency or other water balance terms.

The siting of raingages plays a key role in target-to-control-area evaluations, as has already been discussed above in connection with Rangno's (1979) reanalysis of the Wolf Creek Pass project. The gage siting problem is related to the problem of long-term instabilities in target–control relationships due to long-term climatic swings. Evaluations of nonrandomized data where project and historical target-to-control relationships are compared are especially subject to this type of uncertainty.

9.2.3. Rawinsondes

The rawinsonde has provided valuable information for interpreting results of experiments. It has been subject to errors of various types through many years, in spite of changes in instrumentation. Humidity errors occurred due to time lag effects; temperature and humidity errors occurred due to inadequate shielding from solar insolation. The instrument could easily ascend through thin clouds (above any lower, thicker layers) without recording ice saturation. Empirical relative humidity correction (or temperature–dewpoint spread) factors were developed by various analysts, e.g., Appleman (1954), based upon proximity soundings, and Elliott and Hovind (1965), based upon Santa Barbara storm cloud observations. The latter also found that weather bureau practice required the smoothing of abrupt changes in temperature and humidity that often occur at cloud exit, and this necessitated reference to the original traces to make corrections. Reference to direct cloud observations (especially of the higher ones) also were employed to correct sounding interpretation.

The first carbon element rawinsonde units contained two sources of error. One was the poor shielding of the temperature-sensing element, which led to temperature excesses in the daytime; the other was poor ventilation of the hygristor (Morrisey and Brousaides, 1970). Test data at Bedford, primarily between May and November, revealed errors of several degrees.

In 1972 the housing design for the carbon element was changed, permitting better ventilation and markedly less insolational effect (Brousaides and Morrissey, 1971). Use of the new units started in 1973 in the Colorado River Basin Pilot Project. Hill's (1980) "reexamination" of the cloud-top temperature stratifications used in the CRBPP evaluation was intended to determine what effect the unshielded carbon units used prior to about 1 February 1973 had on this evaluation. For this purpose he used Salt Lake City soundings with unshielded units, which he compared to Hill AFB K-band radar observations. He made no corrections for temperature–dewpoint spread and hence found that daytime soundings often showed no cloud at all. Hill did examine the CRBPP "higher layer present" category (separated by more than 50 mb from the main deck), which appears in the CRBPP final report tables (Elliott et al., 1973), and found that daytime soundings showed a dearth of these high clouds. These higher layer clouds were, however, not part of the sample of clouds used in the cloud-top temperature stratifications. The latter contained only a 1.1°C average daytime–nighttime temperature difference. His surrogate cloud sample failed to represent the deep main cloud deck of concern to the microphysics of nucleation and precipitation production. As in the case of the precipitation gage, any error in cloud-top temperature is unbiased with respect to seeding if the seeded and nonseeded samples are equal.

9.2.4. Microphysical data

Microphysical data are collected on the ground and in the air. Equipment has included a variety of impaction devices for collecting particles, some involving continuous collection on coated film. The latter has been described by MacCready and Todd (1964). Particle measurement in more recent projects has been greatly improved through the use of optical systems. In particular, the Particle Measuring System (PMS) probes are now widely used in winter orographic projects. These include the optical array 1D-P, the 2D-C, the 2D-P, and the FSSP. The basic principle of the probes is the automatic and continuous sizing and counting of particle images passing through a sampling volume. Two-dimensional images of the particles are also displayed. The 2 D-C is ordinarily used to detect smaller-size crystals and the 2 D-P to detect larger, "precipitation" size particles. The FSSP is a forward-scattering spectrometer used for determining drop spectra but also for cloud drop water content (Knollenberg, 1972; Knollenberg, 1976). In addition to the FSSP, the older Johnson–Williams (JW) hot wire device is used for liquid cloud water content and the Turner–Radke instrument (Turner and Radke, 1973) for ice crystal counting.

In the SCPP (see section 9.4) a ground-based microphysical two-dimensional probe is employed. The air is aspirated into the unit through a shaped orifice, thus simulating aircraft probing. Often the probe has a much higher concentration of small (<0.1 mm) particles than are collected by conventional photomicrophysical techniques. This ground-level microphysical facility is regarded as an important supplement to the conventional precipitation gage.

The existence of airplane-produced ice particles (APIPS) of up to 500 L^{-1}, found when the sampling aircraft was carefully maneuvered in return flights through clouds previously sampled (Rangno and Hobbs, 1983), brings into question the validity of observations of high concentrations of particles where repetitive sampling is performed. In particular, Rangno and Hobbs cite an article by Mossop et al. (1968) where high ice crystal concentrations were reported in clouds having a summit temperature of −4°C and were attributed to a natural ice multiplication process. Mossop (1984), in his comments, discusses his own reexamination of flight data and concludes that the procedures employed could not have resulted in APIPS. In their reply Hobbs and Rangno (1984) note that the laboratory evidence of Hallett and Mossop (1974) for ice multiplication used moving metal rods that could as readily simulate structures on aircraft.

9.2.5. Remote sensing

Remote sensing devices used in orographic experiments (radar, radiometers, lidars) focus on a variety of microphysical particles [ice particles, snowflakes (wet or dry), graupel, hail, cloud droplets, water molecules, etc.] and all possess large sampling volumes that can be scanned repeatedly, thus providing certain advantages over the small sampling volume of in situ aircraft observations. The remote sensors are also able to probe cloud volumes where the aircraft cannot travel safely, e.g., near the ground, beneath the minimum operating clearance altitude (MOCA).

Snider and Rottner (1982) discuss the use of microwave radiometry in weather modification experiments. Hogg et al. (1983) explain the functioning of a steerable dual-channel microwave radiometer. Feng and Grant (1982) find a consistent pattern in their comparison of COSE radiometer data with degree of riming and aggregation in ground-based observations. In both COSE and SCPP, interpretation of the radiometric-path-integrated data by use of concurrent Ku band radar has proven useful.

9.2.6. Other systems

The Rosemont icing rate meter has been adapted for use at ground level (Henderson and Solak, 1983). In SCPP the device was placed on Squaw Peak to measure quantitatively the rate of riming, a frequent occurrence at higher levels on the Sierra Nevada barrier. Observations during the past two years match in general those made by the radiometer, about 10 km west of Squaw Peak. Some reservations concerning the representativeness of the measurement at temperatures near 0°C have been raised by Super and Heimbach (1983), based upon tests in the Colorado Rockies. Hindman and Grant (1981) discuss the utility of a different type of riming measurement system.

The counting of natural and artificial ice-forming nuclei (IN) was entrusted to the Bigg–Warner counters in a network of ground sites in the CRBPP. This device is too bulky and observations too manpower greedy for aircraft use. A number of IN measuring instruments have been investigated at three international workshops (Lannemazan, France, 1967; Fort Collins, Colorado, 1970; and Laramie, Wyoming, 1975). The various techniques differ in their measurements by three orders of magnitude for AgI aerosols. The most commonly used airborne instrument for measuring IN (natural or artificial) is the NCAR counter, described by Langer (1973a,b). It processes the sampled air through a chamber containing supercooled liquid droplets, which are nucleated with the resultant ice particles being counted by a sonic device at the base of the chamber. In conventional use a typical correction factor of 10 is applied (Super and Heimbach, 1983).

The polarization lidar for the continuous recording of liquid water and ice particle distribution has found use in both COSE and the Utah Federal–State program (Sassen, 1984).

9.2.7. Seeding systems

The character of seeding agents and their generation and dispersal have been the subject of continuous study since the experiments of Schaefer (1946) and Vonnegut (1949) indicated some of the effects of artificial nucleation. Dry ice was employed by Schaefer in his original field experiment, whereas Vonnegut established the basis for the use of silver iodide smoke. Other chemical agents (mostly in the form of smoke particles) were tested, but silver iodide has remained the paramount agent in use. It became the rule during the 1950s and 1960s to generate silver iodide smoke by forcing an acetone AgI–NaI solution through a nozzle-type burner head, depending upon rapid quenching to crystallize the vapor into submicron-size smoke particles. Organic nuclei have been employed in laboratory and field tests for a number of years and, although not used extensively, do offer considerable potential for practical use (Fukuta, 1963, 1967).

Early work at the Naval Weapons Test Center, China Lake, California, led to the replacement of the 1960s AgI/NaI generators, which had produced mainly condensation–freezing nuclei (hygroscopic ice-forming nuclei), with AgI/NH$_4$I generators that produced mainly contact nuclei. This greatly enhanced the effectivity of AgI at warmer temperatures, considered to be a highly desirable feature at the time. Later, various types of pyrotechnic generators were produced that permitted dispersion of nuclei from aircraft wing-mounted units. Practice now includes the use of wing-mounted pyrotechnic racks holding flares that can be dropped in series into or near the top of clouds, thus generating a "curtain" of nuclei extending downward about 1 km.

Activity centered at the Colorado State University cloud chamber facility has resulted in better laboratory simulations of atmospheric conditions (Garvey, 1975) and in a number of new formulations of AgI nuclei. The facility has been used as a standard test facility by all engaged in weather modification activities. Kinetic theory has been applied to clarify properties of nucleating agents and has provided new insights that may transform seeding methodology. Silver iodide and silver iodide–silver chloride aerosols were compared (DeMott et al., 1983) in a large cloud chamber held at water saturation. Ice nucleation effectiveness for the nonhygroscopic AgI–AgCl aerosols is up to three orders of magnitude higher than that for the nonhygroscopic AgI aerosols. Blumenstein et al. (1983) have reevaluated the ice nucleation by silver iodide–sodium iodide, which has some desirable features under some circumstances.

Dry ice, the seeding agent used by Schaefer in his famous 1946 seeding test, continued to be used extensively in early field tests in this and other countries, but its use declined through the years. However, there has recently been an upsurge in its use in experiments (e.g., Hobbs and Radke, 1975; Marwitz and Stewart, 1981; Stewart and Marwitz, 1982b). Estimates of the effectivity of CO$_2$ have undergone considerable variation over the past years. Project Cirrus estimates were 10^{12} g^{-1}. More recent measurements made in a cloud chamber (Morrison et al., 1984) show a dependence upon temperature not shown by Fukuta et al. (1971). Effectiveness ranges from 10^{12} g^{-1} at -3°C to almost 10^{13} g^{-1} at -20°C. Stith (1984) calculated effectiveness using actual seeding signature data collected in SCPP and found values ranging from 9.8 × 10^{11} to 1.5 × 10^{13} g^{-1}. Dry ice pellets of 1 cm in size dropped from a hopper may fall over a kilometer before evaporating and, when dropped along a line, produce a curtain.

Due largely to the uncertainties associated with the use of ground-based generators, the trend has been to employ aerial seeding in the more recent experiments. Curtains of dry ice pellets or of silver iodide flares are laid within the target cloud volume. The curtains generally are effective through a 1 km depth. Overseeding usually occurs immediately, producing a seeding "signature," i.e., a region of high concentration of small ice particles that are easily detected by a suitably instrumented aircraft penetrating them. While the immediate effect is undesirable, the curtain-signature mixes with its environment and eventually ingests enough water to grow precipitation-size particles.

Field studies of seeding effects using this latter seeding mode have provided important information about the spreading (dispersion) of the seeding signature and changes in its microphysical characteristics with time, including the development of precipitation particles. Marwitz and Stewart (1981) report on airborne calibration seeding tests for convective clouds in SCPP, using different dosage rates and materials. A position reference subroutine allowed the aircraft to be flown repeatedly through moving air columns containing the seeding materials. The usual sequence of events observed was a rapid increase in ice crystal concentration and a decrease in liquid water content during about 10 minutes.

In winter stratiform clouds, the seeding signature has different characteristics. Comparisons were made (Stewart and Marwitz, 1982b) among randomly selected cases of light CO_2 (0.1 g m^{-1}), AgI (1–20 g flare per 250 m), and placebo. Wind shear effects were strong in the stratiform clouds, and this added to the effective dispersion rate (Stewart and Marwitz, 1982a) even though in situ turbulence values were low. In recognition of shear effects and particle size distribution the analysis distinguished changes across the curtain from upwind edge to downwind edge. 2-DC concentrations of 50 L^{-1} were found in one AgI case for up to 15 minutes, in another for up to nearly 50 minutes.

9.2.8. Tracers

For some time, bulk snow samples have been collected in and around various target areas to be used in the analysis of their silver content. The detection methodology employed at the Desert Research Institute in the late 1960s involved neutron activation of the sample, whereas later, flameless spectroscopy and other detection methods were used. Collection has since been improved through use of a snow profiling technique. Collection and analysis for individual storms is possible. The Truckee–Tahoe basin, the target area for an ongoing operational seeding project using ground-based generators, has been sampled (Warburton et al., 1979) and concentrations of up to 35 times background found. The "seeding silver" is present in only about 10%–20% of the seeded period's samples. It often appears in narrow time and spatial zones, suggesting that the plumes are finite and affect one point for periods of 30 minutes to 4 hours.

SCPP field experiments used indium oxide tracer (a gas) emitted by a generator near the regular ground-based

AgI generator (operated by a cooperator). Similar experiments were performed in Utah. Cesium and rubidium salts, which complex with silver iodide, have also been tracked across the Sierra crest in connection with the SCPP project (Warburton et al., 1979). Cesium concentrations in the snowpack were well above background. These tracers move with the silver iodide and therefore follow the nucleation and fallout path. Indium oxide gas, on the other hand, moves with the drift of nucleant prior to nucleation. The possibility of removal of the silver iodide complex by scavenging is being investigated by comparison of indium oxide samples to silver samples.

9.3. Models
9.3.1. Ludlam's model

Ludlam's (1955) conceptual model for seeding of mountain clouds contains, in retrospect, the essence of present day models. For a mountainous region in Sweden, he was able to estimate the difference between conversion of condensate to ice particles through diffusion and accretion (although he considered the latter unimportant) and to arrive at a potential for seeding. From dimensions and wind speeds typical of the area, he developed a seeding mode that could exploit the potential to enhance snowfall in the crest region. To estimate how frequently favorable conditions could be expected, he examined data from a mountain station (Blauhammaren). He assumed conditions would be favorable when the station reported fog, the temperature was −5°C or below, and snowfall was less than 0.3 mm h^{-1}. He noted that rime is frequently reported in snow as well as in fog, indicating an inefficient natural removal mechanism. With reasonable assumptions, he computed the potential silver iodide seeding contribution to removal. In the year analyzed, the total precipitation was below normal, and he estimated seeding would have added 80 mm to the total. Support for this estimate is given in a picture of a heap of rime that had fallen during winter at the feet of a precipitation gage. He calculated it would represent 300 mm of precipitation.

Ludlam discussed the problem of evaluating results, mentioning that target-to-control-area precipitation comparisons would suffer from seasonal climatic trends. He anticipated downwind effects, suggesting that seeding would reduce leeside precipitation only if precipitation was extensive. The evaporation there would increase in proportion to the water removed upstream, and seeding would then transfer to the mountain the water that would otherwise fall on lower ground over a wide area to the lee. If no snow was falling, then seeding would produce no downwind losses.

9.3.2. Quantitative orographic models

Early quantitative models for predicting orographic precipitation used rather simplified versions of the cloud physics (assumed precipitation efficiencies, etc.) but handled the airflow in a more adequate manner. Myers (1962) and Colton (1976) both studied precipitation distribution in the Sierra Nevada of California by such models. Rhea (1978) developed a precipitation model, originally applied in the Rocky Mountains of Colorado and later to the Sierra Nevada in connection with the SCPP. A key element of such models, and also of more recent ones, is the extrapolation of airflow over the entire barrier using as inputs upwind sounding data plus the terrain profile. Models including more adequate cloud physics were developed by Fraser et al. (1973), Young (1974a,b), and Plooster and Fukuta (1975). Dirks (1973) applied a two-dimensional airflow to Elk Mountain in Wyoming to assess precipitation efficiency.

The two-dimensional character of such models made difficult any attempt to track the fallout of particles resulting from seeding. This was because the blocking flow, i.e., the flow parallel to the barrier, was not well represented although the effects of blocking on the flow normal to the barrier were handled. Observations around the Sierra Nevada barrier and other barriers show the presence of complete blocking (flow around the barrier) in the lower levels when the air mass is stable enough relative to the kinetic energy of the basic normal flow. A barrier-parallel jet in low levels extends with diminishing strength up along the barrier slope. Early microbarograph measurements had shown that a strong pressure gradient plays a role in this. To model effectively the blocking jet, the effects of the Coriolis term must be incorporated. Parish (1982) has modeled the flow over the SCPP experimental area, starting with the primitive equations, and has produced a good simulation of the flow parallel and normal to the barrier. At present an operational model employed on the SCPP introduces the barrier jet empirically, using a climatology of observations over the barrier as they relate to the upwind sounding indications of the low-level jet (Elliott, 1981).

A three-dimensional dynamic model, which handles dynamic effects but not convection itself and which handles some cloud physics factors (not ice multiplication), has been developed and tested on a few cases in the San Juan Mountains and the Sierra Nevada (Nickerson et al., 1979).

The CSU 3-D nonlinear time dependent cloud/mesoscale model (Cotton et al., 1982) has been applied to the COSE winter orographic cloud system. Their simulation underpredicted precipitation, and this was believed to be due to its inability to take advantage of transitory "precipitation generating" zones that had been found in this area. Models have been applied in a diagnostic fashion to explain observations. Rutledge and Hobbs (1983) have applied one such model to CYCLES data. It takes into account the seeding from high seeder clouds of lower feeder clouds.

9.4. New generation projects

9.4.1. General

The early randomized experiments have been characterized as "black box" experiments, where only the input (nuclei) and the output (precipitation) were measured. What had occurred along the links in the chain of events from the nucleation source to the ground-level precipitation was not being observed, so that any results, positive, negative, or null, would be difficult to explain on a physical basis. This assessment is harsh since most of those experiments did incorporate extensive data collection beyond ground-level precipitation, and analyses did indeed dwell on links in the chain. Seeding windows that were established, however, tended to be based upon various indices (usually upwind sounding parameters) rather than on direct physical observations.

The current approach seeks to rectify the situation by basing the evaluation of seeding effects directly on physical observation of events along the links in the chain. The observations are to be transformed into a set of response variates representing key physical processes expected to be modified according to the seeding hypothesis. Implicit in this approach is the use of shorter period experimental units, which allow for a better resolution of the physical processes within the changing cloud system. Randomization on this short period counters the adverse effects of serial correlation in treated and control comparisons. Disadvantages include the need to allocate time (and space) to buffer zones and a concomitant reduction in the value of the control area precipitation and other observations as covariates.

SCPP, COSE, and the Utah Federal–State project as discussed herein represent the new generation of orographic experiments. However, prior to the inception of these new projects, the links-in-the-chain approach had been anticipated. These "physical evaluation" projects (they were not randomized) include the Great Lakes project in the winters of 1968, 1969, 1971, and 1972 (for a good resume, see Hess, 1974, Chap. 8). Although this was not a mountain project, there are similarities because of the control of air motions by an effective heat mountain over the Great Lakes. Another example is the field project carried out over the Cascade Mountains from 1969 to 1974. Physical evaluations were made of the modification of winter clouds and precipitation by artificial seeding (Hobbs, 1975a,b; Hobbs and Radke, 1975). In both projects, aircraft were employed for collecting cloud physics data and for seeding. Radar, ground-level microphysics, and precipitation data were also collected. Various trajectory models were employed as aids in linking seeding signatures to ground-level data.

9.4.2. SCPP

The SCPP area covers the American River Basin in the central Sierra Nevada of California and the region immediately to the lee. The north-northwest–south-southeast oriented barrier rises from the Sacramento Valley to crest heights of generally under 3 km over a long slope of 100 km. To the lee is the Tahoe Basin at around 2 km elevation, and beyond this a second barrier. Preliminary field studies, started in the winter of 1976/77, provided the background data (three-hourly upwind rawinsondes, a 5 cm surveillance radar, and special aircraft cloud-top observations) needed to develop a climatology of radar echo types. During subsequent years, observations from other systems were included: cloud physics aircraft, doppler and K-band radar, ground microphysics, precipitation gage network, radiometer, ground-based riming rate, weather satellite, special tracer studies, and numerous calibration seeding tests (many randomized).

The 5-cm surveillance radar precipitation echo types (PETs) continued to be used to identify mesoscale structures of importance to weather modification. The earliest documentation of PETs appeared in Sutherland and Kidd (1978). Descriptive definitions were presented by Heggli et al. (1983). Satellite imagery was used to link PETs to visible forms, and aircraft and Doppler radar data provided more detailed evidence as to their kinematical and microphysical features. An example is the detailed analysis of two prefrontal bands by Gordon and Marwitz (1985). Preliminary conceptual seeding models were next developed for several of these cloud forms, including postfrontal cells, major bands, and the stable/neutral orographic cloud. The potential for seeding of each cloud type was investigated and strategies for exploiting the potential explored. The deep areawide and associated embedded band types present in the advance portion of the frontal system offered little potential. The most favorable types occurred following passage of an upper-level front, identified by the passage of a sharp backside of the deep cloud, topped by cirrostratus or altostratus (Rhea et al., 1982).

Based upon detailed analysis of SCPP aircraft microphysical data (Heggli et al., 1983), it was decided that, assuming a high incidence of supercooled liquid water and low ice-crystal concentrations represented seeding potential, the postfrontal cells exhibited the greatest potential in the Sierra Nevada. On the basis of these estimates and more detailed project analyses, a design for a fully randomized exploratory project on postfrontal cellular cloud was developed, and the "floating-target" exploratory experiment was commenced in the 1982/83 season. The conceptual model for treatment of these clouds includes these features. The condensate, produced in the cells moving up the slope, is partly converted to ice particles that grow first by deposition, then by riming, and fall earthward. At the same time, condensate is lost due to entrainment of dry air from aloft. The object of seeding (by means of dry ice and AgI curtains) is to accelerate conversion to ice, thus forestalling some of the entrainment loss. The seeding-produced precipitation (and first radar echo) would lie upwind from that for natural precipitation. The seeding is carried out in the "formation zone," an elevation zone varying with case but usually over the foothills, where the cells first form and have a maximum of supercooled liquid water and a minimum ice-crystal concentration. Aircraft observations made in the formation zone measure the ratio of liquid-water content to ice-crystal concentration, with seedability declared when the ratio exceeds a certain threshold.

Although flood threat suspensions limited the number of experimental units, some results have been evaluated on a physical basis. The evidence of tracking of the seeding product to the ground by radar and other physical observations under these conditions has been discussed by Huggins and Rodi (1985). In only a small fraction of their cases could the products of seeding be tracked to ground level.

Background data continued to be collected on the shallow orographic cloud type, and also on major bands, but primarily at times when the floating target experiment could not be performed due to suspensions. This provided a rather biased sample since major storms over the SCPP contain an unusually warm and wet air mass. However, a microwave radiometer (see section 9.2), added to the instrumentation in 1980 but not located at an effective elevation (Kingvale) until the 1983/84 season, has provided crucial evidence concerning the presence of liquid water in the shallow orographic clouds that had not been sampled before. The radiometer, although it provides only integrated values, enjoys the distinct advantage of producing a continuous record along one sight line. Accordingly, a second seeding experiment was designed to be carried out on these shallow orographic clouds. The conceptual model was based upon the observation that the supercooled liquid water accumulated to peak values high on the barrier as a result of an imbalance between the condensation and its removal by the natural precipitation process. Kingvale is well sited to measure this accumulated water prior to its departure to the leeside evaporation zone beyond the crest. Other instrumentation located at Kingvale includes a vertical-pointing K-band radar for measuring the depth of cloud and a ground-based aspirated PMS probe (2D-C). The objective is to capture some or all of the excess supercooled liquid water in the form of snow ahead of the crestline by seeding at times when the radiometer observations and coordinated aircraft observations indicate the presence of supercooled liquid water. The seeding curtain would be laid along a line located so that the fallout of seeding products would impact the center of a "fixed target." The economic value of any extra snow produced at these higher elevations in the fixed-target experiment is greater than for the floating-target experiment, where the effect would occur 30 to 65 km upwind from the crest. Therefore, the fixed-target experiment was accorded first priority for experimentation for the 1984/85 season.

It had been noted that different types of storms possessed different distributions of cloud types. In pursuit of the type of storm (and location within it) of high supercooled liquid water content as measured at Kingvale, it was discovered that the katafront, as discussed by Browning and Monk (1982), provided the best and longest enduring cases. Figure 9.2 (from Reynolds and Dennis, 1984) represents a cross section through this type of storm. The special feature of interest in this time section is designated "shallow moist zone," where cloud persists over the barrier (and also over the upwind valley) for many hours, giving a "split-front" character to the vertical structure. In other storm types, the surface frontal zone may occur much closer to the upper cold front, and although an orographic cloud continues over the barrier, it is quite convective in nature.

A special problem arises with curtain seeding because of the high wind shear in this type of cloud. The curtain bends over rapidly, often becoming almost parallel to the ground on impact therewith. Another wind factor is the low-level barrier-parallel jet, which extends far up the barrier. The kinematic structure of this type of flow has been discussed by Marwitz (1983).

As in the case of the floating-target experiment, it was found in preliminary tests that tracking the products of seeding to the ground was successful in only a fraction of the cases. For this reason an operational adaptation of a quantitative orographical model is used to predict where the seeding curtain is to be laid in order to target a region centered at Kingvale. The Kingvale site radiometer, K-band radar, and microphysical laboratory provide the basic data for many of the response variates employed in the formal experimental design.

9.4.3. COSE

The Colorado Orographic Seeding Experiment (COSE) was initiated in 1979. The project area is in the vicinity of Steamboat Springs, Colorado (see Fig. 9.1), and the main barrier is oriented essentially north–south. The emphasis of the program has been on natural cloud system studies, although limited seeding experiments have been conducted. Observation systems included a dual-channel scanning radiometer, a vertically pointing radar, aircraft, sequential soundings, surface meteorological measurements, and ground microphysical measurements, both at mountain base and at a mountaintop laboratory. The purpose of the program is to determine the spatial and temporal distribution of supercooled water and to describe

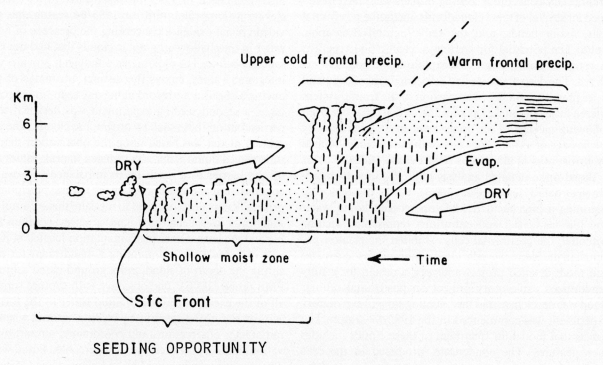

FIG. 9.2. Cross section through a split front. Most seedable regions determined from radiometer and aircraft observations within the storm are annotated.

the microphysical processes leading to ice phase precipitation in northern Colorado River Basin cloud systems.

Recent publications include Rauber et al. (1986) and Rauber and Grant (1986). Rauber et al. described the temporal evolution of supercooled water in Park Range cloud systems. They discussed three general cloud system types, including prefrontal, postfrontal, and orographic cloud systems. They found three common features concerning the evolution of the liquid water field in the prefrontal cloud systems: 1) an inverse relationship between precipitation rate and liquid water content occurred; 2) a direct relationship between cloud-top temperature and liquid water content was observed; and 3) the magnitude of the liquid water content was consistently higher over the mountain slopes.

In the postfrontal cloud systems studied, the liquid water content exhibited little variability upwind of the mountain base but varied considerably in the vicinity of the mountain. In these three storms, the magnitude of the liquid water content over the ridge was inversely related to the precipitation rate at mountain base. Liquid water production near the ridgeline was associated with both orographic and convective forcing.

The three orographic systems discussed in Rauber et al. were shallow, had tops warmer than $-22°C$, and had limited horizontal extent. In these systems, the changes in the liquid water field were inversely associated with changes in precipitation rate. In one case, a decrease in liquid water content was also associated with a decrease in cloud-top temperature.

Based on their studies, it appeared that the highest liquid water contents were most likely to be associated with periods when 1) the precipitation rate at the base of the mountain was low, and 2) the cloud-top temperature was warm. Rauber et al. concluded from these case studies that a weather modification program to enhance wintertime precipitation in the Park Range region should concentrate primarily on the shallow cloud systems.

Rauber and Grant (1986) identified three cloud regions in stratiform cloud systems where liquid water production can occur: cloud top, between cloud base and $-10°C$, and in regions of strong orographic forcing. Together, these papers showed that the structure and evolution of liquid water fields in northern Colorado Rocky Mountain storms are highly variable. The characteristics of the liquid water field depend on processes that occur at all scales. From their data, it appears that supercooled liquid water is present most consistently in shallow cloud systems with warm ($> -22°C$) cloud-top temperatures and low precipitation rates.

9.4.4. Utah Federal–State Project

The Utah Federal–State program carried out field studies during the winter seasons of 1980/81, 1982/83, and 1984/85. The experimental area, over the Tushar Mountains in southern Utah, east of Beaver, is about one-fifth of the operational project area covering the southern two-thirds of the central massif of Utah. The Tushar range lies along the western border of this mountainous area and rises from a 2000 m plain to about 3200 m above sea level. Its orientation is essentially north–south. The operational project, for enhancement of winter precipitation, primarily snow, has been carried out every year since the winter of 1973/74, except 1983/84 when water storage reached an all time peak.

Heavy reliance for nucleation has been placed upon ground-based AgI/NH_4I generators located in the valleys and on mountain slopes, supplemented by aerial seeding during four seasons, 1975/76–1978/79, and thereafter by high-level remotely controlled generators. These supplementary modes were intended for use when low-level stability and blocked flow made the valley-based generators less effective. In applying the "piggy-back" principle enunciated in the Weather Modification Advisory Board's recommendations for federal–state programs (Weather Modification Advisory Board, 1978; Gabriel and Chagnon, 1982), the emphasis was placed upon discovering the distribution of liquid water and the tracking of nucleant in the experimental area.

During the first experimental season (1980/81), the observational system included a fixed-angle radiometer, a vertical-pointing radar, a recording-gage network over the barrier, serial rawinsondes, and a new type of liquid water sounder (Hill and Woofinden, 1980b). A cloud physics aircraft was employed during a portion of the two-month (January–February) study period. In addition, snow samples were collected for silver content analysis. Due to an unfortunate late start and a poor season, only three storms were satisfactorily sampled.

The second experimental season (1982/83), reported on by Long (1984), Rauber and Grant (1983), and Griffith et al. (1983), ran from 15 January to 15 March 1983 and obtained generous samples from 20 storms. The expanded observational system included a dual-channel radiometer in both azimuth and zenith scanning modes, a vertical-pointing Ku-band radar, a C-band surveillance radar, a vertical-pointing polarization diversity lidar, a ground-based recording icing-rate meter, a transbarrier gage network, and a cloud physics aircraft. An indium oxide generator was colocated with a high-level remote-controlled AgI generator to trace particles that might be precipitated through scavenging rather than by nucleation. Extensive snowpack sampling was made for the detection of silver. The generators affecting the experimental area were randomized on a 2 seed to 1 no-seed basis. The storm units were periods of significant cloud cover and weather, typically 18–24 hours long. The spatial and temporal distributions of supercooled liquid water, the delivery of seeding materials to these clouds, and the trajectory of cloud and precipitation particles were primary questions addressed in the analysis.

In both COSE and the Utah Federal–State program, a polarization lidar provided a continuous record of liquid water and ice particle distribution in the vertical. This instrument appears to have considerable potential for resolving the fine-scale distributions (Sassen, 1984).

The key mesoscale analysis feature was the identification of precipitation echo types using the C-band search radar. Bands were much less frequent than in SCPP. Areawide and cellular types were most frequent. A cloud type called NP, precipitation with no echo, was also frequent. In this case a cloud existed over the barrier, as was evidenced by the Ku-band radar return, but was not visible to C-band radar. These types provided a means for stratifying liquid water and other observations in a physically significant way. Median estimates of radiometrically determined liquid water values for areawide, cellular, and NP were 0.085, 0.083, and 0.18 mm, respectively. Mean cloud depths were 3.4, 1.12, and 1.45 km, respectively.

Higher contents can be expected in the vicinity of cells, as shown by aircraft observations, and pulses in the zenith-pointing radiometer trace. The icing-rate meter showed that NP clouds were responsible for the highest LWC values. The liquid-water flux across the crest, as measured directly by the icing-rate meter, was compared to the earthward flux of water as expressed by the precipitation rates at the gages over the barrier.

9.5. Basic problems

This final section summarizes some basic problems that have arisen in the conduct of recent projects. A few suggestions for their amelioration are offered.

The present status of wintertime orographic weather modification mandates detailed observations along the links in the chain from seeding to precipitation on the ground. But tracking the various links becomes more problematical as one proceeds from nucleation and initial growth to fallout, and beyond to impact with the ground. The black boxes have been replaced only by gray ones. The tracking of the links is especially weak in the placebo case, which is supposed to mimic a seeded track. An expanded effort at tagging the links with tracer materials (chemicals, chaff, etc.) might be worthwhile. Lagrangian-type modeling of the chain of events and new and improved seeding modes would support this effort.

The recognition of seedability simply by reference to the real-time observations of liquid and ice water distributions evades the crucial issue of how a given mode of seeding would alter the orographic barrier's water balance in a beneficial way in a particular cloud type. Continuous seeding could impose an immediate perturbation that would be followed by development of a new steady state condition involving complex precipitation redistributions. This extension of viewpoint should also embrace dynamic effects. A remedy would be the use of water balance diagnostic models, improving their performance with improved flow and microphysical components and making provision for incorporating randomly received observations, including remotely sensed data of various types.

The vigorous pursuit of case studies and of calibration seeding tests can focus attention too narrowly on the most easily observed link in the chain. High expectations of success develop for a limited set of circumstances. Eventually, the window of seedability narrows in the minds of the investigators to a few unique situations. We have the case of the "vanishing seeding window." A parallel loss of nerve precludes treatment of the more numerous marginal cases, which is where the greatest hidden potential may exist. The remedy is to return to the true spirit of the exploratory experiment. Deficiencies in the observational chain can be compensated by building a dataset sufficiently large that statistical inference techniques can play a full role. A wide range of marginal cases should be included so as to better define the limits of seeding responses and to add new understanding of the mechanisms involved.

A difficult aspect of planning for any experiment is the balanced allocation of observational resources. What density of raingages is required? Where should the radiometer(s), the radar(s), and the rawinsonde(s) be located? How frequently should observations be made? A crucial problem is that the most expensive units, timewise and moneywise, are the observing and seeding aircraft, which are operated about six hours per storm day, although the other units could be operated around the clock. The limitations are due to the risky and exhausting nature of the work.

Problems of data interpretation arouse basic concerns about both observational and evaluation procedures. The expertise of the experimentalist is needed for quality control. Recognition of faulty data resulting from unusual aircraft maneuvers, natural events (such as lightning strikes), or subtle equipment malfunctions and decisions about the exclusion of outlier data points are examples of situations that cannot be covered by completely automated editing and analysis procedures, no matter how carefully planned in advance. A few blunders could vitiate results for an entire project. It seems important to permit a posteriori data reduction and analyses, relying on blindness with regard to seeding decisions for protection against intrusion of biases.

Target/control area precipitation evaluation methods have particular problems, especially in reanalysis of older projects. The problems result mainly from instability with time of target–control relationships. A review of these problems and how to cope with them in a physical as well as a statistical sense is in order.

A certain lack of appropriate communication exists between various areas of expertise that are involved directly or indirectly within the scope of weather modification. This has lead to some misunderstanding regarding the types of data appropriate to the support of arguments. "Apples, lemons, and limes" are mixed indiscriminately in comparing data and analyses. Inadequate reporting of original experiments, concepts, and conclusions and careless reading of such reports all play a role.

Acknowledgments. The author is grateful for the many suggestions and comments received from various sources concerning the preparation of this invited review paper. In this regard, special mention should be made of Dr. Arnold Court, California State University—Northridge; David Reynolds, Field Director of SCPP, Auburn, California; Dr. Alex Long, Desert Research Institute, Reno, Nevada; Dr. John Marwitz, University of Wyoming; Dr. John Flueck, NOAA; Don Griffith, North American Weather Consultants, Salt Lake City, Utah; and Professor L. O. Grant, Colorado State University, Ft. Collins, Colorado. Special thanks are offered to Gloria Evans and Kristeen Swart for their work on manuscript preparation.

REFERENCES

Advisory Committee on Weather Control, 1957: Vol. I and Vol. II, *Final Report of the Advisory Committee on Weather Control.* Washington, DC.

Appleman, H. S., 1954: Design of a cloud-phase chart. *Bull. Amer. Meteor. Soc.,* **35,** 223–225.

Bergeron, T., 1935: On the physics of clouds and precipitation. *Proc. Fifth Assembly UGGI,* Vol. 2, Lisbon, 156–178.

Blanchard, D. C., 1981: The lighter side of life with Project Cirrus. *J. Wea. Mod.,* **13,** 5–8.

Blumenstein, R. R., W. G. Finnegan and L. O. Grant, 1983: Ice nucleation by silver iodide–sodium iodide: A re-evaluation. *J. Wea. Mod.,* **15,** 11–15.

Brousaides, F. J., and J. F. Morrissey, 1971: Improved humidity measurements with a redesigned radiosonde humidity duct. *Bull. Amer. Meteor. Soc.,* **52,** 870–875.

Brown, K. J., and R. D. Elliott, 1972: Mesoscale changes in the atmosphere due to convective band seeding. *Preprints Third Conf. on Weather Modification,* Boston, Amer. Meteor. Soc., 313–320.

——, ——, J. R. Thompson, P. St. Amand and S. D. Elliott, Jr., 1974: The seeding of convective bands. *Fourth Conf. on Weather Modification,* Ft. Lauderdale, Amer. Meteor. Soc., 7–12.

——, —— and M. W. Edelstein, 1978: *Transactions of a Workshop on Total Area Effects of Weather Modification.* North American Weather Consultants report to NSF.

Browning, K. A., and G. A. Monk, 1982: A simple model for the synoptic analysis of cold fronts. *Quart. J. Roy. Meteor. Soc.,* 435–452.

Chappell, C. F., and F. L. Johnson, 1974: Potential for snow augmentation in cold orographic clouds. *J. Appl. Meteor.,* **13,** 374–382.

——, L. O. Grant and P. W. Mielke, 1971: Cloud seeding effects on precipitation intensity and duration of wintertime orographic clouds. *J. Appl. Meteor.,* **10,** 1006–1010.

Colton, P. E., 1976: Numerical simulation of the orographically induced precipitation distribution for use in hydrologic analysis. *J. Appl. Meteor.,* **15,** 1241–1251.

Cooper, W. A., and J. D. Marwitz, 1980: Winter storms over the San Juan Mountains. Part III: Seeding potential. *J. Appl. Meteor.,* **19,** 942–949.

——, and C. P. R. Saunders, 1980: Winter storms over the San Juan Mountains. Part II: Microphysical processes. *J. Appl. Meteor.,* **19,** 927–941.

Cotton, W. R., G. J. Tripoli and R. Blumenstein, 1982: The simulation of orographic clouds with a nonlinear, time dependent model. *Preprints Conf. on Cloud Physics,* Chicago, Amer. Meteor. Soc., 322–324.

DeMott, P. J., W. G. Finnegan and L. O. Grant, 1983: An application of chemical kinetic theory and methodology to characterize the ice nucleating properties of aerosols used for weather modification. *J. Climate Appl. Meteor.,* **22,** 1190–1203.

Dirks, R. A., 1973: The precipitation efficiency of orographic clouds. *J. Rech. Atmos.,* **2,** 177–184.

Elliott, R. D., 1961: Note on cloud seeding evaluation with hourly precipitation data. *J. Appl. Meteor.,* **1,** 578–580.

——, 1981: A seeding effect targeting model. *Eighth Conf. on Inadvertent and Planned Weather Modification,* Reno, Amer. Meteor. Soc., 28–29.

——, 1984: Seeding effects in the Colorado River Basin Pilot Project. *J. Wea. Mod.,* **16,** 30–33.

——, and E. L. Hovind, 1964: On convection bands within Pacific Coast storms and their relation to storm structure. *J. Appl. Meteor.,* **3,** 143–154.

——, and ——, 1965: Heat, water, and vorticity balance in frontal zones. *J. Appl. Meteor.,* **4,** 196–211.

——, and W. A. Lang, 1967: Weather modification in the southern Sierras. *J. Irrigation Drainage Div., Proc. Amer. Soc. Civil Engrs.,* **5644 IR 4,** 45–59.

——, and K. J. Brown, 1971: The Santa Barbara II project—downwind effects. *Proc. Int. Conf. on Weather Modification,* Canberra, Amer. Meteor. Soc. 179–184.

——, P. St. Amand and J. R. Thompson, 1971: Santa Barbara pyrotechnic cloud seeding test results, 1967–70. *J. Appl. Meteor.,* **10,** 785–795.

——, J. F. Hannaford and R. W. Shaffer, 1973: Twelve basin investigation, Vols. 1 and 2. Report to U.S. Dept. of Interior, Bur. of Recl., Div. of Atmos. Water Res. Mgmt. by North Amer. Weather Consultants. [NTIS PB232-131-3 (Vol. I) and 132-1 (Vol. II).]

——, R. W. Shaffer, A. Court and J. F. Hannaford, 1978: Randomized cloud seeding in the San Juan Mountains, Colorado. *J. Appl. Meteor.,* **17,** 1298–1318.

——, ——, and ——, 1980: Reply. *J. Appl. Meteor.,* **19,** 350–355.

Falconer, R. E., 1981: From Mt. Washington, N.H., to Schenectady, N.Y. and Project Cirrus. *J. Wea. Mod.,* **13,** 12–13.

Feng, D., and L. O. Grant, 1982: Correlation of snow crystal habits, number flux and snowfall intensity from ground observations. *Preprints Conf. on Cloud Physics,* Chicago, Amer. Meteor. Soc. 485–487.

Findeisen, W., 1938: Die Kolloidmeteorologischen Vorgange bei der Niedeerschlagsbildung. *Meteor. Z.,* **55,** 121–133.

Fraser, A. B., R. C. Easter and P. V. Hobbs, 1973: A theoretical study of the flow of air and fallout of solid precipitation over mountainous terrain. Part I: Airflow model. *J. Atmos. Sci.,* **30,** 801–817.

Fukuta, N., 1963: Ice nucleation by metaldehyde. *Nature,* **199,** 475–476.

——, 1967: An airborne generator of metaldehyde smoke. *J. Appl. Meteor.,* **6,** 948–951.

——, W. A. Schmeling and L. F. Evans, 1971: Experimental determination of ice nucleation by falling dry ice particles. *J. Appl. Meteor.,* **10,** 1174–1179.

Gabriel, K. R., and S. A. Changnon, 1982: Piggyback weather experimentation: Superimposing randomized treatment comparisons on commercial seeding operations. *J. Wea. Mod.,* **14,** 7–10.

Garvey, D. M., 1975: Testing of cloud seeding materials at the cloud simulation and aerosol laboratory, 1971–1973. *J. Appl. Meteor.,* **14,** 883–890.

Gordon, G. L., and J. D. Marwitz, 1985: Hydrometeor evolution in rainbands over the California Valley. *J. Atmos. Sci.,* **43,** 1087–1100.

Grant, L. O., and R. D. Elliott, 1974: The cloud seeding temperature window. *J. Appl. Meteor.,* **13,** 355–363.

——, C. F. Chappell, L. W. Crow, P. W. Mielke, Jr., J. L. Rasmussen, W. E. Shobe, H. Stockwell and R. A. Wykstra, 1969: An operational adaptation program of weather modification for the Colorado River Basin. Interim report to Bur. of Recl. from Dept. of Atmos. Sci., Colorado State University.

——, J. O. Rhea, G. T. Meltesen, G. J. Mulvey and P. W. Mielke, 1979: Continuing analysis of the Climax weather modification experiments. *Preprints Seventh Conf. on Inadvertent and Planned Weather Modification.* Banff, Amer. Meteor. Soc., 343–345.

——, J. G. Medina and P. W. Mielke, Jr., 1983: Reply to Comments on "A statistical reanalysis of the replicated Climax I and II wintertime orographic cloud seeding experiments." *J. Appl. Meteor.,* **22,** 1482–1484.

Griffith, D. A., R. D. Elliott, J. L. Sutherland, H. R. Swart and R. L. Atkins, 1983: Analysis of data collected during a NOAA/Utah cooperative research program. North Amer. Weather Consultants report SLWM 83-4 prepared for NOAA and Utah Div. of Water Res.

Hallett, J., and S. C. Mossop, 1974: Production of secondary ice particles during the riming process. *Nature,* **249,** 26–28.

Havens, B. S., J. E. Juisto and B. Vonnegut, 1981: Early history of cloud seeding. *J. Wea. Mod.,* **13,** 14–88.

Heggli, M. F., L. Vardiman, R. E. Stewart and A. Huggins, 1983: Supercooled liquid water and ice crystal distributions within Sierra Nevada winter storms. *J. Climate Appl. Meteor.,* **22,** 1875–1886.

Henderson, T. J., 1966: A ten year non-randomized cloud seeding program on the Kings River in California. *J. Appl. Meteor.,* **5,** 697–702.

——, and M. Solak, 1983: Supercooled liquid water concentrations in winter orographic clouds from ground-based ice accretion measurements. *J. Wea. Mod.,* **15,** 64–70.

Hess, W., 1974: *Weather and Climate Modification.* Wiley and Sons, 842 pp.

Hill, G. E., 1980: Re-examination of cloud top temperature used as criteria for stratification of cloud seeding effects in experiments on winter orographic clouds. *J. Appl. Meteor.,* **19,** 1167–1175.

——, and D. S. Woffinden, 1980: A balloon-borne instrument for the measurement of vertical profiles of supercooled liquid water concentration. *J. Appl. Meteor.,* **19,** 1285–1292.

Hindman, E. D., and L. O. Grant, 1981: Utility of mountain top rime-ice measurements. *Preprints Second Conf. on Mountain Meteorology,* Steamboat Springs, Amer. Meteor. Soc., 404–408.

Hobbs, P. V., 1975a: The nature of winter clouds and precipitation in the Cascade Mountains and their modification by artificial seeding. Part I: Natural conditions. *J. Appl. Meteor.,* **14,** 783–804.

——, 1975b: The nature of winter clouds and precipitation in the Cascade Mountains and their modification by artificial seeding. Part III: Case studies of the effects of seeding. *J. Appl. Meteor.,* **14,** 819–858.

——, and L. R. Radke, 1975: The nature of winter clouds and precipitation in the Cascade Mountains and their modification by artificial seeding. Part II: Techniques for the physical evaluation of seeding. *J. Appl. Meteor.,* **14,** 805–818.

——, and A. L. Rangno, 1979: Comments on the Climax and Wolf Creek Pass cloud seeding experiment. *J. Appl. Meteor.,* **18,** 1233–1236.

——, and ——, 1984: Reply to Comments on "Production of ice particles in clouds due to aircraft penetrations." *J. Climate Appl. Meteor.,* **23,** p. 346.

Hogg, D. C., F. O. Guiraud, J. B. Snider, M. T. Decker and E. R. Westwater, 1983: A steerable dual-channel microwave radiometer for measurement of water vapor and liquid in the troposphere. *J. Climate Appl. Meteor.,* **22,** 789–806.

Huggins, A. W., and A. R. Rodi, 1985: Physical response of convective clouds over the Sierra Nevada to seeding with Dry Ice. *J. Climate Appl. Meteor.,* **24,** 1082–1098.

Knollenberg, R. G., 1972: Comparative liquid water content measurements of convenient instruments with an optical array spectrometer. *J. Appl. Meteor.,* **11,** 501–508.

——, 1976: Three new instruments for cloud physics measurements, the 2-D spectrometer, the forward scattering spectrometer probe, and the active scattering aerosol spectrometer. *Preprints Int. Conf. on Cloud Physics,* Boulder, Amer. Meteor. Soc., 554–561.

Langer, G., 1973a: Analysis of results from second international ice nucleus workshop with emphasis on expansion chambers, NCAR counters, and membrane filters. *J. Appl. Meteor.,* **12,** 991–999.

——, 1973b: Evaluation of NCAR ice nucleus counter: Part 1. Basic operation. *J. Appl. Meteor.,* **12,** 1000–1011.

Long, A. B., 1984: Physical investigations of winter orographic clouds in Utah. Final Report to Utah Dept. of Resources, Div. of Water Resources to NOAA under Contract NA82RAH00001.

Ludlam, F. H., 1955: Artificial snowfall from mountain clouds. *Tellus,* **7,** 277–290.

MacCready, P. B., and C. J. Todd, 1964: Continuous particle sampler. *J. Appl. Meteor.,* **3,** 450–460.

Marwitz, J. D., 1980: Winter storms over the San Juan Mountains. Part I: Dynamical processes. *J. Appl. Meteor.,* **19,** 913–926.

——, 1983: The kinematics of orographic flow during Sierra storms. *J. Atmos. Sci.,* **40,** 1218–1227.

——, and R. E. Stewart, 1981: Some seeding signatures in Sierra storms. *J. Appl. Meteor.,* **20,** 1129–1144.

Meltesen, G. T., O. Rhea, G. J. Mulvey and L. O. Grant, 1977: Certain problems in post hoc analysis of samples from heterogeneous populations and skewed distributions. *Preprints Sixth Conf. on Planned and Inadvertent Weather Modification,* Urbana, Amer. Meteor. Soc. 388–391.

Mielke, P. W., Jr., and L. O. Grant, 1967: Cloud seeding experiment at Climax, Colorado, 1960–65. *Proc. Fifth Berkeley Symp. on Mathematical Statistics and Probability, Vol. 5,* University of California Press, 115–131.

——, and J. G. Medina, 1983: A new covariate for estimating treatment differences with applications to Climax I and II experiments. *J. Climate Appl. Meteor.,* **22,** 1290–1295.

——, —— and C. F. Chappell, 1970: Elevation and spatial variation effects of wintertime orographic cloud seeding. *J. Appl. Meteor.,* **9,** 476–488; *Corrigendum,* **10,** p. 842.

——, L. O. Grant and C. F. Chappell, 1971: An independent replication of the Climax wintertime orographic cloud seeding experiments. *J. Appl. Meteor.,* **10,** 1198–1212; *Corrigendum,* **15,** p. 801.

——, G. W. Brier, L. O. Grant, G. J. Mulvey and P. N. Rosenzweig, 1981: A statistical reanalysis of the replicated Climax I and II wintertime orographic cloud seeding experiments. *J. Appl. Meteor.,* **20,** 643–659.

——, K. J. Berry and J. G. Medina, 1982: Climax I and II: Distortion resistant residual analyses. *J. Appl. Meteor.,* **21,** 788–792.

Mooney, M. L., and G. W. Lunn, 1968: The area of maximum effect resulting from Lake Almanor randomized cloud seeding experiment. *J. Appl. Meteor.,* **8,** 68–74.

Morel-Seytoux, H. J., and F. Saheli, 1973: Test of runoff increase due to precipitation management for the Colorado River Basin Project. *J. Appl. Meteor.,* **12,** 322–337.

Morrison, B. J., W. G. Finnegan, R. D. Horn and L. O. Grant, 1984: A laboratory characterization of Dry Ice as a glaciogenic agent. *Extended Abstracts. Ninth Conf. on Weather Modification,* Park City, Amer. Meteor. Soc., 8–9.

Morrissey, J. F., and F. J. Brousaides, 1970: Temperature induced errors in the ML-476 humidity data. *J. Appl. Meteor.*, **9**, 805–808.

Mossop, S. C., 1984: Comments on production of ice particles in clouds due to aircraft penetrations. *J. Climate Appl. Meteor.*, **23**, p. 345.

——, R. E. Ruskin and K. J. Heferman, 1968: Glaciation of a cumulus at approximately −4°C. *J. Atmos. Sci.*, **25**, 889–899.

Myers, V., 1962: Airflow on the windward side of a large ridge. *J. Geophys. Res.*, **67**, p. 4267.

National Academy of Sciences–National Research Council, 1966: *Weather and Climate Modification, Problems and Prospects.* Publ. 1350, Washington, DC.

Neyman, J., E. L. Scott and M. Vasilevskis, 1960: Statistical evaluation of the Santa Barbara randomized cloud seeding project. *Bull. Amer. Meteor. Soc.*, **41**, 531–547.

Nickerson, E. C., J. M. Fritsch, C. F. Chappell and D. R. Smith, 1979: Numerical simulation of orographic and convective cloud systems. Annual report, U.S. Dept. of Interior, Bur. of Recl., Contract 8-07-83V0017.

Parish, T. R., 1982: Barrier winds along the Sierra Nevada Mountains. *J. Appl. Meteor.*, **21**, 925–930.

Plooster, M. N., and N. Fukuta, 1975: A numerical model of precipitation from seeded and unseeded cold orographic clouds. *J. Appl. Meteor.*, **14**, 859–867.

Rangno, A. L., 1979: A reanalysis of the Wolf Creek Pass cloud seeding experiment. *J. Appl. Meteor.*, **18**, 579–605.

——, and P. V. Hobbs, 1980a: Comments on "Randomized cloud seeding in the San Juan Mountains, Colorado." *J. Appl. Meteor.*, **19**, 342–346.

——, and ——, 1980b: Comments on "Generalized criteria for seeding winter orographic clouds." *J. Appl. Meteor.*, **19**, 906–907.

——, and ——, 1981: Comments on "Reanalysis of generalized criteria for seeding winter orographic clouds." *J. Appl. Meteor.*, **20**, 216–217.

——, and ——, 1983: Production of ice particles in clouds due to aircraft penetrations. *J. Climate Appl. Meteor.*, **22**, 214–232.

Rauber, R. M., and L. O. Grant, 1983: Preliminary analysis of the hypothesis used in the Utah operational weather modification program. Final report to Utah Dept. of Nat. Resources by Dept. of Atmos. Science, Colorado State University.

——, and ——, 1986: The characteristics and distribution of cloud water over the mountains of northern Colorado during wintertime storms. Part II: Spatial distribution and microphysical characteristics. *J. Climate Appl. Meteor.*, **25**, 468–488.

——, D. Feng, L. O. Grant and J. B. Snider, 1986: The characteristics and distribution of cloud water over the mountains of northern Colorado during wintertime storms. Part I: Temporal variations. *J. Climate Appl. Meteor.*, **25**, 468–488.

Reynolds, D. W., and A. S. Dennis, 1984: Fall 1984 status report. *Project Skywater Sierra Cooperative Pilot Project* (SCPP). Div. of Atmos. Resources Research, Bur. of Recl., Denver Engrg. and Research Center.

Rhea, J. O., 1978: Orographic precipitation model for hydro-meteorological use. Atmos. Sci. Rep. No. 287, Dept. of Atmos. Sci., Colorado State University.

——, 1983: Comments on "A statistical reanalysis of the replicated Climax I and II wintertime orographic cloud seeding experiments." *J. Climate Appl. Meteor.*, **22**, 1475–1481.

——, A. W. Huggins, J. L. LeCompte, G. L. Hemmer, A. P. Kuciawskas and C. J. Wilcox, 1982: *Int. Progress Report SCPP Forecasting Support for Period 1 July 1981–30 June 1982.* Office of Atmos. Water Res. Management, Bur. of Recl., Contract 9-07-85-V0021, 228 pp.

Rottner, D., L. Vardiman and J. A. Moore, 1980: Reanalysis of "Generalized criteria for seeding winter orographic clouds." *J. Appl. Meteor.*, **19**, 622–626.

Rutledge, S. A., and P. V. Hobbs, 1983: The mesoscale structure of clouds and precipitation in midlatitude cyclones. VIII: A model for the "seeder–feeder" process in warm, frontal rainbands. *J. Atmos. Sci.*, **40**, 1185–1206.

Sassen, K., 1984: Deep orographic cloud structure and composition derived from comprehensive remote sensing measurements. *J. Climate Appl. Meteor.*, **23**, 568–582.

Schaefer, V. J., 1946: The production of ice crystals in a cloud of supercooled water droplets. *Science*, **104**, 457–459.

——, 1981: The serendipitous happenings which lead to weather modification. *J. Wea. Mod.*, **13**, 1–4.

Shaffer, R. W., 1983: Seeding agent threshold activation temperature height, and important seedability criterion for ground-based seeding. *J. Wea. Mod.*, **15**, 16–20.

Smith, E. J., E. E. Adderley and D. T. Walsh, 1963: A cloud seeding experiment in the Snowy Mountains, Australia. *J. Appl. Meteor.*, **2**, p. 324.

——, L. G. Veitch, D. E. Shaw and A. J. Mitter, 1979: A cloud seeding experiment in Tasmania. *J. Appl. Meteor.*, **18**, 804–815.

Smith, R. B., 1979: The influence of mountains on the atmosphere. *Advances in Geophysics*, Vol. 21, Academic Press, 87–127.

Snider, J. B., and D. Rottner, 1982: The use of microwave radiometry to determine a cloud seeding opportunity. *J. Appl. Meteor.*, **21**, 1286–1291.

Stewart, R. E., and J. D. Marwitz, 1982a: The downwind spread of an initially vertical column of particles in a sheared environment. *J. Appl. Meteor.*, **21**, 1191–1193.

——, and ——, 1982b: Microphysical effects of seeding wintertime stratiform clouds near the Sierra Nevada Mountains. *J. Appl. Meteor.*, **21**, 874–880.

Stith, J. L., 1984: The effectiveness of dry ice as a seeding agent in the Sierra Cooperative Pilot Project. *Extended Abstracts. Ninth Conf. on Weather Modification*, Park City, Amer. Meteor. Soc.

Super, A. B., and J. A. Heimbach, Jr., 1983: Evaluation of the Bridger Range winter cloud seeding experiment using control gages. *J. Climate Appl. Meteor.*, **22**, 1989–2011.

Sutherland, J. L., and J. W. Kidd, 1978: A radar climatology of winter storms in northern California for the Sierra Cooperative Pilot Project. Final report to Bur. of Recl.

Tukey, J. W., D. R. Brillinger and L. V. Jones, 1978: The management of weather resources. Vol. II: The role of statistics in weather resources management. Statistical Task Force report to Wea. Mod. Advisory Board. [U.S. Govt. Printing Office, Stock No. 003-018-00091-1.]

Turner, F. M., and L. F. Radke, 1973: The design and evaluation of an airborne optical ice particle counter. *J. Appl. Meteor.*, **12**, 1309–1318.

Vardiman, L., and J. A. Moore, 1978: Generalized criteria for seeding winter orographic clouds. *J. Appl. Meteor.*, **17**, 1769–1777.

Vonnegut, B., 1949: Nucleation of supercooled water by silver iodide smokes. *Chem. Rev.*, **44**, 277–289.

Warburton, J. A., M. S. Owens, A. V. Anderson and R. Stone, 1979: Temporal and spatial distribution of seeding material in a ground-based seeded target area. *Seventh Conf. on Inadvertent and Planned Weather Modification*, Banff, Amer. Meteor. Soc. 43–44.

Weather Modification Advisory Board, 1978: *The Management of Weather Resources. Vol. 1, Proposals for a National Policy and Program.* U.S. Dept. of Commerce.

Young, K. C., 1974a: A numerical simulation of wintertime orographic precipitation. Part I: Description of model microphysics and numerical techniques. *J. Atmos. Sci.*, **31**, 1735–1748.

——, 1974b: A numerical simulation of wintertime orographic precipitation. Part II: Comparison of natural and AgI seeding conditions. *J. Atmos. Sci.*, **31**, 1749–1767.

CHAPTER 10

Hypotheses for the Climax Wintertime Orographic Cloud Seeding Experiments

LEWIS O. GRANT

Colorado State University, Fort Collins, Colorado

ABSTRACT

The hypothesis used for the initial Climax wintertime cloud seeding experiment and for subsequent Climax replication-type experiments are described and briefly discussed. More recent physical studies of Colorado orographic clouds and seeding hypotheses are briefly summarized. These later tests and studies of orographic cloud seeding hypotheses emphasized direct and remotely sensed cloud and precipitation measurements utilizing instrumentation and modeling capabilities not available during the Climax statistical experiments. The conclusions suggested from the hypothesis testing, considering both the statistical experiments and the later physical studies, are summarized.

10.1. Introduction

This paper presents a brief summary of the hypotheses employed and some of the measurements made during the Colorado State University studies of Colorado orographic clouds. As requested by the organizers of the May 1984 Workshop on Precipitation Enhancement, the emphasis of this summary is on the hypotheses (referred to as "conceptual models" by the workshop organizers) and observations of the randomized experiments carried out near Climax, Colorado, during the 1960s. Physical studies of the Climax hypotheses, carried out subsequent to the Climax I and Climax II statistical experiments, are briefly described. Specific results of the Climax experiments are included only to the extent that they affected modifications in the hypotheses in subsequent phases of the experiments. General conclusions relative to the validity of the hypotheses are presented.

The Colorado State University program of weather modification studies of orographic clouds basically involved four phases. Phase I included the original Climax randomized experiment from 1960 to 1965. This is generally referred to as Climax I. The second phase, Climax II, extended from the fall of 1965 through January 1970. Phase III was embedded within phase II and extended from 1968 to 1970. Phase IV has included the period of increasingly intensive measurements and hypotheses development from 1978 until the present.

The basic design of the Climax experiments has been described in a number of proposals, papers, and reports such as Grant and Schleusener (1960), Grant (1961), Grant and Mielke (1967), Mielke et al. (1970), and Mielke et al. (1971). The abstract from Grant (1961) stated that "the proposed study (involving a continuation effort) would investigate the various physical factors considered by Bergeron and Ludlam for cold-cloud orographic pro-

cesses under wintertime conditions in the high mountainous area of central Colorado." This and the original proposal to NSF (Grant and Schleusener, 1960) describe both the importance of the Ludlam hypothesis to the program and the basic aspects of the experiment. The Climax experiments involved randomized silver iodide seeding with ground generators from the upwind slopes of the Colorado Rocky Mountains near Climax, Colorado. The randomization was restricted only to the extent that large blocks (20–40) had the same number of nonseeded and seeded experimental units. The experimental unit was a 24-hour interval keyed to a distinct climatological diurnal minimum in precipitation during most of the experiment. Experimental units were declared by National Weather Service (NWS) forecasters in Denver, and randomized seeding envelopes were then processed according to the experimental design by observers at the Colorado State University Weather Station. Neither the NWS forecasters nor the weather station observers were involved with the conduct of the research itself.

10.2. Hypotheses for the Climax experiments

The hypothesis for the initial Climax randomized experiment was more specific than those used for most seeding experiments initiated in that era. The hypothesis formulation followed from Bergeron's (1949) early assumption of a steady state supply of condensate for extended periods of time. He assumed that this would result in "releaseable but unreleased" cloud condensate. After Bergeron developed initial concepts concerning the condensate supply that could lead to a potential for "unreleased" cloud water, Ludlam (1955) addressed the microphysical processes that would be required for successful removal of the additional cloud water. This Ludlam formulation served as the basic hypothesis for the Climax experiments. Other experimental criteria were specified and various aspects of the hypothesis were made more specific as the experiments progressed and new information became available. The initial hypotheses and subsequent evaluation of these hypotheses are briefly summarized in the following sections.

10.3. Phase I: Climax I (1960–65)

The basic concepts of the Ludlam theory for initiating snowfall from simple orographic clouds served as the hypothesis for the initial Climax experiment, which started in 1960. The main features of this hypothesis were as follows:

A. Shallow orographic clouds should be extensive. Ludlam referred to these as "extensive low clouds." Ludlam excluded "clouds (that) contain persistent vertical motions of a magnitude sufficient to sustain any considerable precipitation, except in localities where the airstream containing the clouds flows over mountains." The Climax experiments were designed and conducted for blanket orographic clouds that form as moisture is advected into the area in association with various weather systems. The experimental hypothesis, the meteorological indicies used, and the analyses carried out, all placed emphasis on this component of weather systems passing through the area. Organized weather systems passing through the region and deep convective clouds forming over the mountains were considered to be already efficient precipitation producers. Consequently, it was assumed that the precipitation from these systems would be "noise" in the randomized experiment and that similar "noise" should occur on both seeded and nonseeded experimental days. The Climax experiment was designed and conducted as an orographic seeding experiment following consideration of orographic clouds as proposed by Ludlam.

B. Orographic clouds are frequently composed of supercooled water that is not used at all or is not used completely for the production of snowfall during its transit over the mountain barrier.

C. Growth and fallout time are critical for artificially nucleated ice crystals during their transit over the mountain barrier. In the initial stages of the Climax experiment, the time requirement used followed the general calculations of Ludlam. This was specifically considered in ground generator placement and for the studies of ice crystals settling on the mountain.

D. Numbers of ice crystals should be similar to the calculated estimates of Ludlam, $\sim 2 \times 10^4$ m^{-3}.

E. The practical and climatological considerations from Ludlam were evaluated for the specific areas around Climax, Colorado. Several other components of the Climax hypothesis followed from the specific theory of Ludlam. These included components such as:

1) The natural ice crystal concentrations present at appropriate simple orographic cloud elevations are too low to optimize the utilization of all cloud supercooled water.

2) Effective artificial nuclei could be produced in numbers adequate to treat a large volume of the atmosphere. Extensive efforts were conducted to test this and to calibrate the seeding generators used. This led to the development of the calibration and testing program for seeding generators at CSU.

3) The hypothesis included the criterion that artificial ice-nucleating aerosols could be delivered to the orographic clouds from ground sites. Extensive efforts concentrated on studies of the movement of the seeding materials after release.

4) The final component of the statistical hypothesis included the criterion that precipitation would be greater on the seeded than on the unseeded days.

Figure 10.1 is a schematic depiction of the Climax I experiment prepared in the early 1960s. It depicts a simple orographic cloud and summarizes the systems approach of research used. This included considerations of 1) nucleation and seeding agents, 2) transport of seeding material, 3) cloud and precipitation physics, 4) natural and artificial precipitation, and 5) hydrologic impacts.

10.4. Phase II: Climax II (1965–70)

Based on analyses of the Climax I experiment (Grant and Mielke, 1967), it was concluded that important as-

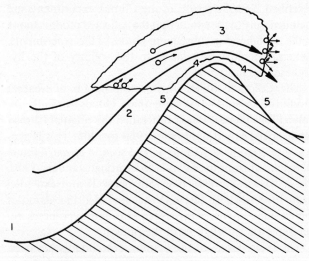

1 NUCLEATION AND SEEDING AGENTS
2 TRANSPORT OF SEEDING MATERIAL
3 CLOUD AND PRECIPITATION
4 NATURAL AND ARTIFICAL PRECIPITATION
5 HYDROLOGY

FIG. 10.1. Systems approach used for the orographic cloud and orographic cloud modification research at Colorado State University.

pects of the Ludlam theory were correct. The Grant and Mielke (1967) paper was presented at the Fifth Berkeley Symposium in December 1965, and with final corrections to comments by reviewers in February 1966. For the Climax II experiment, the hypotheses still followed Ludlam, but were made more specific, and the expectations for precipitation increases were specified only for the warmer cases of orographic clouds. While the precipitation for the seeded cases during Climax I was some 10 percent greater than for the nonseeded cases, the variance of the data was large and the difference was not statistically significant. Post hoc analysis showed that only with orographic cloud systems with temperatures warmer than that for which one would expect $\sim$10 pristine ice crystals per liter was there significantly more precipitation on the seeded days than on the nonseeded days. With colder clouds, the precipitation on the seeded days was the same or less than on the nonseeded days. The Ludlam hypothesis thus still formed the hypothesis for Climax II, but with the qualification that for precipitation increases it applied only to orographic clouds with cloud-top temperatures $\geq -20°C$.

At the start of the Climax II experiment in 1965, the experimenters were not able to routinely measure cloud-top temperatures. Consequently, the 50 kPa temperatures were specified for separating the clouds between warmer and colder cloud temperature categories. It was assumed for statistical analyses that the top of orographic blanket-type clouds were typically at or near (generally at a lower elevation) the 50 kPa level. Individual convective cells or cloud lines that varied greatly from this were assumed to be similar on both seeded and nonseeded days and thus to constitute "noise" in the randomized experiment.

Categories of wind direction and speed also showed substantial differences between the seeded and nonseeded precipitation categories from the post hoc analyses of the Climax I experiment. This was a reasonable expectation since wind direction relative to a mountain barrier and wind speed control both the rate of orographic lifting and the time available for precipitation growth and fallout. The Climax I analyses showed that only modest and statistically insignificant differences between seeded and nonseeded events occurred with weak orographic components of wind flow (less than $\sim$11 m s^{-1} at 70 kPa). The precipitation amounts on seeded days were considerably greater than on nonseeded for intermediate wind speeds ($\sim$12–14 m s^{-1} at 70 kPa). Precipitation amounts for cases with strong winds ($>$15 m s^{-1} at 70 kPa), which adversely impact on time for particle growth and fallout, were less for the seeded cases. These post hoc findings from the Climax I experiment were incorporated as expectation or hypotheses for the Climax II experiment.

10.5. Phase III: Latter part of Climax II (1968–70)

Continuing analyses during the mid- and late 1960s showed the importance of both stability and low-level moisture on the characteristics of the orographic blanket clouds and their seedability. These considerations led to analyses by Chappell (1967), Grant et al. (1968), and Chappell (1970) that placed emphasis on seedability as defined by the comparative rates of cloud water consumption for diffusional ice crystal growth and the condensate supply rate for the orographic clouds being formed. The relationship between condensate supply rate and rate of ice particle vapor consumption was not specified in advance as an analysis criterion when Climax II began. This relationship did, however, become a fixed part of various studies and analyses starting in 1968.

10.6. Phase IV: Physical studies of Colorado orographic clouds and seeding hypotheses (1978–present)

Hypothesis development and testing for the artificial seeding of orographic clouds has continued as a sequel to the Climax I and Climax II experiments. These studies have placed further emphasis on physical evaluation of the basic tenets of the Ludlam model and special conditions placed on it during the Climax statistical experiments. These more detailed physical studies of the hypotheses for the artificial treatment of Colorado orographic clouds have been made possible by recent developments in instrumentation for both in situ and remotely sensed observations of the orographic cloud processes. This program has placed emphasis on physical measurements and modeling efforts to help understand the physical structure and weather modification potential of Colorado orographic clouds. The field measurements for these studies have been designed and planned to test, with improved physical measurements, the hypothesis used for the Climax statistical experiments.

The field program has included measurements to describe the amount and distribution of cloud supercooled liquid water, the concentrations of ice particles, the characteristics of riming and aggregation, the trajectories of both cloud and ice water, cloud system precipitation efficiency, and the distribution of precipitation over an orographic barrier for shallow orographic clouds (and other cloud types) that form over the Park Range of northwest Colorado (Grant and Rauber, 1984; Rauber et al., 1986; Rauber and Grant, 1986). Specific conceptual models based on these studies are described by Rauber (1985). These models place emphasis on the importance of the frequently unreleased liquid water in the lower, warmer portions of the orographic clouds and on a supercooled liquid water cloud layer near cloud top that can be important in permitting more efficient nucleation mechanisms.

10.7. Basic conclusions from tests of these hypotheses

The following are the generalized conclusions reached with respect to the validity of the hypotheses used for the

Climax winter experiments and studies. These are based on both physical and statistical studies and analyses.

1) Shallow orographic clouds present the main opportunity for augmenting winter mountain precipitation in the Colorado Rockies.

2) The main weather modification potential from the shallow orographic clouds occurs with clouds that are below freezing, but with cloud-top temperatures equal to or warmer than about −20°C. In the statistical analyses, 50 kPa was used to specify the warm and cold cloud temperatures.

3) Deep and/or cold clouds (temperature < −20°C) have little or no potential for precipitation augmentation over the Colorado Rockies. These deep clouds include those whose primary forcing is from strong, large-scale dynamic forcing, deep convection, or deep, cold orographic forcing.

4) Many warmer regions of orographic clouds have below-optimal ice crystal concentrations and contain significant amounts of subcooled cloud water that passes over the Rocky Mountain barrier unused for the formation of mountain snowfall.

5) The strength of the orthogonal component of airflow is important both to the rate of formation of cloud condensate and as a control on the time available for the formation of precipitation. Depending on the orientation and slopes of a particular mountain range and the moisture available, airflow can occur in a velocity range that maximizes the production of cloud water and still allows time for artificially nucleated ice crystals to grow and reach the surface in the time available. At lower velocities, the production of cloud condensate is limited, and at higher velocities, time for growth and fallout of ice particles can be insufficient, even with artificial seeding.

6) Large numbers of ice nuclei that are sufficient to adequately treat large volumes of the atmosphere can be generated artificially.

7) Artificial ice nuclei are sometimes carried to appropriate cloud regions when released from upwind surface sites. The delivery, however, is inefficient and constitutes the most serious constraint for weather modification research and operations.

8) Precipitation is frequently increased from seeding clouds that fit the hypothesized criteria (Mielke et al.,

1981). It is likely that the precipitation increases can be of a magnitude to have significant economic, social, and political value.

REFERENCES

Bergeron, R., 1949: The problem of artificial control of rainfall on the Globe I: General effects of ice nuclei in clouds. *Tellus,* **1,** 32–50.

Chappell, C. F., 1967: Cloud seeding opportunity recognition. Atmos. Sci. Paper 118, Dept. of Atmos. Sci., Colorado State University, 87 pp.

——, 1970: Modification of cold orographic clouds. Ph.D. Dissertation, Dept. of Atmos. Sci., Colorado State University, 196 pp.

Grant, L. O., 1961: Study of cloud and snowfall, and changes resulting from the addition of artificial ice nuclei, over the high elevations of the Rocky Mountains. Funded proposal to National Science Foundation, 12 pp.

——, and R. A. Schleusener, 1960: An investigation of snowfall and the effects of cloud seeding on snowfall in the Colorado Rockies. Funded proposal to National Science Foundation, 11 pp.

——, and P. W. Mielke, Jr., 1967: A randomized cloud seeding experiment at Climax, Colorado, 1960–1965. *Proc. Fifth Berkeley Symp. on Mathematical Statistics and Probability,* 115–131.

——, and R. M. Rauber, 1984: Hypothesis evaluation and development for seeding continental wintertime mountain cloud systems. *Extended Abstracts Ninth Conf. on Weather Modification,* Park City, Amer. Meteor. Soc., 33–34.

——, C. F. Chappell and P. W. Mielke, Jr., 1968: The recognition of cloud seeding opportunity. *Proc. First Natl. Conf. on Weather Modification,* Albany, Amer. Meteor. Soc., 372–385.

Ludlam, F. H., 1955: Artificial snowfall from mountain clouds. *Tellus,* **7,** 277–289.

Mielke, P. W., Jr., L. O. Grant and C. F. Chappell, 1970: Elevation and spatial variation effects of wintertime orographic cloud seeding. *J. Appl. Meteor.,* **9,** 476–488.

——, —— and ——, 1971: An independent replication of the Climax wintertime orographic cloud seeding experiment. *J. Appl. Meteor.,* **10,** 1198–1212.

——, G. W. Brier, L. O. Grant, G. J. Mulvey and P. N. Rosenzweig, 1981: A statistical reanalysis of the replicated Climax I and II wintertime orographic cloud seeding experiments. *J. Appl. Meteor.,* **20,** 643–660.

Rauber, R. M., 1985: Physical structure of northern Colorado River basin cloud systems. Atmos. Sci. Paper 390, Dept. of Atmos. Sci., Colorado State University, 151 pp.

——, and L. O. Grant, 1986: The characteristics and distribution of cloud water over the mountains of northern Colorado during wintertime storms. Part II: Spatial distribution and microphysical structure. *J. Climate Appl. Meteor.,* **25,** 489–504.

——, ——, D. Feng and J. B. Snider, 1986: The characteristics and distribution of cloud water over the mountains of northern Colorado during wintertime storms. Part I: Temporal variations. *J. Climate Appl. Meteor.,* **25,** 468–488.

CHAPTER 11

A Comparison of Winter Orographic Storms over the San Juan Mountains and the Sierra Nevada

JOHN MARWITZ

University of Wyoming, Laramie, Wyoming

ABSTRACT

The winter orographic storms over the San Juan Mountains and the Sierra Nevada are compared. The topography of the San Juans is complex while the Sierra barrier is comparatively simple. The barrier jet is well developed upwind of the Sierra Nevada and its development is restricted upwind of the San Juans. The major difference between the storms on the two barriers is that the Sierra Nevada storms are typically maritime while the San Juan storms are continental. The implications for seeding are discussed.

11.1. Introduction

The Colorado River Basin Pilot Project was a randomized seeding experiment designed to determine if it was feasible to increase precipitation in the Colorado River Basin by seeding wintertime orographic storm systems over the San Juan Mountains. The experiment ran for five winters (1970/71 to 1974/75) under the direction of the Bureau of Reclamation. Based on a preliminary analysis of the precipitation data following the third season, it was obvious that a physical evaluation of the natural and modified cloud systems was needed to complement the statistical evaluation. Marwitz (1974) presented a case study of a storm in 1973, Hobbs et al. (1975) presented airborne studies from flights during the 1973/74 season, and Marwitz et al. (1976), Marwitz (1980), Cooper and Saunders (1980) and Cooper and Marwitz (1980) presented their results based on airborne studies in the 1974/75 season.

The Sierra Cooperative Pilot Project (SCPP) represents a rather different approach to cloud seeding experiments. Seeding and physical studies were designed to emphasize understanding and documenting of the chain of events involved in both natural and artificially stimulated precipitation processes. The initial planning and design of SCPP began in 1972. The first field studies began in the 1976/77 season and consisted of radar, rawinsondes, precipitation gages, and an instrumented aircraft. The University of Wyoming instrumented its King Air and began a series of annual field studies in SCPP in December 1977. The first calibration seeding experiments began in 1979, and the first randomized seeding experiment began in the 1982/83 field season. The availability of several new technologies made this new approach possible. In recent field seasons the observational system has been expanded to include Doppler radar, satellite data, microwave radi-

ometer, zenith-pointing radar and detailed microphysical data at the surface.

The objective of this article will be to compare and contrast the storm evolution, airflow characteristics and precipitation processes over these two orographic barriers. The implications for cloud seeding will be discussed. In most cases the results discussed have been published in the refereed literature. In other cases, the results should only be considered preliminary since they only appear in project reports. The opinions in all cases represent those of the author and not necessarily that of the other participants or sponsoring agency.

11.2. Comparison of San Juan and Sierra Nevada storms

The San Juan Mountains are located in southwestern Colorado. The Continental Divide runs along the crest of the San Juans. The crest is oriented north–south in New Mexico and curves to the west for about 100 km in southwestern Colorado before turning toward the northeast into central Colorado (see Fig. 11.1). The mean height of the crest is ~3 km where it enters southern Colorado and rises to ~4 km at its westernmost extent. The foothills are about 100 km upwind and at a height of ~1.5 km. The San Juans rise abruptly in the last ~25 km upwind of the crest. The upwind valley is rather rugged terrain. The Pacific Ocean is ~1000 km toward the southwest. The center of the target area (Wolf Creek Pass) is located at 37°38′N, 106°48′W.

The Sierra Nevada, by comparison, is a much simpler mountain range. They are oriented north-northwest–south-southeast and extend for ~400 km in length through much of eastern California (see Fig. 11.2). The target area for the SCPP is the American River Basin, which lies between Sacramento, California, and Lake Tahoe. The center of the target area (Donner Pass) is lo-

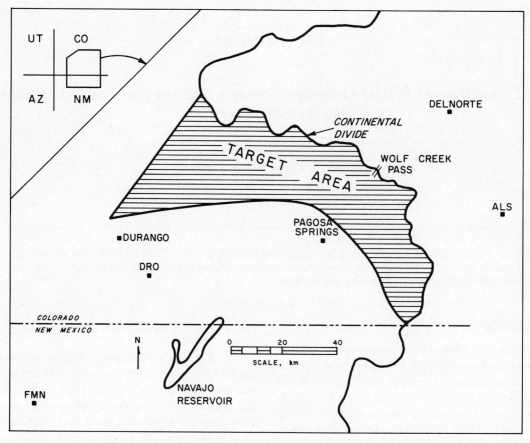

FIG. 11.1. Map of the San Juan research area. FMN is Farmington, NM,
DRO is Durango, CO, and ALS is Alamosa, CO.

cated at 39°20′N, 120°18′W. Both areas are, therefore, in the midlatitudes, with the SCPP area being slightly farther north. The highest region along the Sierra Crest is ∼4 km and is located about 150 km south-southeast of the target. The mean height decreases gradually down to ∼2 km at ∼100 km north-northwest of the target. The crest line in the target area is ∼3.0 km. The upwind foothills are distinct and are ∼100 km from the crest. The California valley is flat and smooth. The Pacific Ocean is ∼200 km west-southwest of the Sierra Crest. The upwind slope of the Sierra barrier has a rather constant slope of ∼3%. The net result is that the Sierra barrier can be viewed as an approximate two-dimensional inclined plane, while the San Juans are a complicated three-dimensional mountain barrier.

Cyclones that produce significant precipitation can approach the Sierra moving from all directions between south through west to north. The cyclone center usually passes to the north of the area but occasionally passes directly overhead or, more rarely, to the south of the area. Extreme flooding situations occur when the subtropical jet extends northward over the Sierra and a series of very warm and very moist storms dump several centimeters of warm rain on a deep snow pack. After a few days, the

accumulation of snow and rain runs off the barrier as a pluvial dump. On the other hand, if a persistent high pressure ridge develops along the west coast, a drought results. In the past ten seasons, a large number of record snowfalls, rainfalls, snow packs, runoffs and, of course, droughts have occurred. In summary, the Sierra precipitation is noted for its extremes. These extremes increase the needs, hazards and difficulties in conducting cloud seeding experiments.

Cyclones that produce significant precipitation on the San Juans are much more restricted in their approach to the area. They must approach from south to west because all other directions cause downslope flow in the target area. The cyclone center typically passes north of the target area, but when the cyclone center passes over the target area these are typically the most intense storms. Precipitation usually begins just prior to passage of the baroclinic zone in the midtroposphere and terminates with the passage of the pressure trough at 500 mb (Marwitz, 1980). The depth of the moisture and amount of forced ascent against the stability are severely restricted prior to the arrival of the midlevel baroclinic zone. Following the passage of the 500 mb pressure trough, the flow comes from north to west and, hence, is a downslope flow. Persistent, shallow

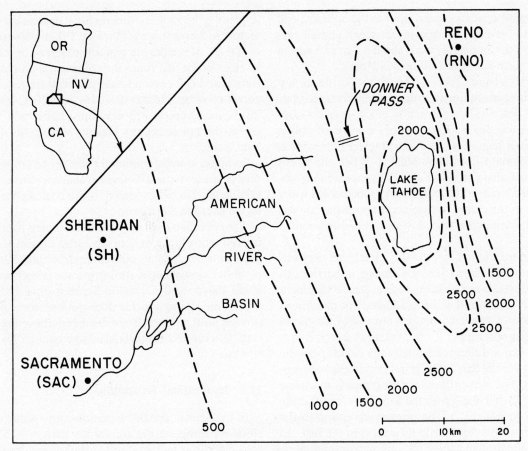

FIG. 11.2. Map of the SCPP research area. Height
contours (dashed) are meters.

clouds with warm cloud-top temperatures are a rare occurrence over the San Juans.

The evolution of most storms as they pass over both
the Sierra and the San Juans is about the same. The storm
stages were identified based on the conditional instability
in the cloud layer below crest level. The front end of the
storms are relatively stable. With the advection of warm,
moist air at the lowest levels and the advection of cold
air at the midlevels, the storms evolve to neutral, then
unstable and finally dissipation. Following passage of the
midlevel baroclinic zone in the mid- and upper levels over
the Sierra, a shallow, warm-topped cloud will persist as
long as upslope flow persists. This typically occurs for 6
to 24 hours. It is these shallow, postfrontal clouds that
are the focus of SCPP seeding experiments.

Rainbands are a common feature in cyclones approaching the Sierra. The rainbands occur in all regions
of the cyclones. Only a few major bands can be tracked
by radar and satellite as they cross the Sierra into Nevada.
These major bands are usually deep, convectively unstable
and are associated with an identifiable baroclinic zone.
The more common occurrence is that the rainbands dissipate as they impinge onto the Sierra barrier. California
rainbands are almost always totally glaciated. It is ex-

tremely rare to observe supercooled water above the $-5°C$
level, and no measurable supercooled water has been detected in rainbands above the $-10°C$ level.

Rainbands have not been reported as occurring upwind
and over the San Juans. This could be because of no radar
data and relatively minimal satellite data. Certainly, it is
clear that if rainbands are present in this area, they are
not the dominant feature they are in California.

The barrier jet is a dominant feature in storms upwind
of the Sierra. Observations (Marwitz, 1983, 1987a) and
numerical simulations (Parish, 1982; Waight, 1984) reveal
that when slightly stable air approaches a 2- to 3-km barrier in the midlatitudes, a barrier jet with speeds of about
twice the approaching winds[1] will occur over the upwind
region and at an altitude of 0.5 to 1.5 km above ground
level. Were it not for surface friction, the barrier jet would
occur at the surface and have speeds of about three times
the approaching winds. In addition to causing substantial
targeting problems, the barrier jet transports large quan-

---

[1] The "approaching winds" are geostrophic and have only a barrier
normal component. Model runs in which the geostrophic winds also
have a barrier parallel component are in progress. Observations suggest
that the initial barrier parallel component is additive to the barrier jet.

tities of sensible heat and water vapor to the northern end of the Sierra Nevada. The maximum mean annual precipitation on the Sierra Nevada is about 2.5 m and occurs ~100 km north-northwest of the SCPP area.

The barrier jet also occurs upwind of the San Juans but by comparison is not nearly as dominant a feature. Marwitz (1980) described the formation of low-level flow upwind of the San Juans. This flow has a speed of only a few meters per second and is diverted toward the west as it enters Colorado from New Mexico. When the storm evolves to an unstable stage, a convergence zone is observed to develop over the foothills and supercooled water is observed in the convective elements above the convergence zone when penetrated aloft with the instrumented aircraft.

Since surface friction and stability clearly are very important factors in barrier jets, the change in barrier orientation plus rough terrain act to supress the development of the barrier jet upwind of the San Juans. The conditions for formation of strong barrier jets upwind of the Sierra are just about optimum.

The last major difference between the clouds over the Sierra and over the San Juans is in their droplet concentrations (N). The N over the Sierra are typically maritime (<100 cm^{-3}), while the N over the San Juans are typically continental (>300 cm^{-3}). The primary exception in the Sierra is convective clouds with their roots in the PBL. In these cases the droplet concentrations are continental.

The droplet concentration is a critical parameter for describing the cloud structure because it is easily measured with modern laser probes and from it one can anticipate the mean volume diameter ($\bar{d}$) from the simple relation LWC = $N \pi \bar{d}^3/6$. When $\bar{d} \sim 20$ μm the processes of accretional and coalescence growth proceed at rather rapid rates. If the temperature is near $-5°$C, the Hallett–Mossop secondary ice crystal production (SICP) process (Hallett and Mossop, 1974; Mossop, 1976; Mossop, 1978) will also act to increase the ice crystal concentrations. Each of these three processes acts to glaciate the cloud, which typically results in an efficient precipitation process. If N is 30 cm^{-3}, then a LWC of 0.2 g m^{-3} results in a $\bar{d}$ of 20 μm. Conversely, if N is 300 cm^{-3}, then a LWC of 2 g m^{-3} is required for $\bar{d}$ to be 20 μm. Obviously, if N is 30 cm^{-3} and LWC is adiabatic, then a parcel need only be displaced upward a few hundred meters above cloud base for $\bar{d}$ to reach 20 μm, where rapid coalescence growth, accretional growth and Hallett–Mossop SICP (near $-5°$C) results in an efficient precipitation process.

The cloud-base temperature in Sierra storms is typically between 10° and 5°C. By comparison, the cloud-base temperature in San Juan storms is typically between 0° and $-5°$C. Given a maritime droplet concentration and warm cloud-base temperature in the Sierra, it is typical for droplets with diameters of 20–30 μm to develop near the $-5°$C level through simple diffusional growth. Con-

sequently, the Hallett–Mossop SICP process is a common feature in Sierra storms (Marwitz, 1987b). Measurements of 100 L^{-1} of needles are typically observed near the $-5°$C level. Over the San Juans the continental droplet concentration and colder cloud-base temperature require that a parcel must be lifted to the $-15°$ to $-20°$C level before 20–30 μm droplets will develop. Cooper and Saunders (1980) did not detect any Hallett–Mossop SICP over the San Juans.

In Sierra storms, millimeter-size drops are sometimes observed above the 0°C level in shallow clouds that contain no ice. No indications of warm coalescence were detected over the San Juans.

The collection efficiency of cloud droplets by ice crystals is exponentially dependent on cloud droplet sizes when the diameters are <30 μm (Pruppacher and Klett, 1980, p. 498). Consequently, riming growth is significant at all levels above the 0°C level in Sierra storms. In San Juan storms the riming process does not become a dominant process until air parcels are displaced above the $-10°$ to $-15°$C level and droplets are large enough for efficient riming.

11.3. Implications for seeding

In the Sierra, seedable situations only occur when the cloud-top temperatures are warmer than $\sim -15°$C. With cold cloud-top temperatures, primary ice develops. Because the cloud droplets are larger at corresponding temperature levels, both the Hallett–Mossop SICP and rapid accretional growth are quite active in Sierra storms. Consequently, the Sierra clouds that develop some initial ice are able to become rather efficient by multiplying their ice and accreting the cloud water. Seedable, warm-topped, shallow clouds primarily occur during the postfrontal period; this is the cloud type and period in which SCPP is presently conducting its seeding experiments.

In the San Juans, long periods of time with shallow, warm-topped clouds are rare. Since the cloud droplet diameters are relatively small at corresponding warm temperatures, both SICP and accretional growth are relatively inefficient. Therefore, the clouds are seedable when the condensation supply rate exceeds the depletion rate by deposition and accretion. This occurs over the crest line during the neutral stability stage because of the strong forced ascent rate. The condensate supply rate especially exceeds the depletion rate in the convection above the convergence zone that develops over the foothills during the unstable stage (Cooper and Marwitz, 1980).

Seeding the Sierra storms using ground generators is, in general, difficult and unreliable. As discussed before, when a stable air mass approaches a mountain barrier, the barrier jet develops. Targeting the seeding agent in both time and space is, therefore, virtually impossible. The operational seeding programs on the Sierra attempt to circumvent this problem by placing their ground generators high on the barrier. It has not been reliably determined whether the seeding agent, when released high on the barrier, is diffused enough in both the horizontal and the vertical directions to effectively seed the clouds. Certainly, the transport/diffusion time is much less for ground generators high on the barrier versus in the foothills region. The modeling by Parish (1982) and Waight (1984) and recent observations by rawinsondes and aircraft indicate that the barrier jet also occurs high on the barrier.

In the San Juans, the spatial targeting problems with ground generators are not as serious. The problems with temporal targeting are significant. When the air mass is stable, the seeding material accumulates near the surface and moves slowly westward with the airflow. Since the clouds contain little supercooled water during stable conditions, the clouds need not be seeded at this time. The only reason to operate the ground generators during stable conditions is to develop a reservoir of well-diffused seeding material so that when the supercooled water in the convection develops, the convection will be seeded from the reservoir of seeding material. Since the Colorado River Basin Pilot Project seeded the storms based on 500 mb temperature ($T < -23°C$) and on a 24-hour experimental unit, the clouds appeared to have been seeded in a carry-over mode in several situations (Marwitz, 1980). If the experimental unit was the storm, then ground seeding could be a viable technique.

Acknowledgments. The author gratefully acknowledges the continuous support of his colleagues at the University of Wyoming, including the scientists, pilots, engineers and technicians. Particularly effective guidance was provided by the Bureau of Reclamation personnel including Drs. B. Silverman, L. Vardiman and D. Reynolds.

This research was conducted under a series of contracts with the Bureau of Reclamation (14-06-D6801, 7-07-83-V0001 and 2-07-81-V0256).

REFERENCES

Cooper, W. A., and J. D. Marwitz, 1980: Winter storms over the San Juan Mountains. Part III: Seeding potential. *J. Appl. Meteor.,* **19,** 942–949.

——, and C. P. R. Saunders, 1980: Winter storms over the San Juan Mountains. Part II: Microphysical processes. *J. Appl. Meteor.,* **19,** 927–941.

Hallett, J., and S. Mossop, 1974: Production of secondary ice particles during the riming process. *Nature,* **249,** 26–28.

Hobbs, P. V., L. F. Radke, J. R. Fleming and D. G. Atkinson, 1975: Airborne ice nucleus and cloud microstructure measurements in natural and artificially seeded situations over the San Juan Mountains of Colorado. Research Report X, Contributions from the Cloud Physics Group, University of Washington, 89 pp.

Marwitz, J. D., 1974: An airflow case study over the San Juan Mountains of Colorado. *J. Appl. Meteor.,* **13,** 450–458.

——, 1980: Winter storms over the San Juan Mountains. Part I: Dynamical processes. *J. Appl. Meteor.,* **19,** 913–926.

——, 1983: The kinematics of orographic airflow during Sierra storms. *J. Atmos. Sci.,* **40,** 1218–1227.

——, 1987a: Wintertime orographic storms over the Sierra. Part I: Thermodynamic and kinematic structure. *J. Atmos. Sci.* (in press).

——, 1987b: Wintertime orographic storms over the Sierra. Part II: The precipitation process. *J. Atmos. Sci.* (in press).

——, W. A. Cooper and C. P. R. Saunders, 1976: Structure and seedability of San Juan storms. Report No. AS 118, Dept. of Atmospheric Science, University of Wyoming, 329 pp.

Mossop, S., 1976: Production of secondary ice particles during the growth of graupel by riming. *Quart. J. Roy. Meteor. Soc.,* **102,** 45–57.

——, 1978: The influence of drop size distribution on the production of secondary ice crystals during graupel growth. *Quart. J. Roy. Meteor. Soc.,* **104,** 323–330.

Parish, T., 1982: Barrier winds along the Sierra Nevada Mountains. *J. Appl. Meteor.,* **21,** 925–930.

Pruppacher, H., and J. Klett, 1980: *Microphysics of Clouds and Precipitation,* D. Reidel, 714 pp.

Waight, K., 1984: A numerical study of the Sierra barrier jet. M.S. thesis, Dept. of Atmospheric Science, University of Wyoming, 119 pp.

CHAPTER 12

How Good Are Our Conceptual Models of Orographic Cloud Seeding?

ARTHUR L. RANGNO

Department of Atmospheric Sciences, University of Washington, Seattle, Washington

ABSTRACT

Some of the complexities of clouds and precipitation that have been encountered in field projects are reviewed. These complexities highlight areas of cloud microstructure and precipitation development that need to be better understood before adequate conceptual or numerical models of orographic cloud seeding can be developed. Some concerns about cloud sampling with regard to the evolutionary behavior of supercooled clouds from water to ice are also discussed.

12.1. Introduction

Recent advances in airborne instrumentation and remote-sensing techniques combined with intensive field programs with dense observation networks are providing much new information on clouds and precipitation, particularly in the mountainous regions of the western United States. These observations have revealed great complexity, but with them comes the promise that many of these complexities will be understood. However, as yet our description of the coupled processes governing air motions and cloud microphysics relating to clouds upwind, over, and downwind of mountains is still inadequate in many ways. This represents a fundamental drawback to the formulation of conceptual and numerical models for use in the prediction of the effects of artificially seeding orographic clouds.

In this paper, the term "orographic" will be used to refer to those cloud systems that form solely as the result of air rising over terrain, which are seen as quasi-stationary clouds of variable coverage on satellite imagery. The term "orographically enhanced" will be used to refer to long-lived cloud systems associated with fronts and troughs that are trackable on satellite imagery prior to impinging on mountain barriers. The discussion will be largely confined to observations made in wintertime clouds in the mountainous regions of the western United States and will attempt to present a holistic view of the seeding potential of these clouds in mountainous regions.

12.2. Brief summary of recent findings

12.2.1. Ice particle concentrations in clouds

Of paramount concern with regard to mounting a cloud seeding effort is knowing what concentrations of ice particles are going to develop naturally in clouds. A logical early expectation in this regard was that ice particle concentrations in clouds, at least in the Rockies, might only be a function of the concentrations of ice nuclei (Grant, 1968) and, therefore, of cloud-top temperature.

However, much of the field work in the West has revealed unexpectedly high ice-particle concentrations in clouds across a wide range of cloud-top temperatures, locations, and types of clouds (Koenig, 1968; Hobbs, 1969; Auer et al., 1969; Cooper and Saunders, 1980; Heymsfield, 1977; Grant et al., 1982). These reports showed that the occurrence of ice at unexpectedly high cloud-top temperatures (some $\geq -10°C$) and in unexpectedly high concentrations (10s to 100s L^{-1} at $> -20°C$) was not restricted to warm-based summertime cumulus clouds (Coons and Gunn, 1951; Braham, 1964), but that it could also occur in wintertime clouds with relatively cold bases.

Several factors have been suggested that can at least partially explain these surpluses of ice particles above those expected based on ice nuclei concentrations (Fletcher, 1962). The two most important of these include the production of "secondary" ice particles, for example, through the breakup of fragile crystals (Hobbs and Farber, 1972; Vardiman, 1978) or through a riming and splintering process (Mossop and Hallett, 1974; Hallett and Mossop, 1974). How these various factors might affect the distribution of ice in orographic clouds will be mentioned in section 12.3.

On the other hand, some studies have appeared to support a temperature-dependent activation of ice nuclei, although not necessarily from cloud top. Cooper and Vali (1981), for example, suggested that a temperature-dependent activation of ice nuclei was evident in their data obtained in two lenticularlike clouds over Elk Mountain, Wyoming, and in the rear portion of a large-scale stratiform cloud system (orographically enhanced) over southwestern Colorado. The temperature effect they reported, however, depended on the accuracy of their estimated origins of the crystals, which were not necessarily from cloud top. In addition, they found that concentrations of ice particles were about an order of magnitude higher in the southwestern Colorado cloud than that at Elk Mountain, despite similar temperatures.

In a further summary of ice particle concentrations in cap clouds in Wyoming, Vali et al. (1982) found that a temperature-dependent effect (again, not necessarily associated with cloud top) was indicated in ice particle concentrations in the upwind-to-crest portion of these clouds. Still higher concentrations were found downwind of the crest, although it was not mentioned whether these higher ice particle concentrations also supported a temperature-dependent effect. Thus, while a temperature-dependent effect was reported in these studies, it is not clear that ice particle concentrations were well predicted by cloud-top temperature.

Mossop (1985) summarized the results of ice particle concentrations in clouds reported by a number of workers. He reported that, with few exceptions, ice particle concentrations in stratiform clouds were dependent on cloud-top temperature when those clouds did not meet the criteria for the riming and splintering production of secondary ice particles. In cumuliform clouds, high ice particle concentrations relative to expected ice nuclei concentrations also appeared to be restricted, with one exception, to those cases where the criteria for riming and splintering were met.

However, there have also been reports of ice particle concentrations in clouds that are *lower* by orders of magnitude than expected ice nucleus concentrations. For example, Stewart (1967) encountered ice-free altocumulus clouds at temperatures below $-25°C$. Heymsfield (1977) stated that in the stratiform clouds that he sampled, ice particle concentrations were, without exception, less than expected ice nuclei concentrations at temperatures below $-20°C$. Measurable liquid water was encountered at temperatures as low as $-35°C$. Grant et al. (1982) reported ice particle concentrations of only about $2 \ L^{-1}$ in a liquid-topped, stratiform cloud in the Rockies at $-31.8°C$. Hobbs and Rangno (1985) encountered altocumulus with tops at $-26°C$ with only a $1 \ L^{-1}$ ice particle concentration. In addition, Sassen (1984) reported liquid water at $-36°C$ over Utah.

In these cases, the presence of liquid water at these temperatures in clouds with characteristically slight updrafts is surprising since Fletcher's (1962) summary of ice nucleus measurements indicates that tens to thousands per liter of ice particles should be present in these clouds, and, if ice nuclei indeed act as sublimation nuclei, they should have formed long before the water phase occurred and therefore been present in more than enough numbers to have prevented the development of liquid water.

Finally, we note that there is no significant correlation

between ice particle concentrations and cloud-top temperature in the large dataset of Hobbs and Rangno (1985). About 25% of their observations of unexpectedly high ice-particle concentrations were in clouds that did not appear to meet the criteria for ice multiplication specified by Hallet and Mossop (1974). A somewhat larger percentage appears not to meet the revised riming–splintering criteria (Mossop, 1978) that specifies the presence of droplets $\leqslant 13 \ \mu m$ diameter in concentrations of $\geqslant 100$ cm^{-3}. Hobbs and Rangno (1985) found a moderate to strong correlation between a measure of the broadness of the cloud-top droplet spectrum and maximum ice particle concentrations over a range of cloud-top temperatures from $-6°$ to about $-26°C$ in both stratiform and cumuliform clouds. This finding may shed light on why cloud-top temperature alone cannot, in general, predict ice particle concentrations, since they may first be dependent on the broadening of the droplet spectrum.

In summary, ice particle concentration surpluses relative to expected ice nucleus concentrations can be found on the warm ($\geqslant -15°C$) cloud-top side of Fletcher's summary curve while large deficits in ice particle concentrations, often combined with the presence of liquid water, can be found on the cold ($\leqslant -25°C$) cloud-top side. These observations suggest a tendency for a leveling of ice particle concentrations in clouds over a wide range of cloud-top temperatures, a phenomenon that was first noted by Weickmann (1957).

12.2.2. Ice nuclei

Some suggested sources for ice nuclei include meteorites, soils, bacteria and spores, and emissions from industrial plants, automobiles, and volcanoes. Kumai (1951, 1961) found that most of the ice particles he examined contained clay particles at their centers. In an example of the potential importance of soil as ice nuclei, Marwitz et al. (1976) reported that the highest natural ice nuclei concentration they measured over southwestern Colorado was on a windy, dusty day associated with a cold front. This maximum ice nuclei count was even greater than the maxima they had measured on all other days above AgI ground seeding sources!

If ice nuclei are important in producing ice particles found in clouds overhead, then numerical models should predict the concentrations of ice nuclei in any given situation. For example, it is conceivable from the Marwitz et al. report that seeding may be unnecessary when it is windy upwind of the region to be seeded but necessary when it is calm or after a regionwide heavy rain or snowfall may have subdued surface sources of ice nuclei. To date, however, and in spite of the potential importance of ice nuclei, ice nuclei concentrations on a day-to-day basis are seemingly unpredictable and their role in ice particle formation is as yet unclear.

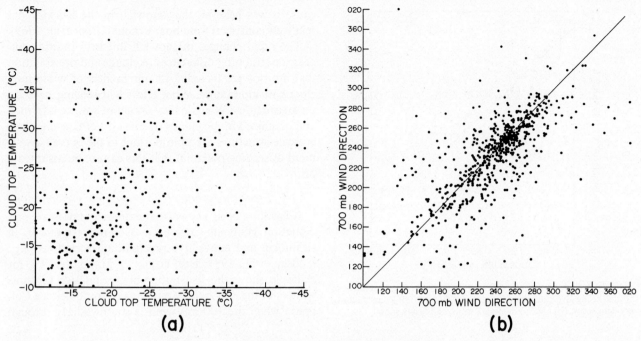

FIG. 12.1. Scatter diagrams of consecutive measurements at Durango, CO, of (a) rawinsonde cloud-top temperatures and (b) 700-mb wind direction for soundings ≤4 h apart. (After Elliott et al., 1976.)

12.2.3. Variability of clouds and winds during storms

While cloud-top temperature as a predictor of ice particle concentrations in clouds is questionable, it is nevertheless useful to examine its variability as a measure of how rapidly project operations may have to be adjusted in order to cope with changing seeding potential. Figure 12.1a is a scatter diagram of cloud-top temperatures (after Elliott et al., 1976) over the Rockies derived from pairs of rawinsondes[1] launched not more than four hours apart during the Colorado River Basin Pilot Project (CRBPP). Elliott et al. used the point where the dewpoint passes ice saturation to deduce cloud tops. The data shown are limited to the period after 1 February 1973, when shielded relative humidity elements apparently came into use (Hill, 1980a). The cloud top of the first sounding is plotted along the ordinate and the cloud top of the second along the abscissa. As can be seen, the serial correlation is poor, illustrating the large, short-term variability in clouds and the forecast and treatment challenge it presents to cloud seeding experiments.

Figure 12.1b is a similar plot of 700-mb wind directions. While the serial correlation is better than in the case of cloud-top temperatures, in about one-third (36%) of the soundings, the 700-mb wind did not remain within 40° of the azimuth value measured ≤4 h earlier.

The purpose of this discussion is to point out that quasi-steady-state situations in mountainous regions (which are often the foundation of conceptual or numerical models) appear to be rare. The cloud variability encountered poses a severe challenge in the forecasting of seeding opportunities and another challenge in the treatment strategy should seeding opportunities be short-lived.

This variability also suggests that experiment strategy may have to be changed in future experiments to better retain homogeneity between control and seeded samples. For example, in an experiment reported by Hobbs et al. (1981) it was shown that long-lived, dry-ice-produced precipitation plumes in stratocumulus clouds, trackable from the cloud to the ground by a vertically pointed 8.6 mm radar that served as the target, can produce a precipitation signal at the ground at or very near forecasted times. Such plume intercepts could be compared directly with the natural precipitation, if any, that immediately preceded and followed the seeding plume intercept while cloud conditions were relatively uniform.

12.2.4. Temporal variations in precipitation

Snowfall intensity in the Rockies varies over periods of minutes (Medenwaldt and Rangno, 1973; Vardiman and Hartzell, 1976). Figure 12.2 shows a plot of visual observations of daytime snowfall intensity over a five-month period (made by the author in Durango, Colorado). It reveals a distribution of intensity changes with time analogous to the relationship between the concentration of precipitating particles and their size reported by Marshall and Palmer (1948). This finding was unexpected be-

---

[1] Rawinsonde-derived cloud-top temperatures can be questioned in some situations, and more comparisons are needed with satellite data and vertically pointed cloud-sensing radars.

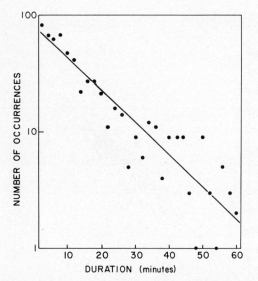

FIG. 12.2. Duration of any initial, visually determined intensity of snowfall using five categories of intensity: flurries (no accumulation), very light snow, light snow, moderate snow, and heavy snow.

cause it was believed that snowfall in the Rockies was relatively consistent from hour to hour (Grant et al., 1969).

The rapid changes in snowfall intensity indicate microscale structural differences in clouds and precipitation that are not yet included in our models of wintertime clouds in mountains. What effect will seeding have on these structures? What do they say about how ice is formed and organized in the clouds overhead? Perhaps these microscale structures are unimportant in the overall assessment of seeding potential, but this cannot be assumed a priori.

12.2.5. *Fine-scale gradients in orographic precipitation*

It has long been known that extremely fine-scale precipitation gradients exist even over modest terrain in stratiform and convective precipitation situations (e.g., Godske et al., 1957; Huff et al., 1975). Figure 12.3, for example, shows the distribution of precipitation for three winters in the San Juan Mountains of southwestern Colorado when the 700-mb wind is southwesterly through

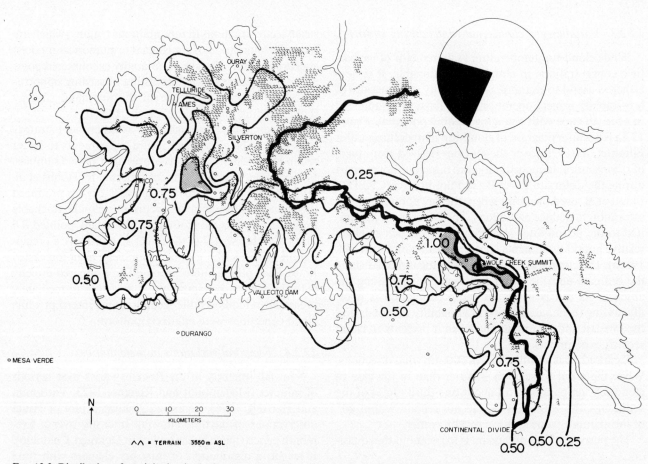

FIG. 12.3. Distribution of precipitation in the San Juan Mountains of southwestern Colorado with the 700-mb wind direction between 206° and 285° (darkened portion of azimuth circle upper right). Isohyets (solid bold lines) are in the percentages of the precipitation observed at Wolf Creek Pass summit. Shading indicates ≥100% of Wolf Creek Pass summit precipitation. Solid light line indicates terrain ≥2670 m (9000 ft) MSL and peaks (⌃) terrain ≥3560 m (12 000 ft) MSL. Gage locations are indicated by small circles.

westerly (Rangno, 1979). An extremely large (>100%) increase in precipitation occurs from the base of Wolf Creek Pass to the summit, a distance of only about 10 km. An examination of the hourly precipitation records of the CRBPP by the author shows that about 70% of this greater total at the summit as compared with the base of Wolf Creek Pass is due to greater intensity (more or heavier ice particles or both) and about 30% is due to greater duration (precipitating cloud was not over the lowest slopes, or the precipitation was not initiated far enough upwind of the lower slopes to fall out on them).

These large gradients in precipitation over extremely small distances, both parallel to and across the barrier, both upwind and downwind of the crest, cannot be produced by changes in cloud-top height resulting from the passage of the cloud system over the barrier. This is because any changes in cloud-top microstructure immediately upwind and over the barrier could not affect snowfall at the ground during the limited time the cloud traverses the barrier. The atmosphere can produce these small-scale gradients in precipitation only through rapid changes in the mass of precipitating particles close to ground level. Bergeron (1950) described a similar situation as a "seeder–feeder" process where ice crystals (or raindrops) fell into and accreted the droplets of a lower-level cloud, and he observed that such a process was capable of producing such fine-scale precipitation gradients in terrain of even modest topography.

What would be the effect of seeding on these fine-scale precipitation patterns? Trajectory changes are likely due to seeding (e.g., Hobbs et al., 1973), and the question of the redistribution of precipitation without an increase is raised. Would some of the heavier precipitation falling onto sharply rising windward slopes be shifted to the next downwind canyon and eventually to the lee of the crest, where some of the precipitation could be lost due to evaporation? This is an old question, but nevertheless one that still needs a firm answer.

12.2.6. Targeting clouds with modification potential

Conceptual or numerical models of clouds for artificial modification purposes should also incorporate the capability to predict if, when, where, and in what concentrations the seeding agent will arrive in targeted clouds. Of the questions that beset cloud seeding, the targeting problem may prove to be the most intractable. Airborne studies of surface-released seeding agents in and near mountains has revealed that they often do not reach the clouds due to intervening stable layers of air (Rhea et al., 1969; Hobbs et al., 1975; Marwitz et al., 1976). Severe targeting problems also develop when there is a large directional wind shear with increasing height (e.g., Elliott et al., 1978), a situation that usually occurs in storms. Or, even when there is strong vertical dispersion, the seeding agent may

nevertheless arrive in the clouds at locations inappropriate to produce an effect in the target (Hobbs et al., 1975). Thus, the opportunities for using ground generators to seed wintertime clouds has been found to be limited in many instances. In addition, no studies have been conducted concerning vertical dispersion of ground-released seeding agents at night over snowy surfaces, a situation that often occurs during cloud seeding operations in mountainous regions.

Even when there is good vertical dispersion in the presence of suitable clouds, however, the nature of turbulence is to produce puffs of higher concentrations interspersed with regions of lower concentrations of the source plume while meandering (e.g., Hobbs et al., 1975). For an update on the complexities of dispersion, the reader is referred to Weill (1985), Wyngaard (1985), and Briggs (1985).

Is it safe to assume, when conducting a cloud seeding experiment, that the actual, chaotic plume will produce the same effects in the clouds as those calculated using a time- or space-averaged plume? Or will some portions of some clouds be continuously underseeded while other portions are overseeded? Should precipitation modification projects or experiments be carried out without firm answers to these questions?

12.3. Conceptual models of clouds over mountains

While uncertainties still exist concerning the structure and evolution of orographic and orographically enhanced clouds, it is useful to summarize some common aspects found in field work. Composites derived from recent observations of orographic and orographically enhanced cloud systems are presented in Figs. 12.4 and 12.5, respectively. These composites are for two generic cloud regimes. The transition between these two regimes is often marked by frontal zones (e.g., Hobbs, 1975). It should be kept in mind that conditions (cloud tops, cloud coverage, stability, and winds) within these regimes change continuously. Observations used to derive these models originate with vertically pointed cloud-sensing radars (e.g., Plank et al., 1955), aircraft observations (e.g., Hobbs and Atkinson, 1976), and satellite and surface observations. Further refinement of these two basic types of meteorological situations affecting mountainous regions has been presented by Heggli et al. (1983) and Heggli and Reynolds (1985).

The class of clouds most pertinent to this discussion is orographic (Fig. 12.4). Cloud coverage over a barrier may range from solid overcast to scattered (less than one-half sky coverage) near the end of a storm episode. Liquid water maxima may vary from $0.1-0.2$ m^{-3} in thin, stable stratiform clouds (Ac and Sc len) over the high elevations of the Rockies (e.g., Cooper and Vali, 1981) to ≥ 1.5 g m^{-3} in postfrontal cumulus and small cumulonimbus near the West Coast (Heggli et al., 1983; Hobbs and Rangno,

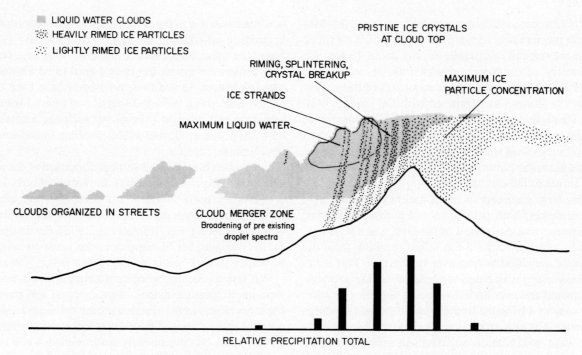

FIG. 12.4. A conceptual model of an orographic cloud system composed of stratocumulus and cumulus.

1985). Due to low areawide cloud coverage, strong diurnal effects are superimposed on orographic clouds that can convert nighttime laminar cloud forms into convective forms during the daytime. Cloud bases rise during the day and descend at night, while cloud tops are likely to do the opposite. While such changes are likely to affect cloud microstructure, no studies of the diurnal effects that are superimposed on orographic clouds have been conducted.

In convective situations, such as that depicted in Fig. 12.4, clouds and showers may align themselves in cloud

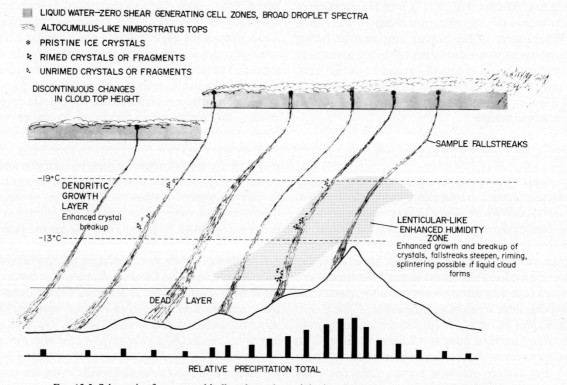

FIG. 12.5. Schematic of an orographically enhanced, precipitating cloud system composed of nimbostratus.

streets emanating from favored spawning points well upwind of the barrier (Heggli et al., 1983). Ice particle concentrations are often highly variable in time and space (e.g., Rauber et al., 1984). Usually, heavily rimed ice particles are found at the ground at upslope and crest locations (e.g., Hobbs, 1975; Vardiman and Hartzell, 1976) with a sharply diminished degree of riming in the lee.

In the stable stratiform case (lenticularlike clouds), ice particle concentrations develop near the upwind edge of the liquid cloud and within a few minutes reach a maximum that remains rather steady (Cooper and Vali, 1981) up to the crest. Since the supply of moisture is continually changing, the upwind edges of these clouds can recede downwind and expand upwind in a matter of minutes. At the same time, layers can be added or subtracted to the bases and tops of these clouds within minutes.

The onset of ice particles in clouds near the West Coast is at cloud-top temperatures between $-4°$ and $-10°C$ (e.g., Hobbs, 1974; Hobbs and Atkinson, 1976; Heggli et al., 1983), but they do not appear until temperatures of about $-10°$ to $-15°C$ are reached in the Rockies (e.g., Auer et al., 1969; Vali et al., 1982). However, data from the CRBPP suggest that only a small fraction (less than 10%) of wintertime clouds in the Rockies have tops that are less than $-15°C$ (Elliott et al., 1976). Hence, most substantial orographic clouds that develop there usually contain some ice particles.

A characteristic of supercooled orographic clouds is their tendency to become increasingly infected with ice particles as they transit a barrier. This facet of orographic clouds has no doubt been visually observed by many, but only on a few occasions has it been mentioned (Hobbs, 1975; Rangno et al., 1977; DeMott and Grant, 1984).

Eight factors, some of which may act in concert, may aid the increase in ice particle concentrations over and downwind of a mountain crest in orographic clouds:

- the ice-forming process is time dependent (e.g., Koenig, 1963) and ice particles would be expected to continue to increase as the orographic cloud progressed across a barrier;
- enhancement of contact nucleation activity (Young, 1974) in the lee side evaporating portions of liquid cloud;
- evaporative cooling, potentially up to several degrees (e.g., Cerni and Cooper, 1980), could lead to an enhancement of temperature-dependent ice-forming mechanisms;
- the lifting and resultant cooling of cloud top as it traverses the barrier also leads to increased activation of temperature-dependent ice nuclei (e.g., Fletcher, 1962);
- the riming and splintering production of small secondary ice particles (Hallett and Mossop, 1974) on the windward slopes that carry to the lee side;
- the mechanical breakup of fragile crystals (e.g., Hobbs and Farber, 1972) whose small fragments could carry to the lee side;

- the blow-off of ice particles from the mountain surface into the downwind cloud (Rogers and Vali, 1981);
- enhanced development of ice in strong uplift zones preceding the crest due to increased values of ice supersaturation (DeMott et al., 1982).

This near-crest and post-crest ice-enhancing region needs to be examined more closely in future research so that it can be incorporated into models of orographic clouds and precipitation and so that what effects, if any, it has on downwind barriers can be discovered.

There is an unambiguous situation, however, where liquid water escapes a mountain barrier and a potential to increase precipitation is present as envisioned by Ludlam (1955). This occurs when there is a flow of nonprecipitating supercooled cloud that satisfies the following criteria: appreciable cloud depth, for the purpose of appreciable crystal growth, say 1 km; cloud base not too far (<1 km) above mountaintop; of such upwind extent that seeding could cause an appreciable crystal fallout on the barrier; and a cloud-top temperature $\leq -5°C$.

To date, however, documentation of this situation is either absent or it has only been inferred to exist (preferentially with 500-mb temperatures $\geq -20°C$) from apparent duration-of-snowfall effects in the evaluation of cloud seeding experiments (Chappell et al., 1971; Mielke et al., 1981). Documentation of this situation would establish a baseline of seeding potential.

Figure 12.5 shows a schematic of orographically enhanced clouds. In these systems, maximum ice particle concentrations are generally a few tens per liter across a wide range of cloud-top temperatures (Hobbs and Atkinson, 1976; Cooper and Saunders, 1980; Hobbs and Rangno, 1985). Liquid water content is modest (<0.5 g m^{-3}) or absent. Often, and perhaps characteristically, the liquid water present is confined to a thin layer at cloud top (e.g., Cunningham, 1957; Walsh, 1980) where small, pristine crystals are encountered (Hobbs and Rangno, 1985). The highest concentrations of ice particles, characteristically, occur at elevations below cloud top and can be due to several factors, including the breakup of fragile crystals, riming-splintering processes, or the direct formation of new crystals.

Lapse rates are often stable in the lowest layers, particularly in the earliest stages of these storm episodes (e.g., Marwitz, 1980), and a "dead" layer a kilometer or more deep in which the flow is light and variable or diverted around the barrier can be present.

Due to the low amounts of supercooled water and to plentiful ice particle concentrations, particularly in and near major bands (e.g., Heymsfield, 1977; Heggli et al., 1983), it appears unlikely in this phase of a storm that precipitation can be increased through seeding, although a possible exception to this conclusion will be considered in the following section.

12.4. Some further comments relating to sampling and seeding

Exciting new measurements in mountainous regions such as in the Sierra Cooperative Pilot Project and in the Colorado Orographic Seeding Experiment are becoming available. These new observations, which include those from scanning radiometers, polarizing lidars, and vertically pointed cloud-sensing radars, will provide much larger and more complete studies of clouds than have been available in the past. If put into climatological perspective, and not just restricted to one location upwind of a mountain, these measuring systems can provide vital information from which decisions can be made on how to conduct, when to conduct, or whether to conduct, cloud seeding operations.

However, the chance for an erroneous assessment of cloud seeding potential still exists, in spite of these new measurements, and it may even be quite high. Some possible scenarios in which erroneous interpretations could be made are outlined below. Wintertime cumuli are included in these discussions because they are being considered as targets for cloud seeding in the Sierra Nevada (Heggli et al., 1983).

Figure 12.6 shows a wintertime cumulonimbus calvus near St. George, Utah. A new cumuli turret, reaching its maximum height at point A, does not outwardly display signs of ice particles and may contain $\geqslant 1.5$ g m^{-3} of liquid water. At point B, a maturing turret that has ceased rising appears to consist only of ice particles. Let us suppose that an aircraft flew only in the small, newly formed cumuli that had not yet produced ice and only in the newest, virtually ice-free portions of the larger cumuli. A researcher examining this data would conclude that there were few ice particles in these supercooled clouds and that there was considerable potential for seeding on this day.

Conversely, the data from another aircraft flown on the same day, but which flew through only the glaciated portions of these cumuli, would lead another researcher to conclude that such clouds were composed solely of ice particles and had no seeding potential.

A third aircraft that flew through both the ice-free and the glaciated portions of these clouds would have collected data that would cause the researcher to choose between two conclusions: a seeding potential was present in some of the clouds, since some regions with large amounts of supercooled water were encountered that had few ice particles, or, there was no seeding potential because it was assumed that all of the liquid water encountered was subsequently consumed by the ice particles that developed later and comprised the glaciated cloud. Clearly, a seeding experiment based on the perception of an unambiguous and appreciable seeding potential on this day would lead to disappointing results. While ice particles could certainly be initiated by seeding earlier in the cloud's lifetime, the ability to produce a strong precipitation signal above the background of natural precipitation at the ground from these clouds is doubtful. A seeding potential would have

FIG. 12.6. Wintertime cumulonimbus calvus near St. George, UT. Point A is a cloud in the early portion of development. Point B is a glaciated portion of cloud.

been unambiguously detected only if it had been shown that ice particle concentrations remained virtually non-existent in these clouds and if the clouds lasted long enough or were wide enough for appreciable-size precipitation particles to develop and fall within them. The problem of cloud lifetime in precipitation development has been recently addressed by Schemenauer and Isaac (1984) and Cooper and Lawson (1984).

A situation similar to that described above for cumuli exists for orographic clouds. Figure 12.7 shows a postfrontal, orographic cloud with modest embedded convection in westerly flow over the La Plata Mountains of southwestern Colorado. The features of interest are the liquid-appearing cloud at point A on the upwind slopes of the La Plata Mountains and the glaciated cloud at the downwind end of the cloud at point B. This behavior mimics that of cumulus life cycles described above, except that the water-to-ice evolution is spread over a greater horizontal extent and usually in thinner clouds.

The risk of an erroneous assessment of cloud seeding potential in orographic situations lies in making incomplete measurements or in making interpretations that do not take into account the evolutionary behavior of cloud microstructure as it traverses a barrier. For example, if orographic cloud observations are confined only to the region of the windward slopes, liquid water and variable, but often low, concentrations of rimed ice particles would be observed. Without additional measurements farther downwind it might be concluded that there exists a po-

tential to increase overall precipitation on the barrier when, in fact, the possibility of redistributing the precipitation without an overall increase exists. Of course, in some cases this may be a desirable goal (e.g., Hobbs et al., 1973).

A seeding scenario within synoptic-scale cloud systems has been suggested by Hobbs and Matejka (1980) based on the banded structure of these precipitation areas. In some cases, the regions between bands incorporate supercooled liquid water (Matejka et al., 1978; Rauber et al., 1984). This supercooled water may appear to present an unambiguous seeding opportunity. However, the presence of differential wind speeds and directions between layers where supercooled water might have been detected and, say, ice-spawning, upper-level generating cells, raises the possibility that supercooled water intercepted at one point could be overtaken and depleted by the fallstreaks from a new crop of generating cells. Also, low-level convergence in and near rainbands can advect supercooled water into the high ice particle rainband regions. Or it might be that these supercooled clouds would eventually transform into high ice-producing clouds that feed the rainband.

These are two possibilities that could make the opportunity to increase precipitation in the "between-band" situation less than it might otherwise be. These regions between rainbands should be tracked for several hours to determine whether the regions of supercooled water always remain liquid and do not contribute significantly to pre-

FIG. 12.7. Orographic clouds with modest convection topping the La Plata Mountains of southwestern Colorado. Point A shows a cloud top ascending the barrier. Point B is the glaciated "exhaust" portion of the cloud.

cipitation at the ground, say, through riming or through a transformation into an ice-producing cloud. Similar comments can be made about the occasionally high values of liquid water that have been encountered in frontal zones (e.g., Hobbs and Persson, 1982).

Also, for the reasons discussed above, aircraft icing reports alone cannot be used to infer seeding opportunities as has been suggested (Hill, 1980b).

Contributing to the problems of sampling and interpretation discussed before is the problem of what days to sample on: Which days would be the most climatologically representative? This is a particularly vexing problem since large field programs are generally planned far in advance and are conducted over miniscule time periods relative to climatological scales. No one, for example, would want to assess the precipitation modification potential in California or Colorado based on the frequency of clouds and storms that occur during an El Niño winter or to deduce the structure of all cumuli based on only those days when the cloud bases were <0°C. Yet rarely is the question of representativeness addressed in field reports—a point made by Johnson (1982). It is recommended that statements relevant to climatological perspective be included in field studies.

12.5. Conclusions

While good progress has been made during the past few years in understanding the nature of the precipitation process in mountainous regions, several outstanding issues remain that hinder the development of more reliable conceptual and numerical models of orographic clouds and precipitation. Those discussed in this paper include the following:

• There is not yet a widely agreed upon parameter that will predict the maximum ice particle concentrations that occur in most stratiform or convective clouds. For example, some studies have indicated that ice particle concentrations in building cumuli and in the upslope portions of lenticularlike, orographic cap clouds are temperature-dependent. In a more recent study, the broadness of the cloud-top droplet spectrum was found to predict ice particle concentrations with no temperature dependence indicated within the cloud-top temperature range of about −6° to −26°C.

• An apparent ice-enhancing process near and downwind of mountain crests exists. Its cause should be elucidated. This will help in answering the question of whether increasing ice particle concentrations over the windward slopes of a mountain range should have a negative or positive effect in the immediate lee.

• Knowledge of the history, evolution, and role of supercooled liquid water regions within storms and over mountain barriers remains scant and needs more attention.

• There is a need for the documentation of the occurrence of substantial, nonprecipitating, supercooled clouds over mountain barriers.

• The microphysical processes that produce microscale temporal and spatial distributions of precipitation need attention.

• Diurnal effects, for example, the rise of cloud bases during the daytime and lowering at night, can produce systematic effects on the microstructure of orographic clouds. The extent of these effects is unknown.

• Considerable uncertainty exists on how to seed clouds. The presence of suitable clouds is useless if ground-released nucleants are inhibited by stable layers, are mistargeted, or arrive in the clouds in inappropriate concentrations, or when a seeding aircraft can provide only a thin strip of treatment in a sea of cloud.

• Statements on the representativeness of field data based on climatological studies should accompany short-term field measurements in order to put them into perspective.

• As a note of practicality, it should be demonstrated prior to field programs that conditions amenable to seeding can be forecast far enough in advance to permit treatment.

Acknowledgments. The author wishes to thank Professor Peter V. Hobbs for his advice, without which this paper would have been impossible. In addition, this paper benefitted greatly from the many comments and suggestions of Professor Roscoe R. Braham, Jr., as well as those of the anonymous reviewers.

REFERENCES

Auer, A. H., D. L. Veal and J. D. Marwitz, 1969: Observations of ice crystals and ice nuclei observations in stable cap clouds. *J. Atmos. Sci.,* **26,** 1342–1343.
Bergeron, T., 1950: Über der mechanismus der ausgiebigen Niederschlage. *Ber. Dtsch. Wetterdienstes,* **12,** 225–232.
Braham, R. R., Jr., 1964: What is the role of ice in summer rain showers? *J. Atmos. Sci.,* **21,** 640–645.
Briggs, G. A., 1985: Analytical parameterizations of diffusion: The convective boundary layer. *J. Climate Appl. Meteor.,* **24,** 1167–1186.
Cerni, T. A., and W. A. Cooper, 1980: Ice crystal concentrations in isolated cumulus clouds of Montana. *Preprints Eighth Int. Conf. on Cloud Physics,* Clermont-Ferrand, WMO, 195–198.
Chappell, C. F., L. O. Grant and P. W. Mielke, Jr., 1971: Cloud seeding effects on precipitation intensity and duration of wintertime orographic clouds. *J. Appl. Meteor.,* **10,** 1006–1010.
Coons, R. D., and R. Gunn, 1951: Relation of artificial cloud modification to the production of precipitation. *Compendium of Meteorology,* Amer. Meteor. Soc., 235–241.
Cooper, W. A., and C. P. R. Saunders, 1980: Winter storms over the San Juan Mountains. Part II: Microphysical processes. *J. Appl. Meteor.,* **19,** 927–941.
——, and G. Vali, 1981: The origin of ice in mountain cap clouds. *J. Atmos. Sci.,* **38,** 1244–1259.
——, and R. P. Lawson, 1984: Physical interpretation of results from the HIPLEX-1 experiment. *J. Climate Appl. Meteor.,* **23,** 523–540.

Cunningham, R. M., 1957: A discussion of generating cell observations with respect to the existence of freezing or sublimation nuclei. *Artificial Stimulation of Rain,* H. Weickmann, Ed., Pergamon, 267–270.

DeMott, P. J., and L. O. Grant, 1984: Development of ice crystal concentrations in stably stratified orographic cloud systems. *Preprints Ninth Int. Conf. on Cloud Physics,* Tallin, USSR, Acad. Sci. USSR, 191–194.

——, W. G. Finnegan and L. O. Grant, 1982: A study of ice crystal formation in a stably stratified orographic cloud. *Preprints Conf. on Cloud Physics,* Chicago, Amer. Meteor. Soc., 488–490.

Elliott, R. D., R. W. Shaffer, A. Court and J. F. Hannaford, 1976: *Colorado River Basin Pilot Project Comprehensive Evaluation Report.* Final Report to the Bureau of Reclamation, Aerometric Research, Inc., Goleta, CA, 641 pp.*

——, ——, —— and ——, 1978: Randomized cloud seeding in the San Juan Mountains, Colorado. *J. Climate Appl. Meteor.,* **17,** 1298–1318.

Fletcher, N. H., 1962: *The Physics of Rainclouds.* Cambridge University Press, 240–241.

Godske, C. L., T. Bergeron, J. Bjerknes and R. C. Bundgaard, 1957: *Dynamic Meteorology and Weather Forecasting.* Amer. Meteor. Soc. and the Carnegie Institution, 608–610.

Grant, L. O., 1968: The role of ice nuclei in the formation of precipitation. *Proc. First Int. Conf. on Cloud Physics,* Toronto, Amer. Meteor. Soc., 305–310.

——, C. F. Chappell, L. F. Crow, P. W. Mielke, Jr., J. L. Rasmussen, W. E. Shobe, H. Stockwell and R. A. Wykstra, 1969: An operational adaptation program of weather modification for the Colorado River basin. Interim Report to the Bureau of Reclamation, Colorado State University, 69 pp.*

——, P. J. DeMott and R. M. Rauber, 1982: An inventory of ice crystal concentrations in a series of stable orographic storms. *Preprints Conf. on Cloud Physics,* Chicago, Amer. Meteor. Soc., 584–587.

Hallett, J., and S. C. Mossop, 1974: Production of secondary particles during the riming process. *Nature,* **249,** 26–28.

Heggli, M. F., and D. W. Reynolds, 1985: Radiometric observations of supercooled liquid water within a split front over the Sierra Nevada. *J. Climate Appl. Meteor.,* **24,** 1258–1261.

——, L. Vardiman, R. Stewart and A. Huggins, 1983: Supercooled liquid water and ice crystal concentrations within Sierra Nevada winter storms. *J. Climate Appl. Meteor.,* **22,** 1875–1886.

Heymsfield, A. J., 1977: Precipitation development in stratiform ice clouds: A microphysical and dynamical study. *J. Atmos. Sci.,* **34,** 367–381.

Hill, G. E., 1980a: Reexamination of cloud-top temperatures used as criteria of cloud seeding effects in experiments on winter orographic clouds. *J. Climate Appl. Meteor.,* **19,** 1167–1175.

——, 1980b: Seeding-opportunity recognition in winter orographic clouds. *J. Climate Appl. Meteor.,* **22,** 1371–1381.

Hobbs, P. V., 1969: Ice multiplication in clouds. *J. Atmos. Sci.,* **26,** 315–318.

——, 1974: High concentrations of ice particles in a layer cloud. *Nature,* **251,** p. 694.

——, 1975: The nature of winter clouds and precipitation in the Cascade Mountains and their modification by artificial seeding. Part I. Natural conditions. *J. Appl. Meteor.,* **14,** 783–804.

——, and R. Farber, 1972: Fragmentation of ice particles in clouds. *J. Rech. Atmos.,* **6,** 245–258.

——, and D. G. Atkinson, 1976: The concentrations of ice particles in orographic clouds and cyclonic storms over the Cascade Mountains. *J. Atmos. Sci.,* **33,** 1362–1374.

——, and T. J. Matejka, 1980: Precipitation efficiencies and the potential for artifically modifying extratropical cyclones. *Preprints Third WMO Conf. on Weather Modification,* Clermont-Ferrand, WMO, 9–15.

——, and P. Ola G. Persson, 1982: The mesoscale and microscale structure and organization of clouds and precipitation in midlatitude cyclones. V. The substructure of narrow cold frontal rainbands. *J. Atmos. Sci.,* **39,** 280–295.

——, and A. L. Rangno, 1985: Ice particle concentrations in clouds. *J. Atmos. Sci.,* **42,** 2523–2549.

——, R. C. Easter and A. B. Fraser, 1973: A theoretical study of the flow of air and fallout of solid precipitation over mountainous terrain. Part II. Microphysics. *J. Atmos. Sci.,* **30,** 813–823.

——, L. F. Radke, J. R. Fleming and D. G. Atkinson, 1975: Airborne ice nucleus and cloud microstructure measurements in natural and artifically seeded situations over the San Juan Mountains of Colorado. Final Report to the Bureau of Reclamation, University of Washington, 89 pp.**

——, J. H. Lyons, J. D. Locatelli, K. R. Biswas, L. F. Radke, R. W. Weiss, Sr. and A. L. Rangno, 1981: Radar detection of cloud-seeding effects. *Science,* **213,** 1250–1252.

Huff, F. A., S. A. Changnon, Jr. and D. M. A. Jones, 1975: Precipitation increases in the low hills of southern Illinois. Part I. Climatic and network studies. *Mon. Wea. Rev.,* **103,** 823–829.

Johnson, D. B., 1982: Geographical variations in cloud-base temperature. *Preprints Conf. on Cloud Physics,* Chicago, Amer. Meteor. Soc., 187–189.

Koenig, L. R., 1963: The glaciating behavior of small cumulonimbus clouds. *J. Atmos. Sci.,* **20,** 29–47.

——, 1968: Some observations suggesting ice multiplication in the atmosphere. *J. Atmos. Sci.,* **25,** 460–463.

Kumai, M., 1951: Electron-microscope study of snow-crystal nuclei. *J. Meteor.,* **8,** 151–156.

——, 1961: Snow crystals and the identification of the nuclei in the northern United States of America. *J. Meteor.,* **18,** 139–150.

Ludlam, F. H., 1955: Artificial snowfall from mountain clouds. *Tellus,* **7,** 277–290.

Marshall, J. S., and W. McK. Palmer, 1948: The distribution of raindrops with size. *J. Meteor.,* **5,** 165–166.

Marwitz, J. D., 1980: Winter storms over the San Juan Mountains. Part I. Dynamical processes. *J. Appl. Meteor.,* **19,** 913–926.

——, W. A. Cooper and C. P. R. Saunders, 1976: *Structure and Seedability of San Juan Storms.* Final Report to the Bureau of Reclamation, University of Wyoming, 324 pp.*

Matejka, T. M., R. A. Houze, Jr. and P. V. Hobbs, 1978: Microphysical and dynamical structure of mesoscale cloud features in extratropical cyclones. *Preprints Conf. on Cloud Physics and Atmospheric Electricity,* Issaquah, Amer. Meteor. Soc., 292–299.

Medenwaldt, R. A., and A. L. Rangno, 1973: *Colorado River Basin Pilot Project Comprehensive Atmospheric Data Report, 1972–1973 Season.* Report to the Bureau of Reclamation, E. G. & G., Inc., Durango, CO, 376 pp.*

Mielke, P. W., Jr., G. W. Brier, L. O. Grant, G. J. Mulvey and P. N. Rosenweig, 1981: A statistical reanalysis of the replicated Climax I and II wintertime orographic cloud seeding experiments. *J. Appl. Meteor.,* **20,** 643–659.

Mossop, S. C., 1978: The influence of drop size distribution on the production of secondary ice particles during graupel growth. *Quart. J. Roy. Meteor. Soc.,* **104,** 323–330.

——, 1985: The origin and concentration of ice crystals in clouds. *Bull. Amer. Meteor. Soc.,* **66,** 264–273.

——, and J. Hallett, 1974: Ice crystal concentration in cumulus clouds: Influence of the drop spectrum. *Science,* **186,** 632–634.

* Available from the Bureau of Reclamation, Denver, CO 80225.

** Available from the Cloud and Aerosol Research Group, University of Washington, Seattle, WA 98195.

Plank, V. G., D. Atlas and W. H. Paulsen, 1955: The nature and detectability of clouds and precipitation by 1.25 cm radar. *J. Meteor.,* **12,** 358–378.

Rangno, A. L., 1979: A reanalysis of the Wolf Creek Pass cloud seeding experiment. *J. Appl. Meteor.,* **18,** 579–604.

——, P. V. Hobbs and L. F. Radke, 1977: Tracer and diffusion and cloud microphysical studies in the American River basin. Final Report to the Bureau of Reclamation, University of Washington, 64 pp.**

Rauber, R. M., L. O. Grant, D. Feng and J. B. Snyder, 1984: The spatial and temporal distribution of supercooled water during wintertime storms over the northern Colorado Rockies. Atmos. Sci. Paper No. 382, Colorado State University, 83 pp.

Rhea, J. O., L. G. Davis and P. T. Willis, 1969: *The Park Range Project.* Final Report to the Bureau of Reclamation, E. G. & G., Inc., Steamboat Springs, 288 pp.*

Rogers, D. C., and G. Vali, 1981: Observational studies of ice crystal production by mountain surfaces inside winter orographic clouds. *Preprints Eighth Conf. on Planned and Inadvertent Weather Modification,* Reno, Amer. Meteor. Soc., 54–55.

Sassen, K., 1984: Deep orographic cloud structure and composition derived from comprehensive remote sensing measurements. *J. Climate Appl. Meteor.,* **23,** 568–583.

Schemenauer, R. S., and G. A. Isaac, 1984: The importance of cloud top lifetime in the description of natural cloud characteristics. *J. Climate Appl. Meteor.,* **23,** 267–279.

Stewart, J. B., 1967: A preliminary study of the occurrence of ice crystals in layer clouds. *Meteor. Mag.,* **96,** 23–27.

Vali, G., D. C. Rogers and T. L. Deshler, 1982: Ice crystal and ice nucleus measurements in cap clouds. *Preprints Conf. on Cloud Physics,* Chicago, Amer. Meteor. Soc., 333–334.

Vardiman, L., 1978: The generation of secondary ice particles in clouds by crystal–crystal collision. *J. Atmos. Sci.,* **35,** 2168–2180.

——, and C. L. Hartzell, 1976: *Investigation of Precipitating Ice Crystals from Natural and Seeded Winter Orographic Clouds.* Final Report to the Bureau of Reclamation, Western Scientific Services, Inc., Fort Collins, 129 pp.*

Walsh, P. A., 1980: Cloud droplet distributions in wintertime Rocky Mountain clouds. *Preprints Eighth Int. Conf. on Cloud Physics,* Clermont-Ferrand, WMO, 183–185.

Weickmann, H., 1957: The snow crystal as aerological sonde. *Artificial Stimulation of Rain,* H. Weickmann, Ed., Pergamon, 315–326.

Weill, J. C., 1985: Updating applied diffusion models. *J. Climate Appl. Meteor.,* **24,** 1111–1130.

Wyngaard, J. C., 1985: Structure of the planetary boundary layer and implications for its modeling. *J. Climate Appl. Meteor.,* **24,** 1131–1142.

Young, K. C., 1974: The role of contact nucleation in ice phase initiation in clouds. *J. Atmos. Sci.,* **31,** 768–776.

CHAPTER 13

Seedability of Winter Orographic Clouds

GEOFFREY E. HILL

Utah State University, Logan, Utah

ABSTRACT

This article is a review of work on the subject of seedability of winter orographic clouds for increasing precipitation. Various aspects of seedability are examined in the review, including definitions, distribution of supercooled liquid water, related meteorological factors, relationship of supercooled liquid water to storm stage, factors governing seedability, and the use of seeding criteria.

Of particular interest is the conclusion that seedability is greatest when supercooled liquid water concentrations are large and at the same time precipitation rates are small. Such a combination of conditions is favored if the cloud-top temperature is warmer than a limiting value and as the cross-barrier wind speed at mountaintop levels increases.

It is also suggested that cloud seeding is best initiated in accordance with direct measurements of supercooled liquid water, precipitation, and cross-barrier wind speed. However, in forecasting these conditions or in continuation of seeding previously initiated, the cloud-top temperature and cross-barrier wind speed are the most useful quantities.

13.1. Introduction

It has been known for virtually a half-century that opportunities for increasing winter orographic precipitation by cloud seeding are primarily related to the presence of supercooled liquid water. As early as 1938, Findeisen discussed the idea that supercooled clouds could be modified by the addition of ice-forming nuclei. Clouds forming over mountains have been observed to occur often at subfreezing temperatures yet to be composed of very small liquid droplets too small to fall as precipitation (e.g., Bergeron, 1949; Ludlam, 1955). However, these cloud droplets may be consumed rapidly by the growth of ice particles if they are present (Wegener, 1911; Bergeron, 1928, 1935). Such ice particles will usually grow large enough to fall out as precipitation. This interactive process between supercooled liquid water and ice particles is known as the Bergeron–Findeisen mechanism.

Ice-forming nuclei were found to be increasingly more active as the temperature becomes lower (Findeisen and Schulz, 1944). This finding was later confirmed and refined by numerous workers (e.g., Rau, 1950; Workman and Reynolds, 1949; Aufm Kampe and Weickmann, 1951). Thus the importance of relatively warm cloud-top temperatures in relation to the presence of supercooled liquid water and possible seeding opportunities was well established by 1950.

When natural ice particles are lacking in sufficient numbers to remove supercooled liquid water, little if any precipitation will develop. Cloud water will subsequently pass to the lee side of a barrier, where warming associated with downflow will cause evaporation of cloud droplets.

Introduction of ice nuclei into such clouds upwind of the barrier may cause the Bergeron–Findeisen process to take place and precipitation to occur. Intentional modification of an orographic cloud to increase precipitation does so by utilizing the cloud water prior to its downwind evaporation. Therefore, cloud seedability for increasing winter orographic precipitation is related to the rate at which cloud water evaporates on the lee side of a mountain.

The first attempts at estimating amounts of supercooled liquid water in orographic clouds were those of Bergeron (1949). Ludlam (1955) used fog frequency and riming deposits to estimate the total potential for increasing winter orographic precipitation.

With the advent of the hot wire device, commonly known as the Johnson–Williams (JW) instrument, airborne measurements of supercooled liquid water were made over mountain barriers. Probably the most comprehensive measurements were made in the 1970s in connection with a search for cloud treatment effects (Hobbs and Radke, 1975; Hobbs et al., 1975). Other studies of supercooled liquid water were made by aircraft (Lamb et al., 1976; Marwitz, 1980; Cooper and Saunders, 1980; Cooper and Marwitz, 1980; Hill, 1980a, 1982b; Rauber et al., 1985). Recently, microwave radiometry developed by Guiraud et al. (1979) and Hogg and Guiraud (1980) has been used to measure remotely supercooled liquid water in winter orographic clouds (Hill, 1982a; Long and Walsh, 1984; Holroyd and Super, 1984; Rauber and Grant, 1986).

Aside from the physical understanding of cloud seeding as applied to winter orographic clouds, important practical difficulties remain. Studies by Auer et al. (1970), Heimbach et al. (1977), Hill (1980b), Hobbs et al. (1975), Rangno et al. (1977), Karacostas (1981), Rottner et al. (1975), and Miller (1984) showed that ice nuclei as they are being delivered often do not reach the intended altitude at the appropriate location and that sufficient diffusion for effective treatment does not take place. Consequently, no attempt is made in this study to include results of field experiments to justify or not justify various cloud seeding hypotheses. On the other hand, a concept of seedability of winter orographic clouds is formulated based upon a variety of physical observations.

Because the presence of supercooled liquid water is of primary importance to the seedability of winter orographic clouds, special attention must be given to a description of when and where supercooled liquid water occurs. On the other hand, the amount of supercooled liquid water passing over a barrier and subjected to evaporation will depend upon the size and distribution of ice particles within the cloud. Consequently, analysis of the distribution of supercooled liquid water must be part of a similar analysis of ice particles and precipitation. In turn, these distributions are related to topography and other meteorological factors, such as cloud-top temperature and airflow.

13.2. Analysis of seedability

13.2.1. Seedability definitions

Cloud seedability for increasing winter orographic precipitation is related to the rate at which cloud water evaporates on the lee side of a mountain. In the absence of precipitation, this rate of evaporation in steady state conditions will be equal to the rate of production of supercooled liquid water. With precipitation occurring, the natural loss of liquid water by evaporation is reduced; the seedability is correspondingly reduced as well.

These considerations may be expressed in terms of a precipitation efficiency, ϵ, which is related to the rates of condensation and formation of nonprecipitation ice, C, and precipitation, P, according to $\epsilon = P/C$. In steady state conditions $C = P + E$, where E is the evaporation rate. Thus, $\epsilon = P/(P + E)$ or $\epsilon = 1/(1 + E/P)$. If it is assumed that evaporation is primarily due to supercooled liquid water passing over the barrier crest, then the supercooled liquid water in a vertical column of air near the barrier crest (where the vertical motion approaches zero) is directly proportional to E. Precipitation from the same vertical column of air occurring as the air passes across the barrier is directly proportional to P.

Thus, the quantity E/P is directly proportional to the quotient of supercooled liquid water crossing the barrier crest and the average precipitation rate over the barrier (Hill, 1980a). This quotient is herein defined as a "seedability factor." This factor is measured in several forms, as will be described later.

The seedability factor does not by itself show the amount of added precipitation that could be realized by cloud seeding, but does describe the efficiency of natural precipitation. When the seedability factor approaches infinity, all of the water will evaporate, whereas with a zero factor there is no supercooled liquid water available for evaporation. The actual potential quantity of precipitation augmentation, or yield, realizable from cloud seeding will depend primarily upon two factors, the precipitation efficiency and the rate of production of the total amount of supercooled liquid water.

Because the precipitation efficiency depends upon the rate at which supercooled liquid water is converted to ice compared to the rate at which it is evaporated, the number and size of ice particles present will be important factors in determining the rate at which supercooled liquid water is consumed. The growth of precipitation particles will, in turn, depend upon their concentration and upon how much time is available prior to evaporation of supercooled liquid water beyond the barrier crest. The concentration of precipitation particles depends in part upon the ice nuclei concentration at higher elevations within the cloud. The time available for the growth of precipitation is largely dependent upon the speed at which air passes from the upwind side of a mountain barrier to the barrier crest, except if clouds are well established upwind of the orographic zone of air flow.

Thus, on general principles, cloud-top temperature and cross-barrier wind speed at cloud levels will be important macrophysical factors governing seedability. The related microphysical factors are the distributions of supercooled liquid water concentration and the ice particle concentration and sizes.

13.2.2. Distribution of supercooled liquid water

The distribution of supercooled liquid water in relation to other factors can be examined in a variety of ways. One way is to separate supercooled liquid water distributions into vertical, horizontal, and temporal categories and look for related factors. Another is to study as completely as possible individual episodes. Each of the measurement systems utilized in field studies offers advantages for particular aspects of the problem. Therefore, an overview of a variety of studies and methods is likely to prove worthwhile.

The vertical distribution of supercooled liquid water has been inferred from airborne measurements such as those by Hobbs (1975). On the other hand, direct measurements have been made by a method developed by Hill and Woffinden (1980). The sensor is a vibrating wire mounted in the humidity duct of a standard U.S. radiosonde. As supercooled liquid water impinges on the wire, the natural frequency of vibration is reduced. The changing frequency yields the concentration of supercooled liq-

uid water (LWC). From such measurements, high resolution vertical profiles of supercooled liquid water are obtained. A sample of one of the soundings is shown in Fig.
13.1. It is noted that the sharp vertical gradients of supercooled liquid water are rather typical of orographic
clouds. Therefore, measurements made from aircraft are
likely to provide a somewhat inaccurate picture on individual flights; with repeated flights a useful statistical
picture emerges.

The altitude at which the indicated concentration of
supercooled liquid water exceeds 0.05 g m^{-3} over 500 m
increments as obtained by vertical soundings over the
Wasatch Mountains is shown in Fig. 13.2 for 57 soundings
over a five-year period (1980–84). These measurements
were made approximately 15 km upwind of the barrier
crest. In many of the soundings, the indicated concentration was less than 0.05 g m^{-3} throughout the entire ascent.
These cases were consistently associated with substantial
precipitation and deep clouds with cold tops.

In a separate study, measurements of vertical profiles
of vertical motion associated with winter orographic
clouds were obtained by a parachute dropsonde system
(Hill, 1980a). The altitude of the maximum updraft for
23 soundings is shown in Fig. 13.3. The most frequent
location of the maximum updraft is at an altitude of 500
to 1000 m above the barrier crest.

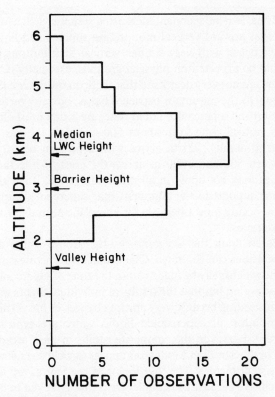

FIG. 13.2. Frequency of observations of supercooled liquid water concentration exceeding 0.05 g m^{-3} averaged for 500 m height intervals over
the upwind base of the Wasatch Mountains, Utah.

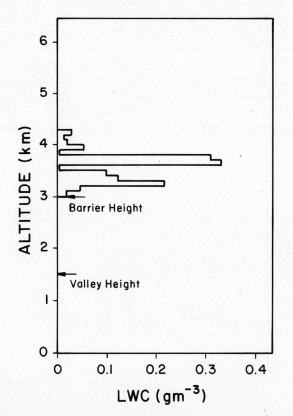

FIG. 13.1. Vertical profile of supercooled liquid water (g m^{-3}) from
cloudsonde release 1910 MST 26 February 1981 from 15 km upwind
of the Tushar Mountains, Utah.

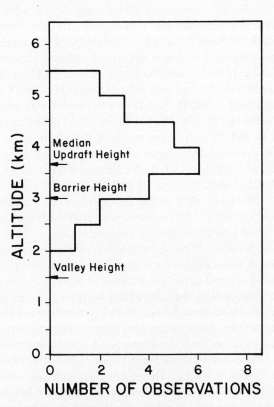

FIG. 13.3. Vertical distribution of updraft maxima in 500 m height
intervals over the upwind base of the Wasatch Mountains, Utah.

Thus, two independent measuring systems yield a finding that upward vertical motion and supercooled liquid water occur with very similar vertical distributions (as would be expected on physical grounds, especially if the layers are not very deep) and that these quantities are tied closely to the mountain barrier. That is, upward vertical motion and supercooled liquid water are found most often in a vertical zone from about 500 m below the barrier crest to about 2500 m above with a median height of between 500 and 1000 m above the crest. While these values may not apply in all mountain ranges, they give strong support to the argument that supercooled liquid water occurs most often not far above the altitude of the barrier crest.

Based upon the foregoing results, several important conclusions can be drawn. One is that airborne measurements will usually be made within the zone of supercooled liquid water, but that the results of individual flights may be misleading because very abrupt changes in the vertical distribution of supercooled liquid water are typically present. However, the composite results in the Cascades and in the San Juan Mountains appear to yield a generally consistent picture. The four flights in the Sierras are less conclusive because the sample size is low, but even in this dataset the general picture is similar to the others.

Another conclusion is that the concern of Vardiman and Moore (1978) regarding precipitation blowover appears overstated because in their criteria for seedability it is inferred that supercooled liquid water occurs often at much higher altitudes than is likely to be the case. If seeding does not result in too great a reduction of supercooled liquid water and riming is not overly reduced, then precipitation blowover need not be a serious problem. Supercooled liquid water converted to precipitation by cloud seeding will generally exist at elevations not very much above the barrier crest; this precipitation will not be carried very far horizontally before reaching the ground.

On the other hand, some of the precipitation induced by cloud seeding will fall on the lee side of a barrier when updrafts exceed the terminal fall velocity of the ice particles. However, with strong downdrafts occurring over or just beyond the barrier crest, precipitation fallout will not be very far downwind of the mountaintop.

The presence of riming will substantially increase the terminal fall velocity of ice particles. Consequently, the presence of small or modest amounts of supercooled liquid water in drop sizes that are collectible by ice particles will generally cause precipitation to fall farther upwind than if supercooled liquid water were absent, other conditions being the same. Only the supercooled liquid water that is not utilized in the natural precipitation process contributes to a seeding potential. Therefore, even with the role of riming taken into account, seedability may still be identified with the rate at which cloud water evaporates on the lee side of a mountain.

The horizontal distribution of supercooled liquid water has been documented for several mountain ranges, mainly by aircraft measurements. Extensive measurements made in the Cascade Mountains (Hobbs, 1975) provide a comprehensive view of the horizontal distribution of LWC. In 22 flights when there was a westerly component to the wind at 3 km elevation, the supercooled liquid water concentration on the average reached a maximum of about 0.4 g m^{-3} 50 km upwind of the crest at an elevation no higher than the altitude of the crest. These data are shown in Fig. 13.4 (after Hobbs, 1975). The actual elevation of the maximum is not known due to flight restrictions at the lower altitudes. Very little supercooled liquid water was found to the lee of the Cascades during these flights. In particular, at an altitude of 2.3 km (where adequate data exist) the average value of liquid water falls from 0.23 g m^{-3} 20 km upwind of the crest to less than 0.05 g m^{-3} over the crest. This distribution apparently reflects the loss of liquid water to precipitation, even though in such conditions liquid water is likely being generated by condensation in the upwind region.

Measurements made over other mountain ranges yield similar results for the horizontal distribution. Such data have been collected in the San Juan Mountains (Marwitz, 1980), the Sierra Nevada (Lamb et al., 1976; Heggli et al., 1983), the Tushar Mountains (Hill, 1982a; Long and Walsh, 1984), and the Park Range (Rauber et al., 1986). However, it is noted that in all of these studies, the horizontal distribution varies greatly from storm to storm and within a storm period. Yet there are preferred zones for the development of LWC. The actual distribution at a given time will depend upon a variety of physical factors, as discussed previously. An example of a particular episode of supercooled liquid water is shown in Fig. 13.5. These data were collected at a constant altitude over the Tushar

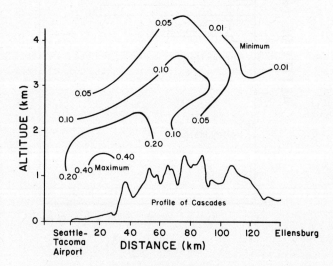

FIG. 13.4. Average distribution of liquid water (g m^{-3}) in clouds over the Cascade Mountains for 22 flights in which there was a westerly component to the wind at 3 km elevation (after Hobbs, 1975).

Mountains of Utah. In this figure LWC reaches a maximum about 10 km upwind of the barrier crest. Even though individual embedded cloud cells are moving from west to east, the maximum LWC remains on the upwind side of the barrier. Intensification of LWC occurs on the upwind slopes of the barrier, similar to the composite for the Cascade Mountains shown in Fig. 13.4.

13.2.3. Relationship of supercooled liquid water to storm stage

Supercooled liquid water forms in the vicinity of an orographic barrier in a manner generally as described in the foregoing section. On the other hand, temporal variations of supercooled liquid water can only be linked to related meteorological factors such as airflow, stability, moisture, and ice particle distributions. In some studies, emphasis has been placed upon storm stage, while in others, the emphasis is on related meteorological parameters.

Clouds associated with storm systems over the Cascades have been categorized as prefrontal, transitional, and postfrontal. Some 90 operational days were so categorized by Hobbs (1975). The prefrontal clouds are noted to occur in a layered form from the surface to 9 km. The transitional clouds occur on the average from the surface to 5.4 km, and the postfrontal clouds have convective tops to 4.6 km.

The principal findings of this extensive research were that 1) ice particles dominate over water droplets in prefrontal clouds, whereas in postfrontal clouds the ratio of water to ice is higher than in prefrontal conditions, the air is unstable, the wind westerly at all levels, the cloud tops are lower, and heavily rimed particles are common; 2) the passage of an occluded or warm front is marked by a change from unrimed to rimed crystals; 3) the concentrations of ice particles are frequently several orders of magnitude higher than the measured ice nuclei concentrations; 4) ice particle growth by riming is particularly fast in the lowest kilometer of fall; 5) the degree of riming of snow crystals reaching the ground and the precipitation rate on the western slopes of the Cascade Mountains generally increase with increasing wind speed.

In the San Juan Mountains, detailed aircraft and supporting observations were made in eight storms during the Colorado River Basin Pilot Project (Marwitz, 1980; Cooper and Saunders, 1980; Cooper and Marwitz, 1980). The categories of storm stages are divided into stable, neutral, unstable, and dissipation stages. The first three of these stages correspond roughly to those of Hobbs.

In the stable stage, practically no supercooled liquid water was found in any region. During the neutral stage the liquid water concentration over the mountain was typically 0.2–0.5 g m^{-3}. Some liquid water was also found well upwind of the barrier. During unstable conditions the liquid water concentration often exceeded 1 g m^{-3} over the mountain. High concentrations were also observed in other regions, but convection far upwind became glaciated before reaching the mountains.

During the stable and early neutral stages, the storms were more widespread than the later neutral and unstable stages when the clouds became more orographic in nature. It was found that supercooled liquid water is associated with regions where the vertical gradient of equivalent potential temperature is weak (unstable).

In the Sierra Nevada, research flights were made in more than 30 storms over a two-year period from 1971 to 1973. Four storms with a total of six flights were selected for detailed analysis on the basis that moderate amounts of precipitation occurred in the area of operations and that adequate airborne data were obtained (Lamb et al., 1976). It was found in the Lamb et al. study that supercooled liquid water was generally located at lower elevations upwind of the barrier, and in some of the storm periods cloud water was found over the crest. It was concluded that the Sierra storms followed much the same pattern as the storms in the Cascades.

Subsequent to these measurements, a series of 45 research flights were made over the Sierra during January–March 1979 and 1980. Analysis of data collected on these flights was made by Heggli et al. (1983). The analysis focused upon in situ concentrations of supercooled liquid water and ice crystal concentrations. These concentrations were then related to radar echo types. It was found that supercooled liquid water and its amount per ice crystal were greatest in convective clouds 40 to 120 km west of the mountain crest and 7 to 10 hours after the passage of

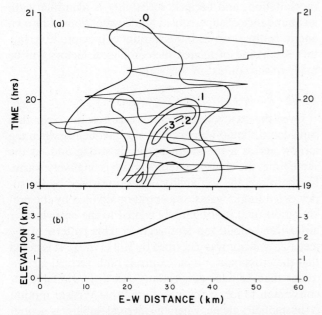

FIG. 13.5. (a) Horizontal distribution of 1 min averages of supercooled liquid water (g m^{-3}) as found by successive traverses on 26 February 1981 over the Tushar Mountains, Utah. (b) The average terrain profile in the vicinity of the flight paths.

the 700 mb trough. It was also noted that the results were in general agreement with the studies of Hobbs (1975) and Lamb et al. (1976).

Over the northern Wasatch Mountains 20 research flights were made during January–March 1978 and 1979. Analysis of these data showed a consistent relationship among supercooled liquid water concentration, cloud-top temperature, and updraft speed.

These results were further supported by analysis of a very large number of aircraft icing reports in 643 six-hour time blocks in northern Utah (Hill, 1982b). In this study the updraft speed was replaced by the cross-barrier wind speed at 700 mb. While the aircraft icing reports are qualitative in nature, the very large number of reports and the absence of substantial sampling bias give added significance to the results.

Over the southern Wasatch Mountains both airborne and radiometric measurements of LWC were made by Hill (1982b). Airborne measurements for ten flights generally supported the idea that supercooled liquid water is found primarily in postfrontal conditions, but occurrence of substantial amounts of LWC were found at other times as well.

13.2.4. Relationship of supercooled liquid water to meteorological factors

It may be inferred from the recent research studies that the storm-stage approach confirms the earlier findings in Europe, i.e., cloud-top temperature is an important factor controlling supercooled liquid water. During prefrontal conditions in the Cascades, when the clouds are deep with cold tops, there is little supercooled liquid water present over the barrier crest. On the other hand, during postfrontal conditions, when cloud tops are relatively lower, there is a general increase in LWC. Similar results are found in the San Juan Mountains. Marwitz (1980) described a representative storm in which LWC was highest during the unstable (postfrontal) conditions when cloud-top temperatures were warmer than in the early storm stage.

The relationship of LWC to cloud-top temperature was further investigated by Hill (1980a). On the basis of aircraft measurements it was found that LWC was highest with warmer cloud-top temperatures and strong vertical motion. The warmer cloud-top temperatures were associated with a lack of precipitation, which if present would deplete the LWC, and the strong vertical motion contributed to the production of liquid water. Extensive analysis of aircraft icing reports have confirmed this finding. In the study by Hill the cross-barrier wind at mountaintop level was used as a measure of vertical motion.

More recently, these results have been strengthened by findings in the Park Range Mountains. Case studies of several storms reveal that highest liquid water concentra-

tions are most likely associated with periods when the precipitation rate at the base of the mountain is low and the cloud-top temperature is warm (Rauber et al., 1986).

While aircraft measurements of supercooled liquid water have yielded a great amount of information, there has been a significant disadvantage in depending upon aircraft as a means of collecting data. Until recently, aircraft measurements were made with little or no prior knowledge as to whether substantial supercooled liquid water would be found during a given flight. Flights were initiated when meteorological disturbances were present. By selecting such conditions for making airborne measurements, more favorable situations for the occurrence of supercooled liquid water may have been systematically excluded.

Since the development of the dual-frequency radiometer (Guiraud et al., 1979; Snider et al., 1980; Westwater, 1978), continuous measurements of the vertically integrated concentration of supercooled liquid water have been made near several mountain barriers (Hill, 1981; Rauber et al., 1986; Long and Walsh, 1984; Holroyd and Super, 1984). Such measurements are particularly useful when made in conjunction with other sensors, such as K- and X-band radar and precipitation gages. A sample of measurements made by this array of instruments is shown in Fig. 13.6. In this example there was a period (0800–1100 MST) when supercooled liquid water near the barrier crest was substantial, yet precipitation was very weak, if not absent. At the same time, cloud tops over the base of the barrier were relatively low with cloud-top temperatures warmer than $-20°C$.

Because the amount of supercooled liquid water is closely related to the same factors as those controlling precipitation, and because seedability is identified with an abundance of supercooled liquid water both of its own and in comparison with the amount of precipitation, further discussion of related meteorological factors will be made in the context of seedability.

13.2.5. Factors governing seedability

Over the past several years it has become clear that supercooled liquid water occurs most frequently when the airflow across a mountain barrier is strong and, at the same time, the cloud-top temperature is relatively warm. Furthermore, the seedability factor (measured by the supercooled liquid water concentration divided by the precipitation rate) is also closely related to the cross-barrier airflow and cloud-top temperature. This pattern of the seedability factor was described by Hill (1980a). It is noted that the cross-barrier wind controls both vertical airflow (other than cellular convection) and time available for conversion to ice. It was also found that vertical motion is substantially greater when the thermal stability is neutral or unstable. Stable airflow with strong cross-barrier flow tends to produce blocking with relatively little vertical motion. However, conditions with low (warm) cloud tops

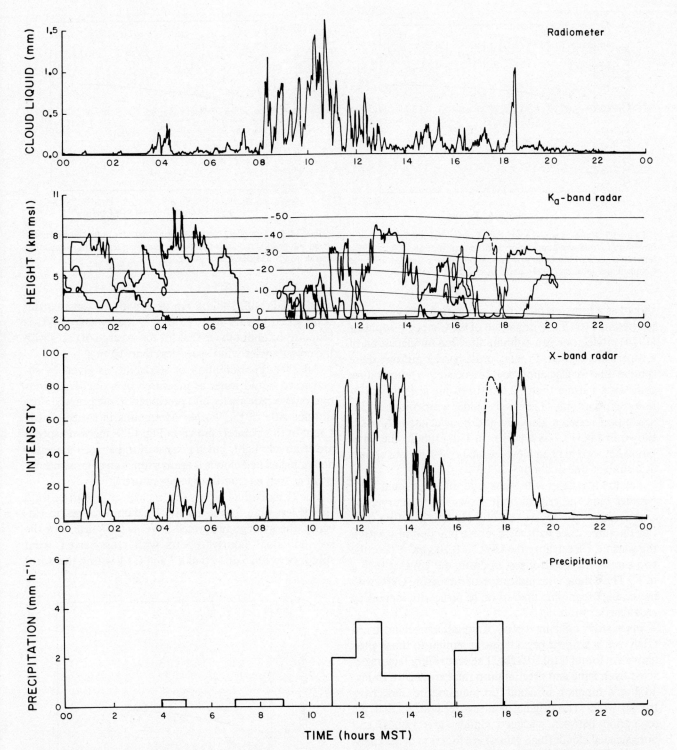

FIG. 13.6. Measurement of cloud structure by dual-frequency microwave radiometer, K_a- and X-band radar, and precipitation in the vicinity of the Tushar Mountains, Utah (radar at the upwind base of the mountain and radiometer and precipitation upwind of the crest). The top panel is derived from 2 min average radiometer values for a vertical path. The second panel is based upon the K_a-band (TPQ-11) radar output at 2 min intervals. The coordinates are height above ground level (1860 m MSL) and time in hours (MST). Clouds are shown by outlined areas, and the temperatures are shown at 10°C intervals. The X-band radar intensity at 2 min intervals as a function of time is shown in the third panel. Hourly precipitation (mm h^{-1}) as a function of time is shown in the bottom panel.

and strong cross-barrier flow are often those associated with postfrontal conditions where the stability is weak. This view is close to that of Marwitz (1980).

In Fig. 13.7 values of the seedability factor derived from airborne measurements of supercooled liquid water and precipitation in the northern Wasatch Mountains are

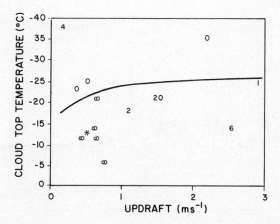

FIG. 13.7. Ratio of supercooled liquid water and precipitation as a function of cloud-top temperature and vertical motion. (The asterisk refers to a nonzero but unknown value.) The curve is derived from the Chappell and Johnson (1974) model to separate the ratios.

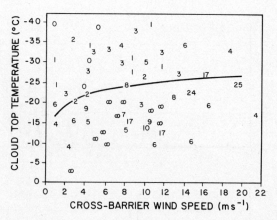

FIG. 13.8. Ratio of supercooled liquid water (as indicated by aircraft icing) and precipitation as a function of cloud-top temperature and cross-barrier wind. The curve is as in Fig. 13.7.

shown as a function of vertical air motion and cloud-top temperature (CTT). With the aid of the Chappell–Johnson (1974) model, one can estimate the time for glaciation of a supercooled cloud. In weak cross-barrier airflow, there is more time for glaciation, so it can occur at a temperature somewhat warmer than −22°C. In strong airflow, glaciation can occur only at relatively colder temperatures. This separation between glaciated and nonglaciated clouds is shown in Fig. 13.7 as a curve. In Hill (1980b) the separation between high and low seedability factors was shown as a straight line at about −22°C.

For the five cases with weak updraft speeds and CTT warmer than the critical limit shown by the curve, the average LWC as reported by Hill (1980a) is 0.10 g m^{-3}. For the three cases with relatively strong updrafts within the warm CTT category the LWC is 0.20 g m^{-3}. (For the two cases with the strongest updrafts the LWC is 0.26 g m^{-3}.) Thus, there is an indication of increasing LWC with increasing orographic updraft or, in turn, with increasing cross-barrier wind.

In the study utilizing reports of aircraft icing in northern Utah over a six-year period, results similar to those cited above are found (Hill, 1982b). The seedability factors derived from icing and precipitation rates are shown in Fig. 13.8 as a function of cloud-top temperature and cross-barrier wind. The original analysis recognized that the separation between glaciated clouds (low values) and nonglaciated clouds (high values) is a function of the cross-barrier wind speed as well as cloud-top temperature. This separation is also reasonably well given by use of the Chappell–Johnson model, which takes into account the increasing number of ice nuclei available at colder temperatures and the time duration of ice crystal growth.

Analysis of 121 individual cases with the cloud-top temperature warmer than the critical limit shows that the LWC increases substantially with increasing cross-barrier

wind speed (Hill, 1982b). For cases with the cross-barrier wind speed greater than 10 m s^{-1}, the average LWC is found to be about twice that for the average of cases with the cross-barrier wind speed less than 10 m s^{-1}.

Another representation of seedability is given by supercooled liquid water as measured by a dual-frequency microwave radiometer and precipitation measured on the upslope side of the Tusher Mountains in south central Utah. In this dataset, shown in Fig. 13.9, the wind speeds are relatively light, but the separation between glaciated and nonglaciated clouds is again a function of cross-barrier wind as well as cloud-top temperature.

For 21 individual hourly reports with cross-barrier wind speeds less than 5 m s^{-1} and CTT within the warm category, the average integrated LWC is 0.07 mm. For the six individual hourly reports with cross-barrier wind speeds between 5 and 10 m s^{-1} and CTT within the warm

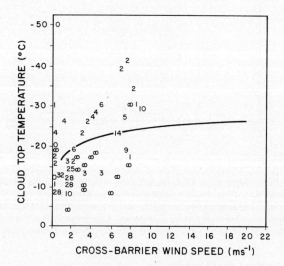

FIG. 13.9. As in Fig. 13.8 as measured by dual-frequency microwave radiometer.

category, the average integrated LWC is 0.27 mm. Once again, the supercooled liquid water is found to increase with increasing cross-barrier wind.

It is noted that if clouds are well formed prior to orographic uplift, then glaciation could well be established at much warmer temperatures than what is indicated by the curve separating glaciated and nonglaciated clouds. Thus, seedability (high seedability factors) is more clearly established when clouds form primarily in response to orographic uplift. Also, natural variations in ice nuclei concentrations from time to time may be present, so that individual instances of seedability may be greater or less than the foregoing criteria imply.

Furthermore, supercooled liquid water may be depleted when ice multiplication processes (Hallett and Mossop, 1974) are occurring. Ice multiplication will generally act to reduce seedability within a cloud-top temperature and cross-barrier wind regime that would otherwise be seedable. Because the combination of temperature and drop size distribution required for ice multiplication is more often found during winter in the far western states compared to the Rocky Mountains, seedability may be reduced more frequently in the coastal states compared to inland mountains.

In summary, it is concluded that supercooled liquid water and orographic precipitation in the vicinity of a mountain barrier are governed primarily by the cloud-top temperature and the cross-barrier wind speed. Seedability is favored when the cloud-top temperature is warmer than a certain value; this threshold temperature decreases with stronger cross-barrier wind.

The amount of yield is evidently related primarily to the strength of the cross-barrier wind. Because the flux of supercooled liquid water during seedable conditions over a mountain barrier is related to the product of the cross-barrier wind speed and the LWC, and because the LWC is, in turn, roughly proportional to the cross-barrier wind speed, the yield will be approximately proportional to the square of the cross-barrier wind speed.

Other evidence supporting these findings is available. The relationship of storm stage to the appearance of supercooled liquid water is well known. During the early storm stage when clouds are deep, there is relatively little supercooled liquid water. During postfrontal conditions when cloud tops are often much lower and warmer, supercooled liquid water is often present over a mountain barrier. The measurements of Rauber and Grant (1986) show a relationship of LWC to cloud-top temperature.

The relationship of LWC to cross-barrier wind has been noted in other studies. Hobbs (1975) reports that the degree of riming of snow crystals reaching the ground on the western slopes of the Cascades generally increases with increasing wind speed. Rauber and Grant (1986) state that their results are in general agreement with those of Hill (1980a), in which it was found that the quantity of supercooled water present in the cloud systems over the Wasatch is related to the cloud-top temperature and cross-barrier wind speed, and inversely related to the precipitation rate.

13.2.6. Application of seeding criteria

Until recently, seedability of winter orographic clouds has been primarily based upon macrophysical quantities such as fixed-level or cloud-top temperatures (Grant and Mielke, 1967; Mielke et al., 1970, 1971; Grant and Elliott, 1974) or barrier trajectory indices (Vardiman and Moore, 1978; Rottner et al., 1980). Obviously, the use of macrophysical properties alone lacks the identification of the essential microphysical properties needed for opportunity recognition.

More recently, microphysical properties have been incorporated into seedability assessments. That is, the concentrations of supercooled liquid water and ice particles (or precipitation) have been utilized (Hill, 1980a, 1982a; Hindman, 1984; Reynolds, 1984). However, if only microphysical properties are included in seeding criteria, there will be uncertainty as to when the opportunity for modifying the cloud by seeding has ended. This uncertainty arises because a successfully treated cloud and naturally precipitating clouds both have lower amounts of supercooled liquid water and higher amounts of ice and precipitation compared to clouds offering a seeding opportunity.

Consequently, it is noted that a combination of macro- and microphysical parameters is needed to identify seedability both before and during seeding operations. The relationship between supercooled liquid water concentration, ice particle concentration (or precipitation), cloud-top temperature, and cross-barrier wind gives such a combination.

Immediately prior to treatment, seedability is identified primarily on the basis of the microphysical properties, i.e., the concentration of supercooled liquid water and the concentration of ice or precipitation. The strength of the cross-barrier wind will also have a bearing on the total yield from the treatment. However, once effective treatment is started, reliance upon the microphysical data is no longer justified. The state of the macrophysical parameters will provide the guidance as to what opportunity would likely be present if the treatment were not being done. In addition, use of the macrophysical parameters will provide guidance for forecasting and preparation for possible seeding opportunities.

Finally, analysis of precipitation trajectories shows that when seedability is identified, treatment of supercooled liquid water in orographic clouds will normally result in precipitation trajectories that will intersect the mountain barrier on the upwind side or near the barrier crest. When the cross-barrier wind speed becomes excessively high, the ice crystal trajectories may reach the ground beyond the high terrain of the mountain. Such a wind speed limit will depend upon the mountain profile as well as the location of supercooled liquid water.

13.3. Conclusions

The seedability of winter orographic clouds for increasing precipitation is related primarily to the supercooled liquid water that evaporates downwind of a mountain barrier. Seedability is reduced by the extent to which ice is formed or precipitation takes place. The amount of supercooled liquid water near the barrier crest divided by the precipitation over the barrier is a measure of the relative abundance of supercooled liquid water available for conversion to precipitation by the addition of ice nuclei. Other measures of ice or precipitation may be used, but whatever measure is used it should be representative of what is occurring over the entire barrier cross section. Also, the measure of LWC should be representative of what water will be subject to evaporation. Therefore, the LWC measurements should be made near the barrier crest.

It is found that high seedability factors (LWC divided by precipitation) occur at cloud-top temperatures warmer than a critical limit. This critical temperature limit is colder for increasing cross-barrier wind speeds. The reason is that there is less time for conversion of supercooled liquid water to ice when the wind is strong. When clouds exist upwind of the orographic zone of uplift, there is an additional time period when conversion to ice may take place. Therefore, not all clouds with tops warmer than the critical temperature limit will be seedable. However, analysis of extensive and variously derived datasets clearly shows a sharp separation between seedable and nonseedable clouds. The reason is that precipitation and supercooled liquid water are somewhat mutually exclusive. While there are frequent observations of supercooled liquid water and precipitation occurring during the same disturbance, the coexistence is transitory.

Strong evidence is found that shows the total amount of supercooled liquid water is related to the cross-barrier wind speed. Evidently, the vertical excursions of saturated airflow are greater with increasing cross-barrier wind. Because the total supercooled liquid water over a barrier crest generally increases with increasing cross-barrier wind speed and the transport of this water is directly proportional to the cross-barrier wind speed, the potential yield from cloud seeding is approximately proportional to the square of the cross-barrier wind speed, provided, of course, that the clouds are within the warm cloud-top temperature category.

In an actual application of cloud seeding technology, identification of seedable clouds is more reliably accomplished by direct measurement of supercooled liquid water and precipitation. When seedability is identified by the microphysical measurements, it is also likely that the guidance criteria described above will be met as well. However, the direct measurements will identify seedability. In that case, the potential yield will depend upon the product of the total supercooled liquid water subject to evaporation and the cross-barrier wind speed.

Once seeding has begun and has effectively modified the clouds being treated, then reliance is made upon the general criteria, particularly cloud-top temperature because the airflow generally changes much more slowly than does the cloud-top temperature. As long as the general criteria remain within a favorable range, then the clouds may be considered seedable even though the microphysical data have been modified by the seeding.

Acknowledgments. This study was supported by the National Science Foundation, Grants ATM-8120416 and ATM-8320330.

REFERENCES

Auer, A. H., D. L. Veal and J. D. Marwitz, 1970: Some observations of silver iodide plumes within the Elk Mountain Water Resource Observatory. *J. Wea. Mod.,* **2,** 122–131.

Aufm Kampe, H. J., and H. K. Weickmann, 1951: The effectiveness of natural and artificial aerosols as freezing nuclei. *J. Meteor.,* **8,** p. 283.

Bergeron, T., 1928: Uber die driedimensional verknupfende wetteranalyse, Teil I. *Geofys. Publ.,* **5** (6).

——, 1935: On the physics of cloud and precipitation. IGGU, Lisbon 1933 (Paris, 1935).

——, 1949: The problem of artificial control of rainfall on the globe. *Tellus,* **1,** 32–43.

Chappell, C. F., and F. L. Johnson, 1974: Potential for snow augmentation in cold orographic clouds. *J. Appl. Meteor.,* **13,** 374–382.

Cooper, W. A., and J. D. Marwitz, 1980: Winter storms over the San Juan Mountains. Part III. Seeding potential. *J. Appl. Meteor.,* **19,** 942–949.

——, and C. Saunders, 1980: Winter storms over the San Juan Mountains. Part II: Microphysical processes. *J. Appl. Meteor.,* **19,** 927–941.

Findeisen, W., 1938: Die kolloidmeteorologischen vorgänge bei der Niederschlagsbildung. *Meteor. Z.,* **55,** 121–131.

——, and G. Schulz, 1944: Experimentelle untersuchungen uber die atmospharisch eistielchenbildung. *Wetterdienst,* **A27.**

Grant, L. O., and P. W. Mielke, 1967: A randomized cloud seeding experiment at Climax, CO, 1960–65. *Proc. Fifth Berkeley Symp. on Mathematical Statistics and Probability,* Vol. 5, University of California Press, 115–132.

——, and R. E. Elliott, 1974: The cloud seeding temperature window. *J. Appl. Meteor.,* **13,** 355–363.

Guiraud, F. O., J. Howard and D. C. Hogg, 1979: A dual-channel microwave radiometer for measurement of precipitable water vapor and liquid. *IEEE Trans. Geosci. Electron.,* **GE-17,** 129–136.

Hallett, J., and S. Mossop, 1974: Production of secondary ice particles during the riming process. *Nature,* **249,** 26–28.

Heggli, M. F., L. Vardiman, R. E. Stewart and A. Huggins, 1983: Supercooled liquid water and ice crystal distributions within Sierra Nevada winter storms. *J. Climate Appl. Meteor.,* **22,** 1875–1886.

Heimbach, J. A., A. B. Super and J. T. McPartland, 1977: A suggested technique for the analysis of airborne continuous ice nucleus data. *J. Appl. Meteor.,* **16,** 255–261.

Hill, G. E., 1980a: Dispersion of airborne-released silver iodide in winter orographic clouds. *J. Appl. Meteor.,* **19,** 978–985.

——, 1980b: Seeding opportunity recognition in winter orographic clouds. *J. Appl. Meteor.,* **19,** 1371–1381.

——, 1981: A case study of a winter orographic cloud seeding system. *Eighth Conf. on Inadvertent and Planned Weather Modification,* Reno, Amer. Meteor. Soc.

——, 1982a: Analysis of precipitation augmentation potential in winter orographic clouds by use of aircraft icing reports. *J. Appl. Meteor.,* **21,** 165–170.

——, 1982b: Evaluation of the Utah operational weather modification program. Final Rep., NOAA Contract NOAA/NA-81-RAC-00023, Utah State University, UWRL/A-82/02, 291 pp.

——, and D. S. Woffinden, 1980: A balloonborne instrument for the measurement of vertical profiles of supercooled liquid water concentration. *J. Appl. Meteor.,* **19,** 1285–1292.

Hindman, E. E., 1984: Colorado winter mountaintop liquid water studies applied to cloud seeding hypotheses. *Ninth Conf. on Weather Modification,* Park City, Amer. Meteor. Soc.

Hobbs, P. V., 1975: The nature of winter clouds and precipitation in the Cascade Mountains and their modification by artificial seeding. Part I: Natural conditions. *J. Appl. Meteor.,* **14,** 783–803.

——, and L. F. Radke, 1975: The nature of winter clouds and precipitation in the Cascade Mountains and their modification by artificial seeding. Part II: Techniques for the physical evaluation of seeding. *J. Appl. Meteor.,* **14,** 805–818.

——, L. F. Radke, J. R. Fleming and D. G. Atkinson, 1975: Airborne ice nucleus and cloud microstructure measurements in naturally and artifically seeded situations over the San Juan Mountains in Colorado. Res. Rep. X, Cloud Physics Group, University of Washington, 89 pp. [Available from Dept. of Atmos. Sci., AK-40, University of Washington, Seattle, WA 98195.]

Hogg, D. C., and F. O. Guiraud, 1980: Simultaneous observation of cool cloud liquid by ground-based microwave radiometry and icing of aircraft. *J. Appl. Meteor.,* **19,** 893–895.

Holroyd, E. W., and A. B. Super, 1984: Winter spatial and temporal variations in supercooled liquid water over the Grand Mesa, Colorado. *Ninth Conf. on Weather Modification,* Park City, Amer. Meteor. Soc.

Karacostas, T. S. 1981: Turbulent diffusion studies of winter storms over the Sierra Nevada Mountains. Ph.D. dissertation, University of Wyoming, 208 pp.

Lamb, D., K. W. Nielsen, H. E. Klieforth and J. Hallett, 1976: Measurements of liquid water content in winter cloud systems over the Sierra Nevada. *J. Appl. Meteor.,* **15,** 763–775.

Long, A. B., and P. A. Walsh, 1984: Radiometric detection and mea-

surement of liquid water in wintertime orographic clouds. *Ninth Conf. on Weather Modification,* Park City, Amer. Meteor. Soc.

Ludlam, F. H., 1955: Artificial snowfall from mountain clouds. *Tellus,* **7,** 277–290.

Marwitz, J. D., 1980: Winter storms over the San Juan Mountains. Part I: Dynamical processes. *J. Appl. Meteor.,* **19,** 913–926.

Mielke, P. W., Jr., L. O. Grant and C. F. Chappell, 1970: Elevation and spatial variation in seeding effects from wintertime orographic cloud seeding. *J. Appl. Meteor.,* **9,** 476–488.

——, —— and ——, 1971: An independent replication of the Climax wintertime orographic seeding experiment. *J. Appl. Meteor.,* **10,** 1198–1212.

Miller, B. D., 1984: Dispersion of AgI released from ground based generators in a winter cloud seeding project. M.S. thesis, Utah State University, 94 pp.

Rangno, A. L., P. V. Hobbs and L. F. Radke, 1977: Tracer and diffusion and cloud microphysical studies in the American River basin. Final Rep., Contract 6-07-DR-20140, University of Washington, 64 pp.

Rau, W., 1950: Uber die wirkungsweise der gefrierkerne in unterkuhlten wasser. *Z. Naturf.,* **5a,** p. 667.

Rauber, R. M., and L. O. Grant, 1986: The characteristics and distribution of cloud water over the mountains of northern Colorado during wintertime storms. Part II: Spatial distribution and microphysical characteristics. *J. Climate Appl. Meteor.,* **25,** 489–504.

——, D. Feng, L. O. Grant and J. B. Snider, 1986: The characteristics and distribution of cloud water over the mountains of northern Colorado during wintertime storms. Part I: Temporal variations. *J. Climate Appl. Meteor.,* **25,** 468–488.

Reynolds, D. W., 1984: Progress and plans of the Sierra Cooperative Pilot Project. *Ninth Conf. on Weather Modification,* Park City, Amer. Meteor. Soc.

Rottner, D., S. R. Brown and O. H. Foehner, 1975: The effect of persistence of AgI on randomized weather modification experiments. *J. Appl. Meteor.,* **14,** 939–945.

——, L. Vardiman and J. A. Moore, 1980: Reanalysis of "Generalized criteria for seeding winter orographic clouds." *J. Appl. Meteor.,* **19,** 622–626.

Snider, J. B., H. M. Burdick and D. C. Hogg, 1980: Cloud liquid measurements with a ground-based microwave instrument. *Radio Sci.,* **15,** 683–693.

Vardiman, L., and J. A. Moore, 1978: Generalized criteria for seeding winter orographic clouds. *J. Appl. Meteor.,* **17,** 1769–1777.

Wegener, A., 1911: *Thermodynamik der Atmosphare.* J. A. Barth, Leipzig.

Westwater, E. R., 1978: The accuracy of water vapor and cloud liquid determination by dual-frequency ground-based microwave radiometry. *Radio Sci.,* **13,** 677–685.

Workman, E. J., and S. E. Reynolds, 1949: Thunderstorm electricity. New Mexico Institute of Mining and Technology Prog. Rep. 6.

CHAPTER 14

Testing, Implementation, and Evolution of Seeding Concepts—A Review

WILLIAM R. COTTON

Colorado State University, Fort Collins, Colorado

ABSTRACT

In this paper, testing, implementation, and evolution of both static and dynamic seeding concepts are reviewed. A brief review of both waterspray and hygroscopic seeding is first presented. This is followed by reviews of static seeding of stable orographic clouds and supercooled cumuli. We conclude with a review of dynamic seeding concepts with particular focus on the Florida studies.

It is concluded that it is encouraging that our testing procedures have evolved from single-response-variable "blackbox" experiments to randomized experiments that attempt to test a number of components in the hypothesized chain of physical responses to seeding. It is cautioned, however, that changes in the seeding strategy to optimize detection of a physical response (in any of the intermediate links in the hypothesized chain of responses) can have an adverse effect upon rainfall on the ground.

14.1. Introduction

In this paper, testing and implementation of warm cloud seeding concepts, then static seeding of supercooled clouds, followed by dynamic seeding of supercooled cumuli will be reviewed. Implementation of a concept will be considered as a part of field testing as well as implementation in an operational system. It is recognized, however, that implementation in a field test may be quite different from the form used in an operational system that must be cost effective.

14.2. Testing, implementation, and evolution of warm cloud seeding concepts

As discussed in an earlier review (Cotton, 1982), the two principal concepts for modifying warm cloud precipitation processes are 1) water droplet seeding and 2) hygroscopic particle seeding.

14.2.1. Water droplet seeding

The concept of warm rain initiation or enhancement as first suggested by Langmuir (see Project Cirrus Summary, Havens et al., 1978) involves the insertion of a spray of water drops (diameters 60 μm to several hundred micrometers), which can then serve as the embryos for droplet collection and can break up to initiate a chain reaction mechanism (Langmuir, 1948). The concept of water drop seeding has been examined and tested in simple mathematical computer models (Bowen, 1950; Ludlam, 1951; Rokicki and Young, 1978). In addition, Coons et al. (1949), Bowen (1952), and Braham et al. (1957) carried out field experiments that tested the concept further. Bra-

ham's experiments provided convincing evidence that waterspray seeding does enhance the rate of formation of precipitation.

Increasing the rate of formation of precipitation, however, does not necessarily mean that rainfall on the ground will be increased. Nelson (1971), for example, simulated Braham's waterspray seeding experiments in a one-dimensional, time-dependent cloud model. He found that the simulated seeded cloud produced rain earlier than the unseeded cloud, but the rain intensity and total rainfall were reduced by 25% and 8%, respectively, in the seeded cloud. Thus, waterspray seeding may only be effective in very marginal clouds where entrainment may destroy the cloud before rainfall commences naturally. Another limitation of the technique is the difficulty in implementing a cost effective method of introducing substantial quantities of waterspray in actual clouds in an operational program.

14.2.2. Hygroscopic particle seeding

The basic concept behind hygroscopic seeding is that giant hygroscopic aerosol particles can be introduced in warm clouds and thereby play an important role in initiating the collision and coalescence process. It is further hypothesized that initiation or speed up of the collision and coalescence process will lead to an enhancement of rainfall on the ground. Early attempts to test hygroscopic seeding concepts were essentially "blackbox" experiments in which salt particles were released primarily from the ground (Roy et al., 1961; Biswas et al., 1967; Murty and Biswas, 1968; Fournier d'Albe and Aleman, 1976). By a "blackbox" experiment I mean that the clouds were seeded in some way and only one output, namely, rainfall on the ground, was measured and analyzed. In spite of the fact that analysis of the Indian experiments suggested a statistically significant 41.9% increase in rainfall on seed days (Murty and Biswas, 1968), many scientists remain skeptical about the results of the study (Mason, 1971; Simpson and Dennis, 1972; Warner, 1973). In retrospect, it is now clear that a "blackbox" experiment does not provide a scientifically acceptable test of a seeding concept.

There have been subsequent experiments that have suggested that hygroscopic seeding changes the cloud droplet distribution, cloud electric field intensity, and cloud chemistry (Kapoor et al., 1976; Ramachandra Murty et al., 1975; Khemani et al., 1981) and lowers radar echo heights (Chatterjee et al., 1969; Dennis and Koscielski, 1972).

The hygroscopic seeding concept has also been examined in several modeling studies (Biswas and Dennis 1971, 1972; Farley and Chen, 1975; Klazura and Todd, 1978; Johnson, 1980; Rokicki and Young, 1978). Johnson's (1980) study, in particular, suggested that if one considers the natural background distribution of salt particles, very large concentrations of seeding particles (on the order of 10^{-3} g m^{-3}) would be required before rainfall initiated by seeding would be predicted to exceed that initiated naturally. Thus, Johnson concludes that "clouds with naturally inefficient warm rain processes will still be inefficient after seeding." The same can be said about clouds that have a very efficient warm rain process. Thus, if a hygroscopic seeding potential exists, the modeling work suggests that it exists only in a class of cloud that lies between those two extremes. Herein lies a problem with the implementation of the hygroscopic seeding concept. Can we assess operationally the natural warm rain potential in order to define a hygroscopic seeding window? When we take into account that turbulence may also influence warm rain potential (see Manton, 1979; Baker et al., 1980), it does not seem that we can currently access the potential for warm rain production with sufficient skill to evaluate a hygroscopic seeding window.

There are a number of other problems associated with the implementation of hygroscopic seeding concepts. One of the reasons the Indian experiments have not received the level of scientific support one might expect from such a statistically significant experiment was that few scientists believe that ground-based hygroscopic seeding can be effective. Unanswered questions such as "Did the seeding material actually enter the clouds and, if so, in what concentrations?" cast considerable uncertainty upon the entire experiment.

Implementation of airborne hygroscopic seeding techniques is by no means straightforward either. It is difficult to grind sodium chloride or a similar salt to a desired and controlled size spectrum and to prevent clumping of the particles. Moreover, most salts are extremely corrosive.

Thus, salt corrosion can result in extremely costly depreciation of an aircraft and/or require very costly preventative measures to isolate the salt from the aircraft. The corrosive nature of salt also imposes an environmental threat in the vicinity of the seeded clouds. Real or perceived concern about salt corrosion of house roofs, automobiles, or electronic equipment can severely jeopardize the implementation of a salt seeding program.

Use of microencapsulated urea (Nelson and Silverman, 1972) can eliminate most of the handling and dispersion problems associated with standard salt and provides a means of very accurate sizing of the particle spectrum. However, unless hygroscopic seeding is implemented on a massive scale, the costs of using this technique may be prohibitive.

Another concern about airborne hygroscopic seeding that is applicable to all forms of cloud seeding is the question of obtaining sufficient area coverage of seeding material in "candidate" clouds to significantly affect the rainfall over a given target watershed or an experimental target area. If only a small fraction of the clouds in a given test area is actually seeded, then the seeding effects are so diluted that even strong responses in individual clouds may not be detectable.

14.3. Testing, implementation, and evolution of "static seeding" concepts in supercooled clouds

According to the "static seeding" concept, some clouds may be naturally deficient in effective ice nuclei and thus inefficient in precipitation efficiency. Consequently, it is hypothesized that if one carefully introduces a selective concentration of nuclei by cloud seeding then precipitation may be increased. However, if the concentration of artificially introduced nuclei is too large, then the clouds can be overseeded, resulting in a reduction of precipitation. Since the primary strategy is to increase the precipitation efficiency of a cloud in a preexisting cloud dynamic framework without intentionally enhancing the circulations of the cloud, the concept is often referred to as "static seeding." This approach has also been called the "optimum nucleus concentration" strategy (Braham, 1979).

The "static seeding" concept has been applied to a variety of cloud types, including supercooled stratus, stable orographic clouds, and convective clouds of all statures. In this review, the application of static seeding concepts to supercooled stable orographic clouds and cumuli will be emphasized.

14.3.1. Stable orographic clouds

The basic concept for the modification of precipitation from orographic clouds was outlined by Ludlam (1955), who predicted the optimum conditions favorable for orographic cloud precipitation enhancement and the optimum concentrations of seeding material needed to cause the desired response. Ludlam applied this concept to clouds over central Sweden and estimated the amount of enhancement of winter seasonal precipitation that could be obtained. A fundamental aspect of this concept is that the natural production of ice crystals obeys an exponential ice nuclei activation model as a function of the degree of supercooling such as described by Fletcher (1962). Early attempts to test this concept were of the "blackbox" type experiment in which seeding material was released from ground-based generators and the primary output, snowfall or precipitation on the ground, was measured. The Climax experiments, which were begun in the late 1950s (Grant and Mielke, 1967; Mielke et al., 1971), were designed to test the Ludlam concept and its applicability to the central Rocky Mountain region of the United States.

Implementation of the concept in the Rocky Mountain region was complicated by several factors. One was that the actual transport dispersion of seeding material from the ground to the cloud systems is not always assured. In some cases the material can be trapped by overlying capping inversions, allowing the accumulation of seeding material in windward mountain valley basins (see Hill, 1984). In other cases the material can be transported away from the mountain barrier in drainage flows and swept aloft upstream of the barrier in convergence established between the drainage flow and prevailing flow (Reid, 1976).

Another problem in the implementation of the Ludlam concept is that orographic clouds rarely behave as in the "pure" orographic cap cloud model. Instead, orographic lifting interacts with varying cyclonic storm structures, including their banded components (Houze et al., 1976a,b), embedded convection, etc. Thus, the conditions suitable for seeding are not as easy to assess as envisaged by Ludlam. In addition, recent studies have clearly shown that estimates of ice nucleus concentrations are not consistent predictors of observed ice crystal concentrations (Hobbs, 1969; Mossop, 1970; Koenig, 1963) and that ice multiplication processes such as the rime–splinter process (Hallett and Mossop, 1974) may prevail in ice crystal production at warm temperatures in some clouds. This further complicates our ability to assess the conditions suitable for seeding.

In spite of these difficulties, the statistical analyses of the "blackbox" Climax I and II experiments have indicated that seeding can significantly increase precipitation in orographic clouds when "estimated" cloud-top temperatures are relatively warm (Mielke et al., 1971; Mielke et al., 1981). Unfortunately, actual cloud-top measurements were not available during the Climax experiments. Instead, the temperature of the 500 mb level over the Climax region was used to represent cloud-top temperature. Grant and Kahan (1974) referred to Furman's (1967) M.S. thesis in which radar, visual, and aircraft observations were used to identify the 500 mb level as being close to cloud tops near Climax. However, Hobbs and Rangno (1979) have questioned the usefulness of the temperature of the 500 mb pressure level as an index of cloud-top temperature. Subsequent reanalysis of the experiment (Mielke et al., 1981) suggested that seeding increases of precipitation were smaller than previous analyses but the statistical significance of the results was stronger.

In parallel with, and subsequent to, the Climax experiments there have been a number of observational and modeling studies designed to evaluate several components of the basic concept or to better quantify Ludlam's static seeding concept. Chappell (1970) extended the Ludlam model to estimate the optimum conditions suitable for cloud seeding. He estimated the rate of supply of condensed water from the rate of adiabatic lifting of air parcels. The rate of removal of water solely by vapor deposition growth of ice crystals was estimated by assuming an exponential increase in crystal concentrations with the degree of supercooling (Fletcher, 1968), where cloud-top temperature was estimated from the 500 mb temperature. He predicted that at temperatures warmer than $-20°C$ the production rate of water exceeded the removal rate, thus yielding a significant potential for precipitation enhancement.

No direct observations of supercooled liquid water content or ice crystal concentrations have been made in the Climax mountain region. However, aircraft observations in the San Juan Mountains of southern Colorado reported by Hobbs et al. (1975) and Marwitz et al. (1976) showed that observed ice crystal concentrations were far in excess of ice nucleus concentrations (Hobbs and Rangno, 1979). Hobbs and Rangno (1979) and Hill (1974) also questioned the reliability of the 500 mb temperature as an index of the cloud-top height. Since there is considerable difference in cloud climatology between central Colorado and the San Juan Mountains, these criticisms certainly cast doubt on the transferability of the Climax results to another region. Recent radiometric and radar observations of orographic clouds over the Park Range (north of Climax) reported by Rauber et al. (1986) indicate that the stratiform component of the cloud systems seldom exceeds the mean height of the 500 mb level, thus the stratiform cloud-top temperature is generally warmer than the 500 mb temperature. This implies that if ice crystal concentrations in the stratiform cloud category are determined by ice nuclei concentrations (i.e., the Fletcher formula) then Chappell's model would be a conservative predictor of seedability.

As can be seen from the preceding discussion, a crucial component of the concept of static seeding of wintertime orographic clouds is the availability of sufficient quantities of supercooled liquid that can be utilized by seeding-produced ice crystals. Aircraft observations in the San Juan Mountains (Cooper and Saunders, 1980) and in the Sierra Nevada (Heggli et al., 1983) indicated that most of the available supercooled water was in cumulus clouds rather than stably stratified orographic clouds. This led Cooper and Marwitz (1980) to conclude that the best opportunity for successful weather modification in the San Juans should be associated with convective clouds because of their high water contents. However, Rauber (1985) noted that aircraft observations are naturally biased toward high liquid water samples in that convective clouds are normally selected for penetration during their active growth stages when liquid water is being produced. Using a ground-based passive microwave radiometer, Rauber et al. (1986) and Rauber (1985) found that cumulus clouds over the Park Range exhibited initially high liquid water contents but natural ice-phase precipitation processes depleted the available liquid water contents in a short time. Thus an ensemble average (e.g., a time and space average over a number of similar cloud systems) of liquid water in wintertime orographic cumuli would exhibit much lower liquid water contents and, hence, seeding potential.

Both Cooper and Marwitz (1980) and Heggli et al. (1983) found little liquid water present in their aircraft observations of stable orographic cloud systems. Rauber (1985) noted that aircraft observations are constrained to fly a minimum flight altitude that is at least 1 km above the highest point on the mountain barrier and often as much as 2 km above cloud base. Radiometric measurements reported by Rauber (1985) along with other supporting data showed that shallow, stable cloud systems with warm cloud tops ($-7°$ to $-20°C$) over Colorado's Park Range frequently exhibit high liquid water both upwind and over the Range, particularly in the vicinity of strong orographic lift. Model calculations reported by Rauber suggested that more than 75% of the liquid water occurs in Park Range cloud at altitudes below minimum aircraft flight altitudes. Rauber also noted that aircraft observations over the Park Range revealed significant liquid water within a few hundred meters of cloud top. Thus, at least over the Park Range, relatively warm-topped, stable orographic clouds exhibit a cloud seeding potential.

What we are seeing today is a rapid evolution of the concept of seeding orographic clouds. First of all, the scientific community is now recognizing that the actual seeding potential occurs only under specific weather regimes or types. This is quite evident in the design of "modern" experiments to test orographic cloud seeding concepts such as the Sierra Cooperative Pilot Project (SCPP; Bureau of Reclamation, 1983). In SCPP a number of precipitation echo types (PETs) are used to identify different meteorological conditions suitable for study in the experimental area and, presumably, for precipitation enhancement. Examples of the PETs are areawide (AW), orographic (O), and embedded bands (EB), which Reynolds (1984) has noted are initially selected types for cloud seeding. Another innovative feature in the SCPP design, which has been transferred from the High Plains Cooperative Program (HIPLEX; see U.S. Dept. of Interior, 1979, design concept), is to test the seeding concept by evaluating various links in the hypothesized chain of events following seeding instead of evaluating only precipitation on the ground as in the "blackbox" approach. In SCPP, liquid water content, ice particles, and ice water content are observed by aircraft during the first 18 minutes following seeding, and radar and raingage observations are used thereafter. The advantage of this approach is that the additional observations allow us to learn more about the entire meteorological response to seeding. Furthermore, we are not putting all of our "eggs in one basket;" if there exists a weak link in the chain of events, all is not lost. We can evaluate where we went wrong and, perhaps, alter the procedure such that the final desired response is obtained. Thus, this evolution of testing procedures for cloud seeding concepts is a healthy one.

Implementation of such a testing procedure has been facilitated by modern aircraft systems allowing fast, accurate computer processing of cloud microphysical data, passive radiometers, and accurate calibrated radars, especially in the K-bands.

Unfortunately, another factor in the SCPP design has prevented the full implementation of such a multiresponse function experiment. That is, whenever there is a likelihood of weather hazard occurrences such as avalanche danger and severe spring runoff, SCPP operations are curtailed. The heavy snowfall in the Sierras in the last few years has prevented the implementation of the seeding experiment.

14.3.2. Supercooled cumuli

The original concept of static seeding of supercooled cumuli differed little from that for static seeding of orographic clouds. As in orographic clouds, it was believed that ice crystal production was controlled by the availability of active ice nuclei, which varied exponentially with the degree of supercooling (i.e., the Fletcher formula). Thus, supercooled cumuli that had relatively warm cloud tops were believed to be inefficient precipitation producers. Seeding was then viewed as a possible means for increasing the efficiency of those relatively warm-topped cumuli. In both orographic clouds and cumuli, timing of the occurrence of precipitation is crucial to the precipitation output of the cloud system. In the case of orographic clouds, the time available for precipitation is controlled by the speed of the winds flowing over a barrier crest. If wind speeds are strong, precipitation must form rapidly before evaporation of condensed water commences as the air warms on its adiabatic decent along the leeward side of the barrier. In the case of cumuli, the time available for the precipitation cycle is bounded by the vigorous updraft stage of the cloud before entrainment of dry environmental air significantly reduces the amount of condensed water.

In this section, testing, evolution, and implementation of seeding concepts is reviewed in two historical experiments, namely, "Whitetop" and the "Israeli" I and II experiments. Some recent results of the analysis of the HIPLEX-I experiments are also reviewed.

Both the core project Whitetop and the Israeli experiments were "blackbox" type experiments. However, both had parallel observational studies that either supported or did not support the basic seeding concept. It is interesting that in both the Whitetop and the Israeli experiments seeding was implemented by broadcast seeding from aircraft flying near the cloud-base level. A randomized crossover design was used in the Israeli I experiment and a fixed target/control with randomized seeding by calendar days was used in the Israeli II experiments. The design in Whitetop, on the other hand, included procedures for determining a floating target and control area based upon measured winds between the surface and 14 000 ft and location of the seeding line. The results of the statistical analysis in the two experiments differed markedly. As summarized by Braham (1979), the overall effect of seeding in Whitetop was a decrease in both rainfall and radar echo cover. Braham noted that the apparent seeding effect varied with the maximum depth of clouds on any given day. If the maximum echo tops were warmer than $-10°C$ or colder than $-40°C$, the inferred seeding effects were negative. On days when the maximum echoes were between $-10°$ and $-40°C$, the target control differences suggested 60% to 100% increases in rainfall. The days with deep convective clouds, however, dominated the overall rainfall during the experiment, leading to a net decrease in rainfall due to seeding. Thus the overall statistical results of Whitetop did not support the original concept, although a possible "window" where the seeding concept may be applicable was elucidated.

Parallel observational studies during Whitetop also raised serious doubt about the validity of the "static seeding" concept. Koenig (1963) and Braham (1964) reported on observations of high concentrations of ice crystals that appeared to be adequate for natural precipitation formation and were in excess of those expected at warmer temperatures by silver iodide seeding. Koenig (1963) also noted that the rapidity of glaciation of the cloud was correlated with the presence of supercooled raindrops. Koenig hypothesized that the high concentrations of ice crystals were a result of the operation of an ice multiplication process. Independent support for these observations was reported by Mossop (1970) and Mossop and Ono (1969). Subsequent laboratory studies reported by Hallett and Mossop (1974) defined a vigorous secondary multiplication process called the rime–splinter mechanism, which operates most effectively at warm temperatures ($-3°$ to $-8°C$) and when a broad droplet spectrum is present.

Thus, the Whitetop experiment and subsequent field and laboratory experiments have shown that the original concept did not stand the "test of time" as far as its general application is concerned.

In contrast to the results of the Whitetop experiment, the Israeli experiments resulted in a confirmation of the original seeding concept. Gagin and Neumann (1981) recently described the results of the statistical analysis of the Israel I and II experiments. They found a statistically significant increase in precipitation for both experiments I and II. Moreover, in both experiments the maximum seeding effect was detected in a belt lying approximately 30 km downwind from the line of seeding. They indicated that transport and diffusion studies (Gagin and Neumann, 1974; Gagin, 1981) suggested that this distance corresponded to the region where diffusion of the silver iodide seeding material resulted in optimum concentrations.

As in the Whitetop experiment, when the results were stratified according to cloud-top temperature, the seeding effect indicated an $\sim$46% increase when convective cloud tops were in the range $-15°$ to $-20°C$. (Remember that in Whitetop 68% to 100% increases were inferred when convective cloud tops were between $-10°$ to $-40°C$.)

A distinct difference between Whitetop and Israeli experiments is that the Israeli experiments were performed in wintertime, postfrontal air masses. Hence, deep, cumulonimbus clouds of the type that overwhelmed the rainfall in Whitetop were typically absent. Moreover, the extreme continentality of the air masses (see Gagin and Neumann, 1974) and the fact that the clouds were cold-based, implied that the clouds are colloidally stable to warm rain processes. As a result, the conditions most favorable for ice multiplication processes were absent. This is consistent with the findings of Gagin (1971) that there is a good correspondence between ice nucleus concentrations and observed ice crystal concentrations.

Thus the statistical experiments and parallel observational studies in Israel have provided a satisfactory test of the original seeding concept. The fact that these experiments were carried out in cold-cloud-based, continental cumuli whose stature did not extend to full cumulonimbi obviously favored the positive outcome of the experiment. The Whitetop experiments suggest that static seeding of naturally heavily raining clouds such as cumulonimbi can lead to significant decreases in rainfall. We thus see that the concept of static seeding of supercooled cumuli has evolved into a far more restrictive form in which the conditions suitable for seeding are dependent upon the air mass properties and the overall intensity of the cloud systems.

Both the Whitetop and the Israeli experiments were "blackbox" type statistical experiments with important parallel physical studies of the cloud systems. While the physical studies were not an integral component of the statistical tests of the basic concept, they helped in the interpretation of the statistical results and placed the physical basis of the concept on a firmer scientific footing.

HIPLEX, however, included many "physical" observations as an integral component of its overall statistical design. It represents the first attempt to move from "blackbox" statistical tests of a hypothesis to a multiresponse statistical experiment in which each step leading to additional precipitation is prespecified and tested by observation during the randomized experiment (see U.S. Dept. of Interior, 1979). Thus the following responses to seeding were hypothesized:

1) Production of an average ice crystal concentration of about 10 L^{-1} in the supercooled water cloud at temperatures higher than $-10°C$. The initial ice crystal concentration in the unmixed seeding plume will be considerably higher to allow for the effects of diffusion. The average ice crystal concentration produced by seeding at these warm temperatures is higher than that found in untreated clouds at comparable times after treatment.

2) Diffusional growth of the ice crystals to a size at which riming occurs, so that higher concentrations of rimed crystals appear in seeded clouds at comparable times after treatment. The seeding-produced crystals tend to develop as columns. Crystals found at these temperatures in unseeded clouds at comparable times after treatment tend to have habits characteristic of growth at lower temperatures.

3) Accretional growth of the rimed ice crystals in the liquid portions of the cloud to graupel on the order of 1 mm in diameter and concentrations of about 0.1 L^{-1}, which then fall through the cloud. Accretional crystal growth is accompanied by a decrease in liquid water content relative to the untreated cloud at comparable times after treatment. The ice crystals produced by seeding have a significant advantage over those that occur in unseeded clouds because they originate earlier in the lifetime of the cloud. This leads to the earlier appearance of precipitating ice particles in the seeded clouds, so greater concentrations and larger sizes of such particles are present at comparable times after treatment.

4) Earlier development of first echoes in seeded clouds as opposed to untreated clouds.

5) Fall of precipitation from the cloud base in the form of graupel and/or rain (melted graupel) earlier in the lifetime of the cloud and in greater volumes than occur in unseeded clouds. In addition, a larger proportion of seeded clouds than unseeded clouds will produce rain.

In many respects the clouds seeded during HIPLEX-1 were similar to the wintertime cumuli seeded in the Israeli experiments. The HIPLEX-1 clouds were generally (but not always) cold-cloud-based, continental cumuli such that warm-rain precipitation broadening was unlikely. Moreover, the HIPLEX-1 design excluded the deep towering cumuli and cumulonimbi that overwhelmed the precipitation statistics in Whitetop. Thus, HIPLEX-1 could be thought of as a more sophisticated statistical/ physical experiment that involved the transfer of technology from the Israeli experiment. There was one important difference, however. In order to facilitate the detection of a clear seeding signal in the hypothesized sequence of events, clouds were individually seeded with dry ice from an aircraft. This should be distinguished from both Whitetop and the Israeli experiments, where seeding was implemented by airborne broadcast seeding of AgI. In HIPLEX, however, an observer first sighted a suitable cloud, then an instrumented aircraft penetrated the cloud, and then a second aircraft commenced seeding the cloud near the $-10°C$ level. Cooper (1984) noted that these maneuvers led to a 3–5 minute delay at the start of the study. During this period, the cloud had typically achieved its maximum liquid water content, the liquid water being eroded by the effects of entrainment. Cooper (1984) concluded that the short lifetimes of the natural clouds provided an important limitation to the precipitation efficiency of the clouds. As noted by Cooper, the statistical and physical evaluations of the seeding experiment indicated that seeding had little effect on precipitation although concentrations of ice and/or rimed crystals had been increased.

One might ask, therefore, why entrainment did not limit the opportunity for precipitation enhancement in the Israeli clouds. One possibility is that the difference between the two experiments was not the dynamic character of the cloud systems, but the method of implementing seeding. The Israeli broadcast seeding strategy allowed some cumuli to be affected by seeding early in the lifetime of cloud before entrainment significantly eroded the liquid water content.

Another possibility suggested by Gagin (personal communication) is that the clouds over the Israeli target area received sustained topographic lifting as the air masses moved inland from the Mediterranean Sea. As a result of this favorable dynamic forcing, it is possible that the Israeli clouds were longer-lived and, therefore, exhibited a larger "seeding window." This is supported by the cloud modeling studies of Tripoli and Cotton (1980) wherein they found that more vigorous subcloud forcing reduced the destructive effects of entrainment quite markedly. In contrast, the clouds seeded over the HIPLEX target were generally over irregular terrain, which does not provide a coherent, sustained subcloud forcing.

In summary, the concept of static seeding of supercooled cumuli has undergone considerable evolution in recent years. The clouds most suitable for "static" seeding are typically cold-cloud-based and continental, where warm-cloud precipitation processes and ice multiplication processes are not very effective. The clouds must also be of moderate depth; clouds of deep, cumulonimbus stature do not appear to be suitable for the application of the static seeding concept. Therefore, we must ask the question: Does a potential exist for enhancement of precipitation by the static seeding concept from warm-based, maritime clouds of moderate depths? We must also keep in mind that in many cumuli a relatively short "time window" exists during which the conditions are optimum for enhancing precipitation by seeding. Thus the resultant response to seeding is strongly dependent upon the actual strategy for implementing the seeding hypothesis.

14.4. Testing, implementation, and evolution of concepts for "dynamic seeding" of cumuli

The "dynamic seeding" concept differs from the "static seeding" concept in that one purposely attempts to seed a cloud in such a way that the updrafts of a cloud intensify. Ultimately, the goal of the concept is to cause an upscale growth of the cloud system such that it becomes a heavily raining cumulonimbus or even a mesoscale convective system. In this section the testing and implementation of the dynamic seeding concept is reviewed as it evolved from a single cloud concept in Simpson's earlier studies (Simpson et al., 1965; Simpson and Woodley, 1971) to the multiple cloud Florida Area Cumulus Experiment (FACE; Woodley et al., 1982; Barnston et al., 1983).

Following some early concept-development experiments (Simpson et al., 1965), a randomized single-cloud experiment was implemented in 1965 in which a number of observational aircraft along with a seeder aircraft were used to evaluate the response of clouds to massive dosages of seeding material. This experiment demonstrated that seeded clouds grew an average of 1.6 km higher than unseeded clouds. An important component in the quantitative description of the concept was the use of a one-dimensional Lagrangian cloud model (see Simpson et al., 1967; Simpson and Wiggert, 1969, 1971). The model was used to define "seedability," which is the difference between the predicted maximum cloud-top height for seeded and unseeded clouds. Moreover, the model became an integral part of the statistical analysis of rainfall from seeded clouds (Simpson and Woodley, 1971; Woodley, 1970). Using radar to assess the rainfall from individual seeded and unseeded clouds, the analysis of this randomized experiment suggested that rainfall from seeded clouds exceeded that from nonseeded clouds by a factor of three or greater.

As noted above, a crucial component in the dynamic seeding concept is a one-dimensional Lagrangian cloud model in which entrainment occurs solely through the cloud's lateral boundaries. Thus, testing the credibility of the model is basic to testing the concept of dynamic seeding. Early experiments with the model showed that it was a good predictor of the maximum heights of seeded and natural cumuli (Simpson et al., 1965; Simpson and Woodley, 1971). However, Warner (1970) claimed that

lateral entrainment models were unable to simultaneously predict the observed cloud-top heights and the sampled liquid water contents. Subsequently, Cotton (1975) confirmed Warner's findings. Evidence was also accumulating that the entrainment process is not principally lateral mixing, but top mixing by penetrative downdrafts (Squires, 1958; Paluch, 1979; Betts, 1982). All of the preceding studies indicate that the lateral entrainment model consistently overpredicts cloud liquid water content while achieving observed heights. They therefore cast considerable doubt upon the fundamental basis of the dynamic seeding concept, since the energy utilized in obtaining invigorated growth of a cloud must come from the freezing of condensed supercooled liquid water. The model predictions of "seedability" would thus be expected to be greatly exaggerated.

However, there is increasing evidence that a small fraction of the area of a growing cumulus (on the order of 10%) remains relatively unmixed along the upshear flank of the cloud. This is supported by observational studies (Heymsfield et al., 1978) and numerical model cloud simulations (Cotton and Tripoli, 1978, 1979; Tripoli and Cotton, 1980). It thus appears that cloud-top height is controlled by the buoyancy in the "protected" portion of the updrafts. Thus, one-dimensional lateral entrainment models may be useful predictors of the behavior of the protected cores even though this is not consistent with the theoretical formulation of the model.

A major deficiency of the one-dimensional Lagrangian cloud model is that it is not able to predict the dynamic response to seeding below the level of artificial glaciation. It is not surprising, therefore, that the stated concept of communication of artificially generated buoyancy aloft to the moist, subcloud layer was particularly vague during the 1960s and early 1970s. Table 14.1 illustrates the hypothesized chain of physical processes linking dynamic

TABLE 14.1. Summary of dynamic seeding hypothesis chain (Simpson, 1980).

1) Silver iodide is introduced at approximately the $-10°C$ level in the cumulus clouds, i.e., in a region where there is believed to be a significant amount of supercooled liquid water.

2) This seeding results in conversion of water to ice, with resultant release of latent heat of fusion (~ 80 cal g^{-1}), producing increased buoyancy. Additional buoyancy is believed to be produced by depositional heating (~ 680 cal g^{-1}) associated with the deposition of water vapor directly onto ice crystals, resulting from the fact that the saturation vapor of ice is less than that of water.

3) This buoyancy produces an increase in the updraft, which is transferred all the way down to the bottom of the cloud.

4) This produces an increase in the inflow of moist air into the bottom of the cloud.

5) This increased inflow of moisture eventually results in more rainfall.

6) By appropriate seeding, neighboring clouds can be caused to merge.

7) The increased size of the merged cloud systems results in increased total rainfall.

seeding to enhanced low-level inflow during this period. During the latter part of the 1970s (Woodley and Sax, 1976; Simpson, 1980), attempts were made to define more explicitly the concept of dynamic seeding in terms of the expected chain of physical responses to artificial glaciation aloft. Table 14.2 shows the elaborate concept of dynamic seeding that still remains to this day.

There has never been an attempt to carry out an experiment (randomized or nonrandomized) designed to test all the links in the hypothesized chain of events. This was largely due to administrative pressures to get on with the business of showing that the concept leads to enhanced rainfall on the ground. In other words, either the main thrust of the experiment was a "blackbox" experiment or

TABLE 14.2. Modified summary of dynamic seeding hypothesis chain (Simpson, 1980).

Stage I: Initial Growth

1) Rapid glaciation of the updraft regions of supercooled convective towers by silver iodide pyrotechnic seeding.

2) Invigoration of the updrafts through the release of latent heats of fusion and deposition, the latter occurring as the cloud air approaches saturation relative to ice.

3) Enhanced tower growth is associated with a pressure fall below cloud, resulting in low-level inflow. At about the same time, strengthened dynamic entrainment (Simpson, 1976) into the cloud occurs just below the invigorated rising tower. The increased inflow of drier air increases evaporation of the liquid water falling from the rising seeded tower, which in turn accelerates and strengthens downdraft processes. This combination of events comprises the initial stage of explosive cloud growth.

Stage II: Enhanced Downdrafts and Secondary Growth
(Duration 30–50 min)

4) Enhanced downdrafts below the invigorated seeded tower as the precipitation and the evaporatively cooled air entrained into the tower move downward. This results in convergence at the interface between the downdraft and the ambient flow, in the growth of secondary towers (which in turn might be seeded), and in the expansion of the cloud system. This is the second stage of explosive cloud growth.

The second stage of explosion involves gust-front forcing of new growth and major explosion on the downshear flank. Location of main expansion/new tower growth may differ depending on the wind profile.

Stage III: Interaction with Neighboring Clouds

5) Seeding of secondary towers in the parent cloud results in their growth, followed by expansion and intensification of the downdraft area, which then moves outward to interact with outflows from neighboring clouds (which also might have been seeded). With the proper ambient conditions, carefully timed seeding might encourage merger by capitalizing on the tendency of the two cumulonimbi of different life cycle stages to approach each other.

6) Accelerated/increased merging, together with larger merged systems, increases the mesoscale convergence, resulting in new cloud growth available for seeding.

Stage IV: Increased Area Rainfall

7) Augmented and more efficient processing of the available moisture from the larger, more organized, seeded cloud systems results in increased rainfall.

8) Increased rainfall over the entire target (assuming the absence of compensatory rainfall decreases in the unseeded portions of the target).

no program at all! This author attempted to carry out an exploratory subexperiment in the 1981 CCOPE to test the various links in the chain. However, it became clear that participants in CCOPE did not want to be "contaminated" by exploratory seeding experiments. In all fairness, it is not obvious today that we have the technology to implement an experiment that can examine Stages I and II outlined in Table 14.2 even with an experiment of the scope of CCOPE dedicated to testing the concept. For one thing, the scanning rates of current Doppler radars are probably too slow to detect direct dynamic communication responses to low levels. As a result, there have been only a few isolated attempts to test the validity of various components in the hypothesized chain of events. Sax et al. (1979) examined the first step in Table 14.2 and showed that pyrotechnic seeding of active cumulus congestus towers over southern Florida does lead to rapid glaciation. As noted previously, the early experiments reported by Simpson et al. (1967) and Simpson and Woodley (1971) did show that seeding leads to enhanced tower growth. However, beyond step 2 (Table 14.2) experiments have not satisfactorily tested the chain of hypothesized responses.

Numerical prediction models have also been used to examine the dynamic seeding concept (Orville and Chen, 1982; Nehrkorn, 1981; Levy, 1982; Levy and Cotton, 1984). Thus far they have not fully verified that the hypothesized chain of events is physically realistic. Orville and Chen (1982) found that the dynamic seeding hypothesis and related chain of events did not hold up in their two-dimensional simulations of seeding. They noted that the onset of precipitation influenced subsequent cloud development in a destructive way. Using a three-dimensional cloud model Nehrkorn (1981) and Levy (1982) also found that locally enhanced rainfall impeded further cloud development. However, Levy (1982) showed that in a slightly different wind field, the enhanced rainfall did not impede further cloud development. Unfortunately, under those conditions the cloud grew so efficiently that a seeding response could not be detected in any significant way. It is thus fair to say that two- and three-dimensional models have not satisfactorily tested the dynamic seeding concept at this time. (For more elaboration on the subject see Cotton, 1984; Levy and Cotton, 1985). This suggests either that the concept is incorrect or that current numerical-prediction cloud models are unable to correctly simulate a seeding response. The latter could be a result of limitations to two dimensions, limited domain and course resolution in three dimensions, and/or the inadequacies of the microphysics and turbulence closure theories used in the models.

As a result we have mainly resorted to "blackbox" type experiments to test the dynamic seeding concept. Based on observational studies that suggested that merged cloud systems were responsible for 86% of the rainfall over an area even though 90% of the clouds were unmerged and that seeded clouds frequently participated in the merger process, the Florida Area Cumulus Experiment (FACE) was initiated to test the concept that dynamic seeding could promote merger and, hence, enhance precipitation over a target area. The results of the statistical analysis of the FACE-1 and -2 experiments have been reported in the literature (Woodley et al., 1982; Barnston et al., 1983; Woodley et al., 1983). Throughout the FACE-1 and -2 experiments, rainfall on seed days exceeded rainfall on no-seed days. However, the magnitude of the enhancement was not large enough to satisfy the criteria that FACE-2 confirm the FACE-1 results. Thus we cannot be sure that the enhancement is real. Herein lies the problem with a pure "blackbox" experiment. If just one weak link exists in the hypothesized chain of physical responses, all is lost without our knowing which was the weak link. Had we observed and tested the various links in the chain, we might have been able to detect the problem and alter our strategies and experimental design to overcome this weakness. As it stands the results of FACE-1 and -2 are particularly unsatisfying!

One additional factor should be considered about implementation of the dynamic seeding concept over a fixed area. In the case of FACE, seeding was performed over a 4000 (n mi)^2 target area. In order to introduce large quantities of seeding material into actively rising cumulus towers, a seeding aircraft had to penetrate the individual cloud towers. Broadcast seeding would not be expected to be effective for this strategy. As a result it became quite clear that a single aircraft or even two or three aircraft could not adequately seed all candidate clouds over the given area. Many clouds over the target area were not seeded, thus their contribution to the fixed target rainfall diluted any real seeding effect if it existed. This was one of the reasons that Woodley et al. (1982) introduced a floating target concept in their analysis of FACE. This clearly illustrates one of the problems of implementing the dynamic seeding concept over a fixed target area. To implement the concept effectively, one needs a fleet of high-performance jet aircraft such that candidate clouds can be selected and seeded before entrainment gets the upper hand on the cloud and enhanced buoyancy simply contributes to turbulent dissipation. The use of such high-performance aircraft would allow more complete coverage of the target area as well. This would be a very expensive proposition, indeed!

14.5. Concluding remarks

In this paper, testing, implementation, and evolution of both static and dynamic seeding concepts have been reviewed. It is encouraging that our seeding concepts have evolved through the years as scientific investigations have elucidated a greater understanding of the physics and dynamics of cloud systems. It is also encouraging that our testing procedure has evolved from a single-response-variable "blackbox" experiment to randomized experiments that attempt to test a number of components in the hypothesized chain of physical responses to seeding. This procedure is not without pitfalls, however. Changes in the seeding strategy to optimize detecting a response in the intermediate links in the chain of responses may have an adverse effect upon the bottom line response, namely, rainfall on the ground. This possibility has to be seriously considered in the design of any future weather modification experiments.

It is also strongly urged that before we invest in major field experiments to test a weather modification concept, we thoroughly test the concept by computer simulation. We have the computer technology today to thoroughly examine the physical consistency of a concept. This may require several years of computer-code development plus the expenditure of large amounts of computer time, but in this author's opinion, we would be in a much better position to design laboratory and field experiments that can definitely test the concept. It should be emphasized that I am not talking about computer models that use simplified physics and dynamics such as we have been forced to use in the past. I am talking about sophisticated models of turbulent transport and diffusion, cloud microphysics, and cloud dynamics that are in the "spirit" of large eddy simulation (LES) models of boundary layer transport processes (i.e., Deardorff, 1978, 1980; Sommeria and Lemone, 1978).

Acknowledgments. The author would like to thank Brenda Thompson for her assistance in obtaining reference material and typing the manuscript.

REFERENCES

Baker, M. B., R. G. Corbin and J. Latham, 1980: The influence on the evolution of cloud droplet spectra: I. A model of inhomogeneous mixing. *Quart. J. Roy. Meteor. Soc.,* **106,** 581–598.

Barnston, A. G., W. L. Woodley, J. A. Flueck and M. H. Brown, 1983: The Florida Area Cumulus Experiment's second phase (FACE-2). Part I: The experimental design, implementation, and basic data. *J. Appl. Meteor.,* **22,** 1504–1528.

Betts, A. K., 1982: Saturation point analysis of moist convective overturning. *J. Atmos. Sci.,* **39,** 1484–1505.

Biswas, K. R., and A. S. Dennis, 1971: Formation of rain shower by salt seeding. *J. Appl. Meteor.,* **10,** 780–784.

——, and ——, 1972: Calculations related to formation of a rain shower by salt seeding. *J. Appl. Meteor.,* **11,** 755–760.

——, R. K. Kapoor and K. K. Kanuga, 1967: Cloud seeding experiment using common salt. *J. Appl. Meteor.,* **10,** 780–784.

Bowen, E. G., 1950: Formation of rain by coalescence. *Aust. J. Sci. Res.,* **3,** 193–213.

——, 1952: A new method of stimulating convective clouds to produce rainfall and hail. *Quart. J. Roy. Meteor. Soc.,* **78,** 37–45.

Braham, R. R., Jr., 1964: What is the role of ice in summer rainshowers? *J. Atmos. Sci.,* **20,** 386–391.

——, 1979: Field experimentation in weather modification. *J. Amer. Stat. Assoc.,* **74,** 57–68.

——, L. J. Battan and H. R. Byers, 1957: Artificial nucleation of cumulus clouds, clouds and weather modification: A group of field experiments. *Cloud and Weather Modification, Meteor. Monogr.,* No. 11, 47–85.

Bureau of Reclamation, 1983: The design of SCPP-1, second revision. Bureau of Reclamation, Division of Atmospheric Resources Research D-1200, Denver, 61 pp.

Chappell, C. F., 1970: Modification of cold orographic clouds. Atmos. Sci. Paper No. 173 (Ph.D. dissertation), Dept. of Atmos. Sci., Colorado State University.

Chatterjee, R. N., K. R. Biswas and Bh. Ramana Murty, 1969: Result of cloud seeding experiments at Delhi as assessed by radar. *Indian J. Meteor. Hydrol. Geophys.,* **22,** 11–16.

Coons, R. D., E. L. Jones and R. Gunn, 1949: Second partial report of the artificial production of precipitation—cumuloform clouds—Ohio, 1948. U.S. Weather Bureau Res. Paper 31, Washington, DC.

Cooper, W. A., 1984: Accretional growth processes in seeded and unseeded clouds of HIPLEX-1. *Preprints Ninth Conf. on Weather Modification,* Park City, Amer. Meteor. Soc.

——, and J. D. Marwitz, 1980: Winter storms over the San Juan Mountains. Part III: Seeding potential. *J. Appl. Meteor.,* **19,** 942–949.

——, and C. P. R. Saunders, 1980: Winter storms over the San Juan Mountains. Part II: Microphysical processes. *J. Appl. Meteor.* **19,** 925–941.

Cotton, W. R., 1975: On parameterization of turbulent transport in cumulus clouds. *J. Atmos. Sci.,* **32,** 548–564.

——, 1982: Modification of precipitation from warm clouds: A review. *Bull. Amer. Meteor. Soc.,* **63,** 146–160.

——, 1984: In search of communication mechanisms linking glaciogenic seeding to the boundary layer. *Preprints Ninth Conf. on Planned and Inadvertent Weather Modification,* Park City, Amer. Meteor. Soc.

——, and G. J. Tripoli, 1978: Cumulus convection in shear flow—three-dimensional numerical experiments. *J. Atmos. Sci.,* **35,** 1503–1521.

——, and ——, 1979: Reply. *J. Atmos. Sci.,* **36,** 1610–1611.

Deardorff, J. W., 1978: Efficient prediction of ground surface temperature and moisture, with inclusion of a layer of vegetation. *J. Geophys. Res.,* **83,** 1889–1093.

——, 1980: Cloud top entrainment instability. *J. Atmos. Sci.,* **37,** 131–147.

Dennis, A. S., and A. Koscielski, 1972: Height and temperature of first echoes in unseeded and seeded convective clouds in South Dakota. *J. Appl. Meteor.,* **11,** 994–1000.

Farley, R. C., and C. S. Chen, 1975: A detailed microphysical simulation of the hygroscopic seeding on the warm rain process. *J. Appl. Meteor.,* **14,** 718–733.

Fletcher, N. H., 1962: *Physics of Rain Clouds.* Cambridge University Press.

——, 1968: Ice nucleation behavior of silver iodide smokes containing a soluble component. *J. Atmos. Sci.,* **25,** 1058–1060.

Fournier d'Albe, E. M., and P. Mosino Aleman, 1976: A large-scale cloud seeding experiment in the Rio Nazas Catchment Area, Mexico. *Proc. Second WMO Scientific Conf. on Weather Modification,* Boulder, WMO 143–149.

Furman, R. W., 1967: Radar characteristics of wintertime storms in the Colorado Rockies. Atmos. Sci. Paper No. 112, Dept. of Atmos. Sci., Colorado State University, 53 pp.

Gagin, A., 1971: Studies of the factors governing the colloidal stability of continental cumulus clouds. *Proc. Canberra Int. Weather Modification Symp.*

——, 1981: The Israeli rainfall enhancement experiments. A physical overview. *J. Wea. Mod.,* **13**(1).

——, and J. Neumann, 1974: Rain stimulation and cloud physics in Israel. *Weather and Climate Modification,* W. N. Hess, Ed., Wiley and Sons, 454–494.

——, and ——, 1981: The second Israeli randomized cloud seeding experiment: Evaluation of the results. *J. Appl. Meteor.,* **20,** 1301–1311.

Grant, L. O., and P. W. Mielke, Jr., 1967: A randomized cloud seeding experiment at Climax, Colorado, 1960–1965. *Proc. Fifth Berkley Symp. on Mathematical Statistics and Probability,* Vol. 5, University of California Press, 115–131.

——, and A. M. Kahan, 1974: Weather modification for augmenting orographic precipitation. *Weather and Climate Modification,* W. N. Hess, Ed., Wiley and Sons, 282–317.

Hallett, J., and S. C. Mossop, 1974: The production of secondary ice particles during the riming process. *Nature,* **249,** 26–28.

Havens, B. S., J. E. Jiusto and B. Vonnegut, 1978: Early history of cloud seeding. New Mexico Tech. Press for the Geophys. Res. Ctr., New Mexico Institute of Mining and Technology.

Heggli, M. F., L. Vardiman, R. E. Stewart and A. Huggins, 1983: Supercooled liquid water and ice crystal distributions within Sierra Nevada winter storms. *J. Climate Appl. Meteor.,* **22,** 1875–1886.

Heymsfield, A. J., D. N. Johnson and J. E. Dye, 1978: Observations of moist adiabatic ascent in northeast Colorado cumulus congestus clouds. *J. Atmos. Sci.,* **35,** 1689–1703.

Hill, G. E., 1974: Factors controlling the size and spacing of cumulus clouds as revealed by numerical experiments. *J. Atmos. Sci.,* **31,** p. 646.

——, 1984: The science and technology of winter mountain cloud modification. *Preprints Ninth Conf. on Weather Modification,* Park City, Amer. Meteor. Soc.

Hobbs, P. V., 1969: Ice multiplication in clouds. *J. Atmos. Sci.,* **26,** 315–318.

——, and A. L. Rangno, 1979: Comments on the Climax and Wolf Creek Pass cloud seeding experiments. *J. Appl. Meteor.,* **18,** 1233–1237.

——, L. F. Radke, J. R. Fleming and D. G. Atkinson, 1975: Airborne ice nucleus and cloud microstructure measurements in natural and artificially seeded situations over the San Juan Mountains of Colorado. Res. Rep. X, Contributions from the Cloud Physics Group, University of Washington, 89 pp.

Houze, R. A., Jr., J. D. Locatelli and P. V. Hobbs, 1976a: Dynamics and cloud microphysics of the rainbands in an occluded frontal system. *J. Atmos. Sci.,* **33,** 1921–1936.

——, P. V. Hobbs, K. R. Biswas and W. M. Davis, 1976b: Mesoscale rainbands in extratropical cyclones. *Mon. Wea. Rev.,* **104,** 868–878.

Johnson, D. B., 1980: Hygroscopic seeding of convective clouds. SWS Contract Rep. 244, Illinois State Water Survey, Meteorology Section, 222 pp.

Kapoor, R. K., K. Krishna, R. N. Chatterjee, A. S. R. Murty, S. K. Sharma and Bh. V. Ramana Murty, 1976: An operational rain simulation experiment using warm technique over Rihand Catchment in northeast India during summer monsoons of 1973 and 1974. *Proc. Second WMO Scientific Conf. on Weather Modification,* Boulder, WMO, 15–20.

Khemani, L. T., G. A. Monin, M. S. Naik and A. S. R. Murty, 1981: Chloride and sodium increases in rain form salt seeded clouds. *J. Wea. Mod.,* **13,** 182–183.

Klazura, G. E., and C. J. Todd, 1978: A model of hygroscopic seeding in cumulus clouds. *J. Appl. Meteor.,* **17,** 1758–1768.

Koenig, L. R., 1963: The glaciating behavior of small cumulonimbus clouds. *J. Atmos. Sci.,* **20,** 29–47.

Langmuir, I., 1948: The growth of particles in smokes and clouds and the production of snow from supercooled clouds. *Proc. Amer. Phil. Soc.,* **92,** p. 167.

Levy, G., 1982: Communication mechanisms in dynamically seeded cumulus clouds. Atmos. Sci. Paper No. 357, Dept. of Atmos. Sci., Colorado State University, 142 pp.

——, and W. R. Cotton, 1984: A numerical investigation of mechanisms linking glaciation of the ice-phase to the boundary layer. *J. Climate Appl. Meteor.*, **23**, 1505–1519.

Ludlam, F. H., 1951: The production of showers by the coalescence of cloud droplets. *Quart. J. Roy. Meteor. Soc.*, **77**, 402–417.

——, 1955: Artificial snowfall from mountain clouds. *Tellus*, **7**, 277–290.

Manton, M. J., 1979: On the broadening of a droplet distribution by turbulence near cloud base. *Quart. J. Roy. Meteor. Soc.*, **105**, 899–914.

Marwitz, J. D., W. A. Cooper and C. P. R. Saunders, 1976: Structure and seedability of San Juan storms. Final Rep. to the Bureau of Reclamation, College of Engineering, University of Wyoming, 326 pp.

Mason, B. J., 1971: *The Physics of Clouds.* 2nd ed., Clarendon Press, 671 pp.

Mielke, P. W., Jr., L. O. Grant and C. F. Chappell, 1971: An independent replication of the Climax wintertime orographic cloud seeding experiment. *J. Appl. Meteor.*, **21**, 788–792.

——, G. W. Brier, L. O. Grant, G. J. Mulvey and P. N. Rosensweig, 1981: A statistical reanalysis of the replicated Climax I and II wintertime orographic cloud seeding experiments. *J. Appl. Meteor.*, **20**, 643–660.

Mossop, S. C., 1970: Concentrations of ice crystals in clouds. *Bull. Amer. Meteor. Soc.*, **51**, 474–478.

——, and A. Ono, 1969: Measurements of ice crystal concentration in clouds. *J. Atmos. Sci.*, **26**, 130–137.

Murty, Bh. V. Ramana, and K. R. Biswas, 1968: Weather modification in India. *Proc. First Natl. Conf. on Weather Modification*, Albany, Amer. Meteor. Soc., 71–80.

Nehrkorn, T., 1981: A three-dimensional simulation of the dynamic response of a Florida cumulus to seeding. M.S. thesis, Colorado State University, 99 pp.

Nelson, L. D., 1971: A numerical study on the initiation of warm rain. *J. Atmos. Sci.*, **28**, 752–762.

——, and B. A. Silverman, 1972: Optimization of warm-cloud seeding agents by microencapsulation techniques. *Mon. Wea. Rev.*, **100**, 153–158.

Orville, H. D., and J.-M. Chen, 1982: Effects of cloud seeding, latent heat of fusion and condensate loading on cloud dynamics and precipitation evolution: A numerical study. *J. Atmos. Sci.*, **39**, 2807–2827.

Paluch, I. R., 1979: The entrainment mechanism in Colorado cumuli. *J. Atmos. Sci.*, **36**, 2462–2478.

Ramachandra Murty, A. S., A. M. Selvam and Bh. V. Ramana Murty, 1975: Summary of observations indicating dynamic effect of salt seeding in warm cumulus clouds. *J. Appl. Meteor.*, **14**, 629–637.

Rauber, R. M., 1985: The physical structure of northern Colorado River basin cloud systems. Ph.D. dissertation, Colorado State University.

——, D. Feng, L. O. Grant and J. B. Snider, 1986: The characteristics of cloud water over the mountains of northern Colorado during wintertime storms. Part I: Temporal variations. *J. Climate Appl. Meteor.*, **25**, 468–488.

Reid, J. D., 1976: Dispersion in a mountain environment. Atmos. Sci. Paper No. 253, Dept. of Atmos. Sci., Colorado State University, 150 pp.

Reynolds, D. W., 1984: Progress and plans of the Sierra Cooperative Pilot Project. *Preprints Ninth Conf. on Weather Modification*, Park City, Amer. Meteor. Soc.

Rokicki, M. L., and K. C. Young, 1978: The initiation of precipitation updrafts. *J. Appl. Meteor.*, **17**, 745–754.

Roy, A. K., Bh. V. Ramana Murty, R. C. Srivastava and L. T. Khemani, 1961: Cloud seeding trials at Delhi during monsoon months, July to Sept. (1957–1959). *Indian J. Meteor. Hydrol. Geophys.*, **12**, 401–412.

Sax, R. I., J. Thomas and M. Bonebrake, 1979: Ice evolution within seeded and nonseeded Florida cumuli. *J. Appl. Meteor.*, **18**, 203–214.

Simpson, J., 1980: Downdrafts as linkages in dynamic cumulus seeding effects. *J. Appl. Meteor.*, **19**, 477–487.

——, and V. Wiggert, 1969: Models of precipitating cumulus tower. *Mon. Wea. Rev.*, **97**, 471–489.

——, and ——, 1971: 1968 Florida Cumulus Seeding Experiment: Numerical model results. *Mon. Wea. Rev.*, **99**, 87–118.

——, and W. L. Woodley, 1971: Seeding cumulus in Florida: New 1970 results. *Science*, **172**, p. 117.

——, and A. S. Dennis, 1972: Cumulus clouds and their modification. NOAA Tech. Memo. ERLOD-14, Washington, DC, 148 pp.

——, R. H. Simpson, D. A. Andrews and M. A. Eaton, 1965: Experimental cumulus dynamics. *Rev. Geophys. Space Phys.*, **3**, 387–431.

——, G. W. Brier and R. H. Simpson, 1967: Stormfury cumulus seeding experiments 1965: Statistical analysis and main results. *J. Atmos. Sci.*, **24**, 508–521.

Sommeria, G., and M. A. Lemone, 1978: Direct testing of a three-dimensional model of the planetary boundary layer against experimental data. *J. Atmos. Sci.*, **35**, 25–39.

Squires, P., 1958: Penetrative downdraft in cumuli. *Tellus*, **10**, 381–389.

Tripoli, G. J., and W. R. Cotton, 1980: A numerical investigation of several factors contributing to the observed variable intensity of deep convection over south Florida. *J. Appl. Meteor.*, **19**, 1037–1063.

U.S. Dept. of Interior, 1979: The design of HIPLEX-1: A randomized rain augmentation experiment on summer cumulus congestus clouds on the Montana High Plains. Office of Atmospheric Resources Management, Div. of Research.

Warner, J., 1970: The microstructure of cumulus clouds. Part III: The nature of the updraft. *J. Atmos. Sci.*, **27**, 682–688.

——, 1973: The microstructure of cumulus clouds. Part V: Changes in droplet size distribution with cloud age. *J. Atmos. Sci.*, **30**, 1724–1726.

Woodley, W. L., 1970: Precipitation results from a pyrotechnic cumulus seeding experiment. *J. Appl. Meteor.*, **9**, 242–257.

——, and R. I. Sax, 1976: The Florida Area Cumulus Experiment: Rationale, design, procedures, results, and future course. NOAA Tech. Rep. ERL 354-WMPO 6, 204 pp.

——, J. Jordan, J. Simpson, R. Biondini, J. Flueck and A. Barnston, 1982: Rainfall results of the Florida Area Cumulus Experiment, 1970–1976. *J. Appl. Meteor.*, **21**, 139–164.

——, A. Barnston, J. A. Flueck and R. Biondini, 1983: The Florida Area Cumulus Experiment's second phase (FACE-2). Part II: Replicated and confirmatory analyses. *J. Climate Appl. Meteor.*, **22**, 1529–1540.

CHAPTER 15

An Engineer's View on the Implementation and Testing of Seeding Concepts

PAUL L. SMITH

South Dakota School of Mines and Technology, Rapid City, South Dakota

ABSTRACT

Comments are made on opportunity recognition, treatment, and evaluation aspects of the implementation and testing of seeding concepts. The main topics include experimental design, experimental units, delivery and dispersion of seeding agents, and statistical evaluation procedures.

15.1. Introduction

Engineers deal mainly with how to do things. The implementation and testing of seeding concepts involve three main activities:

(i) Opportunity recognition
(ii) Treatment
(iii) Evaluation.

The comments herein concern a variety of "how to" topics related to selected aspects of each activity. However, they involve broad procedural matters rather than items of the nuts and bolts type.

15.2. Opportunity recognition

In weather modification experiments, opportunity recognition is roughly equivalent to the selection of experimental units. This selection is normally made on the basis of guidelines developed from prior exploratory studies. Operational seeding requires similar selection guidelines, but they generally have to be applied on the basis of much less information. Opportunity recognition is an acknowledged problem in winter orographic seeding, but it is also a difficult issue for cumulus cloud modification.

Kempthorne (1980), thinking of the complex behavior of clouds, spoke rather wistfully of the idea of a "pig in the sky" experimental unit. In comparing biomedical and weather modification experiments, people may argue about whether pigs are more alike than clouds, but at least a pig is an entity with fairly well defined boundaries. One can tell what's pig and not pig, and pigs don't merge with other pigs, or dissipate. They also change slowly in comparison to the duration of many of the experiments one might wish to perform on them. Clouds should grant us such luxury!

There seem to be three possibilities for the experimental units in weather modification:

(a) They should be essentially alike; or
(b) We should be able to measure the important differences; or
(c) We should be prepared to wait a long time for the results of black-box-type randomized experiments.

Experimental approaches involving each of the three possibilities have been employed, generally with limited success. Experiments treating "matched pairs" of clouds are based on possibility (a), but assurances that the paired clouds are *essentially* alike are hard to come by. Stratification procedures or cloud selection criteria like those used in HIPLEX-1 (Smith et al., 1984) are other ways of attempting to obtain similar experimental units. Attempts to develop covariates or predictor variables are based on possibility (b), but usually one either does not know or is unable to measure all of the important differences. Experiments based on possibility (c) run afoul of the long time required to get meaningful results and the difficulty of determining what may have gone wrong (or right) when the results are confusing or implausible.

Several remarks seem in order here. First, one's ideas about establishing similarity among experimental units are necessarily tied to some conceptual model. For example, suppose one seeks clouds that have similar liquid water concentration (LWC). Apart from such questions as whether to use "point" or average values and what range of values may constitute "similar" clouds, the matter of where (and when) to measure the LWC depends largely on one's conceptual model. Consider the following possible viewpoints concerning aircraft measurements of LWC:

1) Temperature is the important factor; therefore, LWC should be measured at, say, the $-5°C$ level.
2) Depth of cloud is the important factor; therefore, LWC should be measured at, say, 1.5 km above cloud base.

3) Particle growth time is the important factor; therefore, LWC should be measured at a level corresponding to, say, 3 min of ascent above cloud base (in the observed updraft).

The procedures for establishing cloud similarity will vary depending upon the particular viewpoint adopted, and clouds that are similar in one respect may not be so similar in others.

Second, our measurements are governed as much by what we know how to measure as by what we ought to measure. Continuing the above example, knowledge of the three-dimensional distribution of cloud liquid water as a function of time would probably be needed to assess fully the similarity in LWC of different clouds. Disregard for the moment the unlikelihood that any two clouds would ever be identical, even in LWC alone, in all four dimensions. With aircraft, we know how to measure one-dimensional distributions of LWC at intervals of time. Therefore, we attempt to assess the similarity among clouds by using a limited sampling strategy based on the existing observing capability.

Third, the similarities and differences among clouds may be multivariate. For example, if coalescence growth is an important factor, a cloud with high LWC and strong updrafts could be equivalent, at least in some sense, to another with lower LWC and weaker updrafts.

Finally, in practice, achieving timely opportunity recognition is a major problem. The basic question is, how does one recognize an opportunity soon enough to apply the treatment in time to have the desired effect? This question is especially important in situations where the duration of any "seeding window" can be short. The matter of timing was a significant problem in HIPLEX-1 (Cooper and Lawson, 1984), and Dr. Gagin presented evidence at the workshop suggesting that rainfall increases occurred in FACE mainly in cases where the seeding was initiated within 5 min of the appearance of the first radar echo.

There is great hope that remote sensing methods will help with the timing as well as with volume sampling, but I sense that too many people use remote sensing without having a sound understanding of what is being measured. A common characteristic of remote sensors, and of electronic instruments in general, is that if one is turned on it will point to, say, 72. (Technology has really advanced, because we now get a digital indication of 72.) What the 72 really means or whether, in fact, it means anything at all, can be a difficult question.

15.3. Treatment

Apart from the matter of timing discussed above, the major issues in applying treatments include what kind of seeding material to use, how much to use, and how to deliver it to the intended places in the clouds. The subject of the kind and amount of agent desirable seems to involve a continuing, and sometimes circular, story. The length of time required for nucleation scientists, having discovered an "improved" formula, to decide after much testing that it really doesn't nucleate so well after all, appears to be comparable to the time needed to plan, execute, and evaluate a seeding experiment using the formula. The history of cloud seeding experiments, especially of the black-box type, has certainly been confused by shifting assessments of the efficacy of the seeding agents employed.

The concept of an optimum ice nucleus concentration ought to provide a useful benchmark against which to evaluate candidate seeding methods, but no one seems very eager to specify the optimum. Recent laboratory and field work shows that dry ice seeding can produce too many ice crystals for efficient production of precipitation. The same would seem to be true for silver iodide, but in view of uncertainties about the mechanisms and rates of nucleation of the different formulas, we may, in practice, not really know how many ice crystals silver iodide seeding will produce or where they will appear. For seeding experiments, good arguments can be made that dry ice seeding may be a better way to produce relatively certain quantities of ice crystals in relatively known places in the clouds.

This, of course, presumes that the problem of delivering the seeding material to the desired parts of the clouds can be solved. Statisticians have emphasized the importance of being sure that the treatments are actually applied to the experimental units as specified in the experimental design (e.g., Kempthorne, 1980). The issue of where the desired points are located is essentially one of opportunity recognition rather than treatment and is especially troublesome with orographic clouds. Dry ice has to be dropped from above, which presents logistical difficulties that are usually surmountable for experimental purposes. They may, however, be a more serious obstacle to large-scale operational dry ice seeding.

If silver iodide is dispensed from ground generators, serious problems arise with respect to the transport and diffusion of the material, especially in winter orographic storm seeding. If the silver iodide is dispensed from aircraft at cloud base into convective clouds, there are questions about the dispersion of the agent. Attempts to associate observed ice crystals with cloud-base seeding in Montana were not very successful (Hobbs and Politovich, 1980). Results from the National Hail Research Experiment (Linkletter and Warburton, 1977) suggested that the seeding agent often ascended through the clouds in a ribbon pattern instead of spreading throughout the updraft region as desired. Diffusion calculations using the turbulence intensities observed inside NHRE storms (Sand et al., 1976) also offer no encouragement about dispersion of the agent. Dropping the seeding material along a flight line through or above the clouds (using pyrotechnics, if the agent is silver iodide) may be more effective. The observations with droppable pyrotechnics discussed in Hobbs and Politovich (1980) are rather encouraging in this respect, as are the HIPLEX investigations of the diffusion of ice crystals produced by dropping dry ice pellets (Holroyd et al., 1979).

For all of these problems, the question of how operators can apply the treatments over large target areas should also be considered. The transfer of the experimental technology to operations will require suitable answers.

15.4. Evaluation

As noted at the workshop by Dr. Flueck, evaluation really begins with the project (or experiment) design. The most difficult problem in evaluation is to establish a design that asks the right questions in the right ways. Black-box-type cloud seeding experiments have, for the most part, proved to be inadequate, although they might be more satisfactory if they gave more conclusive results.

One important aspect of the formulation of an experimental design is the distinction between a conceptual model and an experimental hypothesis. A conceptual model is needed to provide a framework for developing the design, but such a model is not equivalent to an experimental hypothesis. The distinction can be illustrated by comparing the hypothesis statements for FACE (e.g., Woodley et al., 1982) and HIPLEX-1 (Smith et al., 1984).

The FACE statement is a conceptual model, containing general statements to the effect that latent heat released by seeding invigorates the updrafts, downdrafts are enhanced, mergers are promoted, and so forth. It does not provide much indication of when, where, or how the hypothesized effects are to be observed. For example, updrafts vary with position and time, even in unseeded clouds, and differ from cloud to cloud; how does one determine whether seeding has invigorated the updrafts? A conceptual model does not address such questions, but they must be dealt with in an experimental hypothesis intended to test the model. Thus, seemingly important elements in the FACE conceptual "chain of events" were not subject to measurement during the experiment.

In contrast, the HIPLEX-1 experimental hypothesis states, rather specifically, the expected differences between the seeded and nonseeded clouds and provides indications of where, when, and (at least generally) how to measure those differences. The response variables and the experimental procedure to investigate the differences then follow rather directly from the statement of the experimental hypothesis. Unmeasured, or unmeasurable, differences are not of much substantive value in an experimental design, even though they may be included in the broad conceptual model.

The response variables for an experiment should be related closely to the experimental hypothesis. If a multistep physical hypothesis is to be tested, several iterations may be needed to develop this relationship within the context of available measurement capabilities. The hypothesis can be reformulated to be compatible with those capabilities, even though the underlying conceptual model remains the same.

Statistical methods play an important role in the evaluation of most weather modification experiments. The nonparametric permutation procedures applied in FACE (Woodley et al., 1983) or in HIPLEX-1 (Mielke et al., 1984) seem to be well adapted for statistical evaluation of suitably designed experiments. One issue that remains to be resolved is how to evaluate chain-of-events experiments with multiple response variables. HIPLEX-1 used Mielke's multiresponse permutation procedures (MRPP) for the primary statistical evaluation, but much of the information actually came from applying those procedures to one response variable at a time (i.e., in a univariate mode). The overall multivariate p-values indicate the significance (if any) of differences between the seeded and nonseeded groups of clouds in a multivariate sense. However, those values appear to be dominated by a small number of the response variables (specifically, the ones for which the seed/no-seed differences were greatest).

One HIPLEX-1 analysis involved moving down the chain of events in the hypothesis and adding one response variable at a time to observe the step-by-step trend of the multivariate p-values (see Mielke et al., 1984). However, it is not clear just how to interpret the results. For example, some people initially thought that adding a variable which, by itself, showed a strong seed/no-seed difference would decrease the composite p-value, but that did not always turn out to be true. It appears from the HIPLEX-1 results that a composite p-value cannot be lower than the lowest univariate p-value among the included response variables. Adding another variable with a given univariate p-value for the seed/no-seed difference caused the composite p-value to increase in some cases and decrease in others. The main point here is that while the mechanics of these multivariate p-value calculations are well established, the way to interpret the results is not.

In closing, it is worth noting that the evaluation of operational cloud seeding projects is also a very difficult problem. Statements of the conceptual models guiding the seeding operations are often vague or nonexistent; measurements of quantities that might fit into a chain-of-events analysis are rarely available; and the absence of randomization limits the kinds of statistical approaches that can be applied. Work is underway to develop better techniques, both statistical and physical, for evaluating operational weather modification projects (e.g., Hsu et al., 1981; Miller et al., 1983). However, much remains to be done in this important area.

REFERENCES

Cooper, W. A., and R. P. Lawson, 1984: Physical interpretation of results from the HIPLEX-1 experiment. *J. Climate Appl. Meteor.,* **23,** 523–540.

Hobbs, P. V., and M. K. Politovich, 1980: The structures of summer convective clouds in eastern Montana. II: Effects of artificial seeding. *J. Appl. Meteor.,* **19,** 664–675.

Holroyd, E., III, J. T. McPartland and A. B. Super, 1979: Treatment system. Appendix D, *The Design of HIPLEX-1,* U.S. Dept. of the Interior, Bureau of Reclamation, Denver, 48 pp. + app.

Hsu, C.-F., K. R. Gabriel and S. A. Changnon, Jr., 1981: Statistical techniques and key issues for the evaluation of operational weather modification. *J. Wea. Mod.,* **13,** 195–199.

Kempthorne, O., 1980: Some statistical aspects of weather modification studies. *Statistical Analysis of Weather Modification Experiments,* E. J. Wegman and D. J. DePriest, Eds., Marcel Dekker, 145 pp.

Linkletter, G. O., and J. A. Warburton, 1977: An assessment of NHRE hail suppression seeding technology based on silver analysis. *J. Appl. Meteor.,* **16,** 1332–1348.

Mielke, P. W., Jr., K. J. Berry, A. S. Dennis, P. L. Smith, J. R. Miller, Jr. and B. A. Silverman, 1984: HIPLEX-1: Statistical evaluation. *J. Climate Appl. Meteor.,* **23,** 513–522.

Miller, J. R., Jr., S. Ionescu-Niscov, D. L. Priegnitz, A. A. Doneaud, J. H. Hirsch and P. L. Smith, 1983: Development of physical evaluation techniques for the North Dakota Cloud Modification Project. *J. Wea. Mod.,* **15,** 34–39.

Sand, W. R., J. L. Halvorson and T. G. Kyle, 1976: Turbulence measurements inside thunderstorms used to determine diffusion characteristics for cloud seeding. *Proc. Second WMO Sci. Conf. on Weather Modification,* Boulder, WMO, 539–545.

Smith, P. L., A. S. Dennis, B. A. Silverman, A. B. Super, E. W. Holroyd III, W. A. Cooper, P. W. Mielke, Jr., K. J. Berry, H. D. Orville and J. R. Miller, Jr., 1984: HIPLEX-1: Experimental design and response variables. *J. Climate Appl. Meteor.,* **23,** 497–512.

Woodley, W. L., J. Jordan, A. Barnston, J. Simpson, R. Biondini and J. Flueck, 1982: Rainfall results of the Florida Area Cumulus Experiment, 1970–76. *J. Appl. Meteor.,* **21,** 139–164.

——, A. G. Barnston, J. A. Flueck and R. Biondini, 1983: The Florida Area Cumulus Experiment's second phase (FACE-2). Part II: Replicated and confirmatory analyses. *J. Climate Appl. Meteor.,* **22,** 1529–1540.

CHAPTER 16

Principles and Prescriptions for Improved Experimentation in Precipitation Augmentation Research

JOHN A. FLUECK

CIRES, University of Colorado, and NOAA/ERL, Environmental Sciences Group, Boulder, Colorado

ABSTRACT

Proper field experimentation in precipitation augmentation, or virtually any other topic, is not an easy task. Some general research considerations, i.e., the objectives of research, the quest for believability, and the two principal types of field studies, are discussed. The anatomy and stages of life of an experiment are presented, and the three levels or classes of an experiment (i.e., preliminary, exploratory, and confirmatory) are depicted. A number of prescriptions for improved experimentation are offered in regard to conceptual models, treatment design, treatment selection and allocation, treatment effect models, and analyses for treatment effects. Lastly, a few comments are appended on the role of statisticians in quality field research efforts.

When the well's dry, we know the worth of water.
Ben Franklin, 1758
Poor Richard's Almanack

16.1. Introduction

The modern origins of the science of precipitation augmentation research began with the laboratory and field work on smoke screens, precipitation static, and aircraft icing conducted by General Electric Research Laboratory scientists during World War II (Byers, 1974). By the end of 1946, Langmuir (1948) and Schaefer (1946) had shown that their laboratory nucleation results could be duplicated in the field for carefully selected cold clouds. Two years later, Vonnegut (1947) also succeeded in showing that his laboratory nucleation results with AgI could be repeated in the atmosphere. The "cloud seeding rush" began.

In the almost 40 years since this historic period, well over 500 operational precipitation projects (Todd and Howell, 1985) and over 40 precipitation augmentation research experiments have been conducted under a variety of meteorological conditions, geographical settings, and experimental expertise (see Neyman, 1967; NRC, 1973; Tukey et al., 1978; Williams and Elliott, 1985, for details). However, the general treatment effect picture that emerges from these experiments is mixed and inconclusive. This situation appears to be due to three factors: 1) the initial heightened expectations for treatment effects (e.g., Langmuir, 1950), 2) the complexity of precipitation processes (e.g., Braham and Squires, 1974), and 3) the lack of proper experimental tools and techniques (e.g., NAS–NRC, 1959; NRC, 1973).

Fortunately, recent years have seen some improvements in all three of these areas. The treatment effect expectations have become more reasonable (e.g., 10% to 15%; AMS,

1984), there is increased awareness of the need to investigate the full physical chain of events in the precipitation process (e.g., Braham, 1981), and researchers in precipitation augmentation are being advised on how to better design, implement, and analyze their experiments (e.g., Flueck, 1978, 1982; Tukey et al., 1978).

This article will focus on this third factor and offer some principles and prescriptions for improving field experimentation in precipitation augmentation research. The general research problem and the desired approach to its solution will be discussed in section 16.2. The morphology of an experiment will be presented in section 16.3 to show its stages, components, and classes. Some prescriptions will be offered in section 16.4 for improving particular components of an experiment's "anatomy." Lastly, section 16.5 will offer some concluding comments on proper field experimentation and on the role of statistics and statisticians in atmospheric science research. Although the emphasis in this paper is on field precipitation experiments, much of the material is equally applicable to other research settings.

16.2. General research considerations

16.2.1. General objectives

The general objectives in any research effort should be, in principle, twofold: address an important question and achieve a "believable" answer. The characteristics of a potentially solvable question include 1) being able to state clearly and concisely the question of interest, including its subquestions, 2) assuring that the question is amenable to quantitative analysis, and 3) having confidence that appropriate evidence and useful results can be obtained within the resource constraints. Hence, the criteria for a solvable question is based partly on expectations, and a priori expectations may differ from a posteriori results. In fact, it has often been said that a successful scientist is one who judiciously picks his research questions.

The characteristics of a "believable" answer can be best viewed as composed of two parts (Flueck, 1982): 1) the plausibility of the proposed theory or conceptual model for the process of interest and 2) the credibility of the offered statistical evidence (it should be noted that statistical evidence includes both case studies and aggregate results). Thus, it is interesting to note that the rigorous theorem–proof approach to problem solving in mathematics has a counterpart of "proposed theory–offered evidence" in science.

16.2.2. Believable answers

It is useful to examine the two parts of believability in order to uncover some of the underlying principles. The plausibility of a proposed theory or conceptual model is a function of at least four factors:

(i) the clarity and succinctness of the proposed theory (e.g., no murky long-winded explanations),

(ii) the level of detail and the completeness of the proposed theory (e.g., all k steps of the process and not just one "big-black-box"),

(iii) the logic of the proposed theory, including internal consistency (e.g., no unexplained jumps or miracles), and

(iv) the consistency of the proposed theory with past accepted evidence (e.g., the theory is not contradictory with other well-accepted evidence or theory).

The weighting and integration of these four factors to produce an overall measure of plausibility of the proposed theory or conceptual model is performed subjectively (perhaps this is where Bayes Theorem applies). Thus, there may be differences of opinion between scientists as to the plausibility of a proposed conceptual model.

The second part of believability, the credibility of the offered statistical evidence, is also a function of at least four factors:

(i) the quality of the design, implementation, and analyses of the experiment (e.g., the care, skill, and methodology utilized in the experiment),

(ii) the basis for the treatment effect inferences (e.g., whether treatments are randomized, the size of the P-value associated with the observed treatment effect, and how multiplicities are accounted for; Tukey, 1977a),

(iii) the quality and internal consistency of the basic data (e.g., reasonable values and few, if any, outliers or missing values), and

(iv) the integrity of the research investigation (e.g., scrupulous researchers with a publicly available design document, operations plan, and final data base).

The combining of these four factors into a probability that the observed results are due to treatment is also done subjectively, and hence there can also be differences of opinion. Consequently, the ex post facto evaluation of how well the experiment performed its tasks should be done formally such that both the "believability" and future improvements have a justifiable basis.

One can illustrate the relation between the two components of "believability" using the traditional 2×2 contingency table. Table 16.1 presents the states of the believability problem cross-classified by model plausibility and evidence credibility.

Clearly, the desired result of a research study is to achieve both a high probability of model plausibility and supporting evidence credibility and thus have the experiment be declared a "success." This joint event is the goal of all scientists, but it is infrequently achieved. The opposite is to have a low probability of both of the two believability characteristics and thus be declared a "failure." This is the joint event we all abhor but too often achieve.

The two remaining states of Table 16.1 represent combinations of the two extreme states (i.e., success and failure). If one achieves credible evidence but not a plausible model, then one is in the situation of being a "number's man." In short, one has some supporting numbers or evidence but no acceptable scientific explanation of how they could have occurred. Alternatively, if one achieves a plausible model but has no supporting credible evidence, then one is in the situation of being a "model's man." This situation could be termed the "Pygmallion problem." To become emotionally attached to one's proposed theory or model before achieving credible supporting statistical evidence can be terminal.

Given that one has achieved the "success" state in an experiment, the task becomes to move the conceptual model from plausible to principle or a scientific law. This typically is accomplished by replicating the "success" result for other locations and conditions. As R. A. Fisher (1926) stated many years ago,

A scientific fact should be regarded as experimentally established only if a properly designed experiment rarely fails to give this [.05] level of significance.

TABLE 16.1. The believability state table.

		Plausible model	
		Yes	No
Credible evidence	Yes	Success	Number's man
	No	Model's man	Failure

In this situation, the conceptual model is often modified to provide a broader domain of explanation and relevance.

16.2.3. Types of field studies

There are two fundamental types of field studies: 1) observational and 2) experimental, and each has its place in furthering our knowledge of our universe. A field experimental study, or experiment, is a situation in which the researcher can at least control what, when, where, and how much "treatment" is to be applied to each treatment situation or unit. Furthermore, the researcher also may be able to partially "control" a number of ancillary conditions (e.g., the relative humidity of the day, the amount of natural ice in the clouds, etc.) by judicious selection of the experimental conditions (e.g., prescreening). Thus, in a field experiment the researcher can select some of the initial conditions, the time of intervention, and the size and place of the treatment.

In a field observational study the researcher has relatively little control over the "treatment" and the ancillary or intervening factors. The event or treatment is due to "Nature" in the past or future (e.g., flooding of the area one million years ago or "local heating" of a mountainside by solar radiation), and the researcher attempts to observe the results of this treatment by collecting data in the proper location and time period presumably affected by the treatment. In short, the researcher does not control the preconditions, the time of intervention, or the size or location of the "treatment." Hence, the observational study is a more difficult method for furthering knowledge.

Where there is a choice between experimental and observational studies, one typically can learn more quickly and more thoroughly from the former than the latter. Hence it is not surprising to find that the more "advanced disciplines" are those that learn by experimentation (e.g., physics, chemistry, microbiology, medicine, agriculture, etc.).

16.3 The stages, components, and classes of an experiment

An experiment is like many living things in that it has a beginning, a middle, and an end. These stages are important to delineate and describe in order to understand the principles of a proper experiment.

16.3.1. The stages

The seven stages that an experiment proceeds through in its "life cycle" are as follows (Flueck, 1982):

(a) *The conception:* An awareness of a problem, some questions, initial ideas, and insights, typically followed by a literature search, a look at some past evidence, and some preliminary investigations of the process of interest.

(b) *The design:* The construction of a general plan for the research effort, including the goals and scope of the experiment, the conceptual model, the treatment effect

parameters and their estimators, the measurement process and systems, the desired analyses, etc.

(c) *The feasibility study:* The "pilot testing" of the proposed experiment and a formal benefit–cost assessment of the project given the present knowledge, the offered design, and the available resources and time.

(d) *The implementation of the experiment:* The field performance of the experiment guided by a detailed "operations plan," which specifies how to select experimental units, apply treatment, collect the data, field check the data, etc.

(e) *The analyses:* The final data checking and reduction, data base construction and documentation, performance of the a priori specified analyses and the a posteriori or exploratory analyses.

(f) *The reporting:* The public presentation(s) of the experiment, including a summary of the design and its implementation, the results of the specified a priori and a posteriori analyses, and the response variable data base cross-classified by at least experimental unit and treatment.

(g) *The ex post facto studies:* Reanalyses of the collected data to sharpen the evidence for a treatment effect, revised conceptual models and their associated analyses, and ex post facto evaluations of the quality of the design, implementation, analyses, and data.

These seven stages, or the "life cycle," of an experiment are linked, and Fig. 16.1 presents the typical flow diagram. The solid lines represent the major flow of activities and the dotted represent minor flows. Of course, an experiment may iterate between stages (e.g., design and feasibility) as it is being performed.

One should note that there are two conditional stopping points before the end of the experiment (i.e., not feasible and major problems). Both situations represent admission of problems that cannot be overcome given the present experiment. Thus one should terminate the present experiment and either return to "start" or quit. The final stopping point produces three possible results: 1) a confirmed effect (i.e., positive, negative, or zero), 2) an indication of an effect, or 3) simply noise (great uncertainty).

16.3.2. The components

If we look closer at any experiment, we note that it has an "anatomy" of its own. This anatomy is composed of a number of components, each of which plays an important role in the life cycle of the experiment. These components and their specific activities are as follows:

(a) *The experiment's problem, objectives, and scope:* A brief statement of the problem of interest should be given with the principal and subobjectives that are to be pursued, the scope or populations that will be investigated, and any other highly distinguishing characteristics of the proposed experiment.

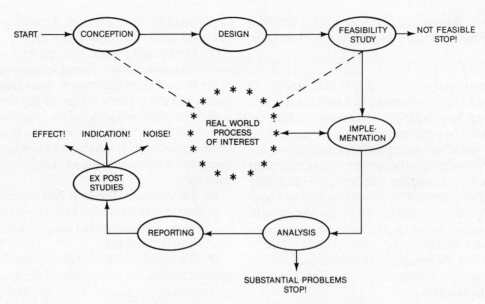

FIG. 16.1. The typical stages, or "life cycle," of an experiment.

As an example, with regard to objectives, one might specify that the principal objective of the experiment is to determine whether a presumed microphysical chain of events can be artificially altered to produce a greater number of precipitation-size particles in wintertime convective clouds in the High Sierras of California.

The subobjectives are more detailed and precise and should attempt to specify each of the research questions that will be pursued.

(b) *The experiment's conceptual model(s):* This is an offered prescription(s) in prose, symbols, or mathematical form of how the overall process of interest might function, including the particular deduced relations that are to be studied in each of the subobjectives.

An example could be a numerical model of the nucleation and growth processes of precipitation particles in a cloud under specific conditions and the particular relations that will be studied.

(c) *The statistical treatment effect model and its estimators:* Precise statistical specification of the conceptual model's selected relations and parameters that are to be evaluated for evidence of treatment (including auxiliary variables). Typically, each specification refers to a particular research question(s) presented in the subobjectives section.

An example is that, based on the conceptual model, we will describe the population of ice crystal concentrations (ICC) by a statistical linear model containing one index variable for the treatment effect and four covariates [e.g., amount of treatment, cloud-top temperature, prior ICC, prior supercooled liquid water concentration (SLWC), and time since initial treatment].

(d) *The dimensions of interest and their measurement:* The specification of the properties that one desires to measure, the translation to what can be measured, and finally, the compromise to what will be measured. Lastly, the equipment, procedures, and accuracy of the measurements must be carefully specified.

As an example, if we desire to measure ICC distributions in the core of the convective cell over time, we compromise and measure ICCs using a Particle Measuring System 2D-C optical imaging probe during two penetrations (e.g., time $t = 0$ and $t = 6$) of the convective cell at the $-10°C$ level by a cloud physics aircraft.

(e) *The treatment design:* A plan indicating what the experimental and other units of the investigation are, what k ($k \geq 2$) treatments and dosages are to be used, and which units are to be treated, how many units (sample size), etc.

As an example, we could specify that a particular volume of a predesignated cumulus cell of given characteristics will be treated at time $t = 0$ by 10 grams of AgI, CO_2, or a placebo following a particular randomized treatment allocation plan. This procedure will be repeated on similar clouds during the day until the group or block of all three treatments is utilized. Further blocks of three treatments will be performed on subsequent days until 20 blocks (i.e., 60 treatment units) are completed.

(f) *The treatment application:* The specification, in an "operations plan" or "manual," of how the treatments are to be applied to the units (i.e., when, where, how much, and in what manner).

An example is that 10-gram flares of either AgI or a placebo (on a random basis) will be ejected at 100-m intervals throughout the entire diameter of the cell at the $-8°C$ level by a seeder aircraft. All field participants will be blind to the treatment being applied.

(g) *The data collection system:* Specification, again in

an operations plan, of how the data are to be collected (when, where, how, etc.) and how the preliminary field "checking" of the collected data is to be performed.

As an example, we might specify that a cloud physics aircraft will penetrate the predesignated convective cell at the $-10°C$ level at zero and six minutes after treatment, and ICC data will be collected by a 2D-C probe and digitized on magnetic tape. The field checking of the data will be performed by visual monitoring in flight and in more detail upon return to the airport.

(h) *The data-base construction and documentation:* Instructions (typically in the design document) are given for data checking, cleaning (editing), coding, and transforming of data observations to experimental unit values. The construction, storage, and documentation of the resulting multivariate data base are also specified.

As an example, one would present instructions for detailed checking, comparing, editing, and coding of the ICC data, and reduction of it to the appropriate experimental unit time periods. Also, plans for the assembling of the resultant data base into a computerized, readily accessible form, with complete documentation, would be given.

(i) *The analyses of the assembled data base:* Specifications (typically in the "design" document) of the a priori case study and aggregate analyses for evidence of treatment effects. Also, guidelines for the performance of the a posteriori analyses, which are directed to uncovering further evidence and understanding of the process of interest, would be given.

As an example, one could fit the previously specified statistical linear treatment effect model to the designated data and perform an analysis of the residuals for the a priori suggested effects. The a posteriori or exploratory analyses could be an enlarged linear modeling of ICC values to assess the previous fit and to explore for other factors that might influence ICC values.

These nine components constitute the basic building blocks or anatomy of a comparative (e.g., treatment compared to no treatment) experiment. Consequently, each component and its attendant activities must be given explicit and careful attention if one desires to produce a proper field experiment. The failure to adequately plan or implement a particular component increases the chance that bias or other problems will enter the experiment and thereby reduce the experiment's "believability." As Hill (1951) has so well reminded us, the hallmark of a quality experiment often is the degree of attention given to the "small details." Some of the recent precipitation augmentation experiments (e.g., the HIPLEX, Bureau of Reclamation, 1979; the SCPP-1, Bureau of Reclamation, 1982; and the FACE-II, Woodley et al., 1983) appear to have explicitly addressed most of the above nine components and typically with considerable care.

16.3.3. The intersection

As a proper comparative experiment moves through its life cycle, the various components are "spotlighted" and required to perform their role. As in a play, these roles are quite precise and each builds upon and must be integrated and coordinated with the previous roles. Each component's performance generates numerous detailed activities for the researchers and supporting staff.

Table 16.2 summarizes the major intersections between the stages and the components of a proper experiment using an activity or "performance" matrix. An "×" in this matrix designates that the particular component performs a major role in the particular stage, and a blank designates a minor role or no role. As the matrix indicates, there is a definite association among stages and components, with the performances typically flowing from the earlier to the later components of an experiment as it moves through its life cycle. It is interesting to note that following the important stages of design, implementation, and reporting are three evaluation stages (i.e., feasibility, analyses, and ex post facto studies). Thus a proper experiment can be viewed as conservative in that "check points" are regularly interspersed throughout its life cycle.

TABLE 16.2. The performance matrix for a proper comparative experiment.

Components	Conception	Design	Feasibility study	Implemen- tation	Analyses	Reporting	Ex post facto studies
1. Problems, objectives, and scope	×	×				×	
2. Conceptual model	×	×				×	×
3. Treatment effect model	×	×	×		×	×	×
4. Variables and measurements		×	×			×	
5. Treatment design		×	×			×	
6. Treatment application		×	×	×		×	
7. Data collection		×	×	×		×	
8. Data-base construction		×		×	×	×	
9. Analysis of data		×			×	×	×

16.3.4. The classes

There actually is a third dimension to the performance matrix of Table 16.2, namely, classes, or tiers, of experimentation. The three classes are 1) preliminary experiments or studies, 2) exploratory experiments, and 3) confirmatory experiments. Each class plays a specific role in the research process and is needed to complete the journey from conception to scientific law.

Preliminary. A preliminary experiment or study is typically a "background study" to a larger issue. Its objectives are to give evidence or insight on particular questions. As such, the study typically is focused, has no formal design document, uses a relatively small number of observations, employs nonrandomized treatment allocation, and relies on simple point estimates (i.e., "indications" in the Mosteller and Tukey, 1977, vernacular). Thus, a preliminary study is usually informal and hence does not allow for formal statistical inference.

In a precipitation augmentation research experiment, numerous preliminary studies are needed. They include cloud climatologies, transport and diffusion investigations, treatment opportunity studies, treatment calibration trials, etc. Clearly, preliminary studies are an important antecedent to the actual experiment. As the NAS–NRC report (1959) states,

> The effectiveness of formal testing programs can be greatly enhanced by carefully conducted preliminary studies of cloud and other atmospheric conditions prevailing in a proposed test area.

Malkus and Simpson (1964) present an example of such a preliminary study in tropical cumulus cloud modification. Furthermore, some of these studies can become exploratory experiments (e.g., Simpson et al., 1966; Simpson et al., 1967).

Exploratory. An exploratory experiment has been described as

> an attempt at "staking a claim" on a planned or unplanned relation(s) among events based on a plausible (though often crude) conceptual model and appropriate scientific evidence (Flueck, 1978).

In short, it can be viewed as a flexible interactive guided search for evidence of a treatment effect.

Some characteristics of a proper exploratory experiment are

(a) An investigation of potential treatment effects guided by a crude but developing conceptual model, a simplified random treatment allocation design, usually a flexible sample size, and often a growing number of covariates (e.g., sounding-derived variables, prior LWC, precipitation variables, etc.);

(b) Flexible interactive exploratory analyses of the accumulating data (emphasizing visual display of response and other values, e.g., Tukey, 1977b) at frequent intervals (typically at the end of each field season) to guide the search for effects of treatment;

(c) Recognition of the presence of different types of "multiplicities" (there is a cost to exploration; Tukey, 1977a) and thus the use of *P*-values and confidence intervals as indications rather than formal "classical" inferences of treatment effects; and

(d) Feedback of the interim results at predesignated times to allow for changes in the experiment (e.g., the conceptual model, the research questions, the treatment design, the implementation, and the analyses).

An exploratory experiment may have a few a priori analyses, and these analyses should be directly related to the conceptual model and to particular research questions. The a posteriori analyses typically are many, and they should lead to a better "picture" of how the particular precipitation process functions. (The results from a priori analyses are typically a higher order of evidence than those from a posteriori analyses.) The Climax-I (Mielke et al., 1970), the Whitetop (Flueck, 1971), and the FACE-1 (Woodley et al., 1982b) projects are examples of exploratory experiments.

Confirmatory. A confirmatory experiment has been described as

> a well defined inflexible process closely focused on replicating (confirming) a result while minimizing sampling and nonsampling errors (i.e., variability and bias). Hence, there should be a small chance that the results are due to anything other than the treatment (Flueck, 1982).

Thus, a confirmatory experiment is typically an independent confirmation of the results of a previous experiment. This confirmation effort can take many forms from a split-sample cross-validation study (i.e., weak confirmation) to a separate experiment by other experimenters (i.e., strong confirmation; Flueck, 1978).

Some characteristics of a proper confirmatory experiment are

(a) The confirmatory experiment should have a few (e.g., one or two) well-defined and focused research questions derived directly from the proposed conceptual model (e.g., Can we expect to achieve a 100% increase in median ice crystal concentration due to treatment at the $-10°C$ level in our particular subset of summertime isolated convective clouds?).

(b) There should be a detailed description of the experiment and its implementation (i.e., a Design Document and an Operations Plan), including the latest version of the conceptual model, the treatment effect model and parameters, the efficient treatment design, etc. These documents should be finalized and made public prior to the data collection stage of the confirmatory experiment.

(c) There should be a small set of predefined confirmatory analyses (i.e., preferably one per research question or hypothesis) to be conducted only at the conclusion of the entire data-gathering task. All multiplicities should be clearly identified, limited, and properly assessed.

(d) No changes should be allowed in the design, implementation, or analyses of the experiment once the experiment is implemented. The presumption is that one simply desires to replicate or confirm an earlier result (e.g., an exploratory result), and thus there is no need to search. The result either will or will not be there.

The confirmatory experiment appears to be the appropriate place for utilization of the sequential analyses methodology (e.g., Armitage, 1975; Pocock, 1977) often recommended for medical research. As discussed elsewhere (i.e., Flueck, 1978), there typically is considerable similarity between a medical–clinical trials study and a precipitation augmentation field study. Hence, both disciplines should benefit from understanding each other's techniques and methodology.

Although confirmatory-type experiments were recommended as early as 1959 for precipitation augmentation research (i.e., NAS–NRC, 1959), it was 1978 before a planned confirmatory experiment consistent with the previous discussion was attempted (i.e., FACE-2; Woodley et al., 1982a). However, the Climax II results (Mielke et al., 1971b, 1976) did contain two analyses based on temperature partitions that were replications of two Climax I analyses (i.e., Mielke et al., 1970, 1971a), and thus a small portion of Climax II can be viewed as an attempt at confirming Climax I. Unfortunately, neither Climax II nor FACE-2 confirmed its prior results at the conventional 5% P-value. [It should be noted that a later exploratory reanalysis of both Climax I and II (Mielke et al., 1981b) produced consistent positive treatment indications at P-values smaller than 5%. The value of ex post facto studies is well illustrated by this example.]

Finally, it should be recognized that an experiment can be both an exploratory and a confirmatory experiment (Flueck, 1978). For a number of reasons, including the cost and time requirements, one often is faced with the need to design a multipurpose experiment (i.e., both confirmatory and exploratory). In this situation, the two parts of the experiment must be clearly identified and properly separated so that the confirmatory part is not compromised by the exploratory effort. Furthermore, a confirmatory experiment can use exploratory analyses after completion of the confirmatory analyses (e.g., FACE-2; Woodley et al. 1983). In fact, it is generally important and effective to explore for further understanding on every experiment's data set.

16.4. Prescriptions for improved experimentation

A number of reports and articles have been published providing suggestions and recommendations for improv-

ing precipitation augmentation field experiments (e.g., NAS–NRC, 1959; NAS–NRC, 1966; Neyman, 1967; Moran, 1970; NRC, 1973; Brier, 1974; Tukey et al., 1978; Flueck, 1978, 1982). However, given the apparent complexities of the precipitation processes and the resulting need for further field experiments, it seems appropriate to review in more depth some of the previously presented components of an experiment and offer some prescriptions for improved experimentation.

16.4.1. Conceptual models

As indicated earlier, a proper precipitation augmentation experiment should have a conceptual model of the precipitation process of interest, and it should include both the natural and modified situations. The conceptual model (sometimes termed a "working" model; Flueck, 1976) should be the researcher's description of how he believes the particular precipitation process functions and should be constructed prior to the implementation stage of the experiment. It typically is based on a combination of intuition, past laboratory and field results, and current preliminary field evidence. Although the conceptual model initially may be rather crude and incomplete, it should become more detailed and refined as further insight and evidence are obtained from case and aggregate studies. ("Case studies" are typically an examination of one event over many variables as opposed to "aggregate studies," which are typically many events over few variables.)

The conceptual model can be presented in any of three different descriptive forms: 1) prose, 2) analog, and 3) mathematical or numerical.

1) *Prose*—A written description of the overall process (e.g., Woodley et al., 1982b). It could be as elementary as the following: Cloud seeding with AgI flares at the −10°C level provides an increase in the number of IN, which grow by vapor deposition and then accretion of supercooled liquid water to become rimed crystals or graupels or rain, depending on the surface temperature.

2) *Analog*—Here emphasis is on the physical deterministic relations, and hence it sometimes is termed a "physical" model of precipitation (e.g., Chappell, 1970). An example is

$$\text{PI} = f(\text{INN}, \text{INA}, \text{SLWC}, \text{CSR}, \text{TMP}, \text{TIME})$$

where

PI	precipitation intensity,
INN	ice nucleus natural concentration,
INA	ice nucleus artificial concentration,
SLWC	supercooled liquid water concentration,
CSR	condensate supply rate,
TMP	temperature at the INA deposition level,
TIME	time to first echo,

and f represents the particular regionalized relation between the response variable PI and the six predictor variables.

3) *Numerical*—Typically, a system of simultaneous nonlinear deterministic (and some stochastic) equations (e.g., Cotton, 1972), which can be represented as follows:

$$y_1 = b_{10} + b_{11}x_{11} + b_{12}x_{12}/x_{13} + \cdots + b_{1m}x_{1m-1}x_{1m}$$
$$\vdots \quad \vdots \qquad \qquad \qquad \qquad \qquad \vdots$$
$$y_k = b_{k0} + b_{k1}x_{k1} + b_{k2}x_{k2}/x_{k3} + \cdots + b_{km}x_{km} + e_k.$$

Furthermore, there are usually zero, one-, two- and three-dimensional (actually four: x, y, z, and t) versions of a numerical model.

In the early days of precipitation augmentation research (e.g., 1940s and 1950s), not much attention was given to the conceptual model. Whatever attention was rendered typically was in prose and in essence was "One Big-Black-Box." The description might mention the start (e.g., artificially creating ICs) but mainly would focus on the finish (e.g., rainfall amount or intensity at the ground) with little said about all of the subprocesses. Correspondingly, only one response variable was identified and measured, typically mean rainfall amount on the ground.

Clearly, this approach to the conceptual model was not very satisfying, and in 1966 the NAS–NRC review report (1966) recommended that the overall precipitation process be partitioned "into a chain of subprocesses" so that the links between each subprocess could be examined for evidence of treatment effects. (In essence, they were advocating a closer tie between the physics and the statistics to improve both evaluation and understanding.) They also believed that the number of subprocesses of the precipitation process were many (i.e., 17), but for purposes of evaluation they should be limited to a few basic subprocesses, "perhaps half a dozen." For static seeding of the cold rain process, this appears to be a useful number, for one can allocate two response variables to the initial nucleation stage, two to the growth of ice particles stage, and two to the final stage of precipitation on the ground.

HIPLEX (Bureau of Reclamation, 1979; Smith et al., 1984) appears to be the first precipitation experiment to follow this partitioning process. The current SCPP-1 winter precipitation augmentation experiment (Bureau of Reclamation, 1982; Reynolds and Dennis, 1986) has followed the full recommendation with an added detail. This detail requires the first response variable in each stage to measure the core of the basic subprocess while the second measures the linkage to the next subprocess. It should be noted that this procedure essentially results in the conducting of three subexperiments and assessing the relations between them. Thus, one can readily see how this "divide and conquer" approach strengthens the evaluation and understanding of the precipitation process.

This chain-of-events approach can be implemented in different ways. One can measure sequentially each of the designation subprocesses in the same cloud or cell to achieve one replication of the full experiment (e.g., a single observation for each of three different response variables). Alternatively, one can measure the same response variable in three different clouds (presumably all treated the same) to produce one replication. In this latter case, a new set of three different clouds would be used for investigating each subprocess. The former approach generally is preferred on both theoretical and empirical grounds (e.g., each cloud serves as its own control), and both HIPLEX and SCPP utilized this sampling method.

In summary, the conceptual model fulfills three roles: 1) a storehouse for developing ideas and theories, 2) a device for predicting precipitation treatment effect results, and 3) a candidate for the desired "scientific law." As such, the conceptual model can be utilized to both describe and guide the research effort.

16.4.2. Treatment design

The treatment design for a precipitation augmentation experiment should at least address the following topics: 1) the populations and units of interest, 2) the design configuration, 3) the methods for reducing variability, 4) the allocation of treatment to the units, and 5) the total number of sample units. Each of these topics will be briefly addressed; further discussion of these topics can be found in applied statistics texts covering the design of experiments (e.g., Fisher, 1935; Snedecor and Cochran, 1967; Winer, 1971).

Populations and units. There are at least four populations that enter a precipitation augmentation experiment. First, there is the population of record or the general group in which you desire to work (e.g., the population of all wintertime clouds in the High Sierras of California). Second, there is the population of particular interest, or the subgroup, that you desire to investigate (e.g., the population of all wintertime orographic clouds in a particular target area in the High Sierras). Third, there is the identified population of the subgroup of interest (e.g., the population of wintertime orographic clouds that occur in daylight hours, in the particular target area, are individually identified, and remain for at least three hours). Lastly, there is the treatment population or the actual subset of clouds you are able to treat. Thus, the sequence of the population of record, interest, identity, and treatment is a continued winnowing of the set of clouds to eventually reach the set that is studied. Of course, each succeeding population is a proper subset of the preceding one.

The units also have a hierarchical sequence: 1) experimental, 2) treatment, and 3) observational. The experimental unit is a physical entity such as a person, animal, agricultural plot, cloud, etc. In a precipitation augmentation experiment it is typically a mass of air with space and time dimensions (e.g., the volume of air that moves through the target area in a three-hour experimental period). The treatment unit can be synonymous with the experimental unit or a subset of it (e.g., the subset of cells that have CTT < −10°C and ICC < 10 L^{-1}). Of course, one should have supplemental information on the proportion of the experimental unit that becomes the treatment unit.

Lastly, the observational unit is what is actually measured, and this unit determines the basic resolution of the response variable. The observational unit may be synonymous with the treatment unit or it may be smaller. Also, the observational unit may change with the response variables over the chain of events [e.g., a 2D-C median ICC value in each of three cells at the nucleation stage, an area or volume of given radar reflectivity (e.g., >20 dBZ) in each of the cells at the growth stage, and individual raingage intensity values in the presumed "floating precipitation target" for the three-hour experimental unit]. Thus, three response variables would be measured within each experimental unit.

Design configurations. An important question that must be faced in the treatment design is, where do I expect, in space and time, to see the effects of treatment? In short, what should be the domain of the response variable? This question usually is answered by selecting a single geographical area (i.e., the so-called target area of the experimental unit) and some period of time (i.e., the time period of the experimental unit) over which the response variable is defined. This design configuration has been termed the target-only design (NRC, 1973) in that the single target area is utilized both for treated and nontreated units on a presumably randomized basis. This was the design frequently used in the 1950s and 1960s precipitation experiments. Note that the designated target area and time are only predicted and may not be congruent with the realized area (or volume) and time of the treatment effect.

If one is attempting to monitor the entire precipitation chain of events, there is a sequence of related target areas or volumes, each dedicated to one of the subprocesses in the chain. Assuming the overall precipitation process is partitioned into the three subprocesses of nucleation, growth, and rain-on-the-ground, one would have a space and time sequence for these three target areas. Thus, the first target area (T_1) could be dedicated to measuring ICC at 5 minutes after treatment initiation, the second (T_2) to the extent of coverage of a given radar reflectivity at 15 minutes, and the third (T_3) to the rain amount on the ground at 30 minutes. Typically, each of these consecutive targets is of increasing size in recognition of the increasing uncertainty.

Given the substantial variability that accompanies most atmospheric processes, it is reasonable to add a control area to the target-only design and hence produce a target–control design. The control area has a variable defined on it, and this covariate, or predictor variable, aids in increasing the "precision" of the design. The control area is not treated and thus is a "baseline" or prediction of what would have happened in the target in lieu of treatment. The control area typically is located relatively near the target area to provide a sizable correlation (e.g., >0.50) but not too near to be affected by the treatment.

One can expand on the target–control design to produce a number of additional designs, including the crossover design (Moran, 1959), the floating target and floating control design (Braham, 1966), and the target–multiple control designs (Flueck and Mielke, 1978). Kulkarni (1968, 1969) has enlarged this set of designs by providing a generalized crossover and a generalized target–control design for fixed and variable treatment effects. Clearly, the set of designs applicable to precipitation experimentation can be further expanded. Thus, there is still a need for increased emphasis on experimental designs in precipitation experiments from the randomized, complete block, and factorial designs, so successfully employed in agricultural research, to the more complex multiresponse designs.

Lastly, it should be noted that the use of a control area in a design is simply a special case of the use of a covariate or predictor variable. The generalization of this approach is the statistical linear model, which will be discussed later.

Confronting variability. It is well known that many atmospheric processes have substantial natural variability, which often makes it difficult to detect treatment effects (e.g., NAS–NRC, 1959). Four principal methods are available to mitigate this "noise": 1) prescreening, 2) blocking, 3) covariating, and 4) replication. Every precipitation augmentation experiment should consider each of them.

1) *Prescreening* is a technique designed to limit the scope of the experiment. Its two goals are to obtain a more homogeneous subset of experimental units (i.e., a less variable set of units) and to provide a "more responsive to treatment" subset (this is sometimes termed the "seedability" subset in precipitation experiments). The prescreening is typically based on meteorological criteria (e.g., only utilize orographic clouds with measurable supercooled liquid water), and the experimental units must be designated prior to treatment. Prescreening has been utilized for many years in precipitation research experiments [e.g., Whitetop (Braham 1966), Climax-1 (Mielke et al., 1970), FACE-1 (Woodley et al., 1982b), etc.], and it continues to be an important method for both mitigating the variability of the response variables and enhancing the potential treatment signal.

2) *Blocking,* or "local control" as termed by Fisher (1935), is the prior-to-treatment skillful grouping of the experimental units so that they, and the observational units, are more alike (homogeneous) within a group (block) than among groups. Hence, blocking uses prior-to-treatment information to secure a block of similar units

and thus "block off" variation among blocks and thereby reduce the variability of the analyzed data.

A simple example of blocking is the randomized block design where, for example, each day is a separate block that contains a pair of experimental units with one treated (T) and one not treated (N) on a randomized basis (Fig. 16.2). Hence, the day-to-day contribution to the experimental error can be removed (the within-day still remains), and the remaining variability should more properly represent chance or random variation. When each treatment is randomly assigned only once to each block, then each block constitutes a single replication of the treatments, and the design is termed a Randomized Complete Block Design (e.g., Snedecor and Cochran, 1967; Winer, 1971). More complex forms of blocking for weather research have been proposed by Tukey et al. (1978).

3) *Concomitant information or covariating* is a third method for reducing the variability of a response variable. The covariate should be well correlated with the response variable, unrelated to the treatment, and have theoretical justification. The covariate information may be utilized in three ways:

    (i) to partition the set of response values into relatively homogeneous subsets (e.g., precipitation under high, medium, and low SLWC),

    (ii) to adjust the response values (Y) by the so-called control values (X) [e.g., the difference ($Y - X$) or ratio (Y/X) of the target and the control area values], and

    (iii) to adjust the response values by a statistically derived linear model (e.g., $Y - bX$, where Y and X are some measure of target and control precipitation and b is the data-derived regression coefficient).

The goal of each of these methods is to reduce the contribution of other factors to the response variability and thereby provide a less noisy background for detection of the potential treatment effects. The concomitant information can take many forms: from precipitation in a control area to sounding-generated quantitites or numerical model predictions.

The first method has been used rather extensively in past precipitation research (it is sometimes termed "stratification"), and it is based on the theory of "divide and conquer." The subsets created by the partitioning are presumed to be more alike within each subset than among subsets. In essence this is ex post facto blocking. The weakness of the approach is the continued reduction in sample size within each subset as one continues to partition. Furthermore, the effects often are conditional on the subset, and thus one cannot pool results to regain the desired strength in numbers.

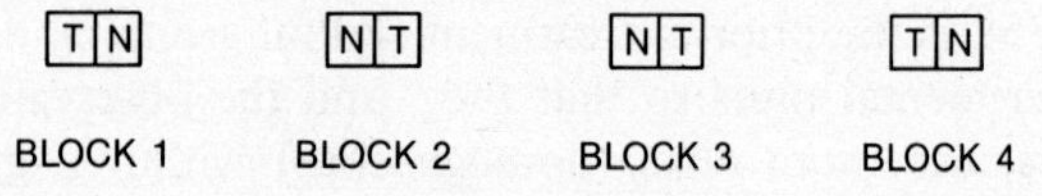

The second method is a traditional adjustment procedure that attempts to take out the natural or background effect. However, it arbitrarily weights the control value the same as the target value (i.e., both have a weight of 1).

The third method of using concomitant information often is termed ANOCOV, and it is an adaptive approach in that the data dictate what weight (b) should be assigned to the control precipitation (e.g., Snedecor and Cochran, 1967). This method usually is preferred to the previous methods because one gets a check on the arbitrary weight of 1.0 and the sample size is not progressively reduced as in the first method.

Mielke et al. (1977) and Tukey et al. (1978) present some interesting views on the use of covariates, and Spar (1957), Flueck et al. (1981), Woodley et al. (1982b, 1983), and others have used this method in analyzing precipitation augmentation experiments.

Cox (1957) has examined the question of covariating versus blocking and suggests that covariating is preferred provided the correlation between the covariate and the response variable is "at least 0.60." Of course, if both can be performed then one secures the most benefit.

4) *Replications* are attempts to overwhelm the variability by brute force (i.e., large numbers of observations). However, the variability typically is reduced only by a function of $(n)^{1/2}$, and this reduction only applies to random error. Furthermore, the expense and time now needed for replications in precipitation augmentation research typically constrains an experiment to relatively small sample sizes (e.g., <60 units). Thus, this method is less preferred than the others.

Most precipitation augmentation experiments have disjoint field periods (e.g., three months in each of four winters). This is generally beneficial for it allows the researchers to examine the most recent preliminary or exploratory results and reconsider the experiment's conceptual model, design, and implementation. To presume that one can obtain a sufficient sample size in one field season to properly complete an exploratory or confirmatory experiment is foolish! Quality precipitation augmentation experiments typically require multiple trips to the field.

Treatments and allocations. The principal issues here are the number of different treatments, their dosage amounts, the allocation procedure, and the number of replications of each treatment.

The number of different treatments employed in an experiment typically is determined by a compromise between the research interests and the funding constraints. Past precipitation experiments typically used two different

FIG. 16.2. A layout for a randomized complete block design
with two treatments and four blocks.

treatments: an active material (e.g., AgI) and a placebo (e.g., sand or nothing). However, too little appears to be known about the competitive advantages of various treatment compounds in the field environment, and hence it seems advisable to compare at least three different treatment materials in future preliminary or exploratory precipitation experiments. The recent pioneering laboratory work by DeMott et al. (1983) on alternative nucleation compounds makes this comparative field testing even more important.

One of the treatments used in a comparative precipitation experiment should be a placebo. The results for the placebo-treated units can serve as a baseline for the treatment effect estimations. This also provides the researcher with an opportunity to investigate both the natural and the artificial (i.e., treated) precipitation processes to better understand the physics of precipitation.

The dosage levels or amounts of a treatment should be based on laboratory results augmented by preliminary studies in the field. These preliminary studies should contain different levels (i.e., amounts) of the active treatments, and they should be conducted with great care. Furthermore, one must become proficient in placing the specified dosages in the desired portion of the cloud at the proper time, and this action should be fully monitored.

Given that a particular experimental design has been judiciously selected, one should determine how the specified treatments are to be allocated to the experimental units. There is now little doubt that the allocation procedure should be randomized (i.e., known probability-based selection of the treatment for each experimental unit). The principal purposes for randomized allocation of treatments to experimental units are 1) to guard against selection bias, 2) to reduce unsuspected (or unaddressed) sources of bias, and 3) to provide the basis for formal classical statistical inference. Selection bias arises, for example, if one picks the more vigorous clouds for the treated group and the less vigorous for the untreated group. The second purpose refers to the attempted balancing (among treatment groups) of other factors that might affect the response variable, and the last purpose refers to the ability to apply probability statements to the outcomes of the analyses (e.g., P-values, confidence intervals, etc.).

In the interest of reducing possibilities for bias at all stages of an experiment, it is recommended that a "blinded" approach be used. Thus, the scientists and staff of an experiment should not be informed which experimental units got which treatments until all relevant data are collected, reduced, and prepared for analysis. Preferably, only the designer of the randomization plan and the treatment material "loader" have advance information of the treatment decisions, and they must be sworn to secrecy. Furthermore, these two people should construct a "verification plan" such that any mistakes in treatment allocation can be discovered and corrected.

The question of what percentage of experimental units should be treated with the placebo is an important one. The preferred allocation plan, from both a meteorological and a statistical viewpoint, is equal frequency of each treatment, including the placebo (e.g., 50% if an active and a placebo treatment are used). However, some deviation from this preferred allocation situation can be tolerated (e.g., perhaps as much as a 34% to 66% allocation), and the theoretical loss of precision due to unequal probabilities can be calculated prior to the selection of the particular treatment plan.

A problem that arises in some experimental designs is the opportunity for runs of the same treatment (e.g., four AgI-treated units in a row). This problem can be removed by the use of constrained or restricted randomization. In short, the actual selected treatment allocation plan can be randomly selected from the subset of all possible plans that obey the specific restrictions. This approach has been known for many years, practiced in medical research for some years (e.g., Zelen, 1974), recently expanded upon by Tukey (1978) for use in weather research, and used in a number of past precipitation research experiments (e.g., Woodley, 1983).

The number of replications of each treatment (i.e., total fixed sample size) is an important decision in all classes of experiments. The basic principle is to select a sample size that provides a "substantial chance" of seeing the expected treatment effect. There are a number of formal methodologies available for estimating the needed sample size (e.g., coefficient of variation, confidence interval, hypothesis testing, and Bayesian methods), and each can be conducted in a parametric or nonparametric manner. Flueck and Mielke (1978) presented details for the confidence-interval and hypothesis-testing parametric approaches to sample size estimation, Twomey and Robertson (1973) presented details for a confidence-interval permutation approach, and Flueck and Boik (1982) presented details and examples for the hypothesis-testing permutation approach.

Although estimates of the total (fixed) sample size needed in preliminary and exploratory experiments may be made, often they are not realized. Thus, one is again in a searching situation, and a variable stopping rule may be appropriate (again, the methodology of sequential analysis may be appropriate; e.g., Armitage, 1975). However, for confirmatory experiments the fixed sample size estimate should be made and strictly followed. Both the standard classical inference procedure and the believability judgment require it.

16.4.3. Treatment effect models and analyses

There appear to be three distinct historical periods in the modeling and analyses of treatment effects in precipitation field experiments. They can be classified as 1) visual response (i.e., 1940s–early 1950s), 2) univariate response (i.e., 1952–79), and 3) multivariate response (i.e., 1980–).

Visual response. The analyses of the early precipitation augmentation field studies (i.e., 1940s and early 1950s) were largely visual or qualitative, and the approach appears to have been borrowed from the pioneering laboratory work of Langmuir, Schaefer, and Vonnegut. These three research chemists conducted a number of laboratory and field experiments on nucleation, and many of the results could be easily replicated and visually demonstrated. Hence, it is not surprising to find that Schaefer's landmark field seeding trial of a cold stratus deck with CO_2 on 13 November 1946 was analyzed visually.

Subsequent field precipitation experiments in New York and New Mexico, under the title of Project Cirrus, brought increased claims by the researchers of seeding effectiveness both locally and many miles downwind while continuing to rely largely on visual evidence (e.g., visual sightings, photographs, and radar pictures) and qualitative analysis (Langmuir, 1948, 1950). The treatment model in all of these experiments appears to have been implicit and of a qualitative nature (e.g., treatment should change the shape or height of the cloud or the flow of water in a stream). The analyses typically relied on looking at the cloud or stream and observing changes that were ascribed to treatment.

This situation culminated in a committee of four prominent scientists being asked by the American Meteorological Society to review the Langmuir claims. The committee acknowledged the high natural variability of the atmosphere and its lack of predictability and reported,

> It is the considered opinion of this committee that the possibility of artificially producing any useful amounts of rain has not been demonstrated so far if the available evidence is interpreted by any acceptable scientific standards (Haurwitz et al., 1950).

Needless to say, this report marked the end of the visual or qualitative approach to analysis of precipitation augmentation research in the United States.

Univariate response variables. A quantitative and more objective approach to analyses of precipitation research experiments appears to have originated with the Artificial Cloud Nucleation (ACN) Project of the early 1950s. For the first time, statistical expertise and methodology were utilized in the design and analysis stages of precipitation experiments. Improvements in experimentation included the a priori designation of experimental units, randomization of treatment assignment, tests of significance, and the formal use of covariates (e.g., Braham et al., 1957; Hall, 1957; Spar, 1957). These three ACN experiments

would now be judged as exploratory, and each utilized univariate response variables in their analyses (e.g., difference in mean adjusted rainfall on seeded and nonseeded storms). Numerous other precipitation experiments utilizing substantial statistical inputs followed in their footsteps (e.g., Arizona, Israel, Climax, Whitetop, etc. in the 1957–70 period).

There have been three general types of treatment effect models used in randomized comparative precipitation experiments: 1) the difference or ratio of two summary statistics (e.g., y^*/x^*), 2) the double difference or geometric mean of four summary statistics (e.g., $[y^*/x^* \cdot x^{**}/y^{**}]^{1/2}$), and 3) the statistical linear regression equation (e.g., $y^* = b_0 + b_1x_1 + \cdots + b_kx_k$). A brief review of each model and its accompanying analyses follows.

(i) Difference or ratio
This model can be written as

$$y^* = x^* + T, \qquad (16.1)$$

where y^* is a summary measure (e.g., median) of the response variable (y) (e.g., precipitation intensity) for the treated (S) experimental units, x^* is the same summary measure of the response variable (x) for the nontreated (N) experimental units, and T is the additive treatment effect. If the ratio form is used, then the log transformation produces Eq. (16.1). In this latter case the estimator of T is biased, but Neyman and Scott (1960) have presented equations for the correction. The extension to three or more treatments is handled by the classical ANOVA (e.g., Snedecor and Cochran, 1967).

As a point estimate of the additive treatment effect, the median or trimmed mean appears preferable. Stigler (1977) has shown for a number of empirical situations that the trimmed mean (e.g., 10% trim in each tail) can perform better than both the mean and the median as a measure of central tendency. Assuming one desires to visually present the empirical distributions of the two response variables (i.e., y and x), the stem-and-leaf plot or the box plot (e.g., Tukey, 1977b) is recommended.

The corresponding statistical estimation and testing techniques that have accompanied this treatment effect model have been varied and include the "t" (e.g., Woodley et al., 1982b), the Wilcoxon–Mann–Whitney (e.g., Mielke et al., 1970; Flueck, 1971), the rerandomization[1] (e.g., Woodley et al., 1982b), and the MRPP (e.g., Mielke et al., 1981a, 1984). Each of these techniques has its strengths and weaknesses, and if the response data are roughly normally distributed (i.e., "bell" shaped and no "outliers"), then all four techniques will give similar results (e.g.,

---

[1] The so-called rerandomization technique was first suggested by Fisher (1935), developed by Kempthorne (1955), popularized by Tukey et al. (1978), and critiqued by Gabriel (1979).

Mielke, 1985). However, if one is concerned about transforming the data and hence working in units different from the original measured units (typically Euclidean), then rerandomization and MRPP are preferred.

Each of these two techniques has its comparative advantages. Rerandomization is more general and can be applied to any statistic and any design, including those using restricted allocation. However, if the rerandomization procedure only samples from the set of all possible randomizations, then it incurs the possibility of a Type I error (i.e., falsely rejecting the null hypothesis when it is true). Fortunately, this problem can be made vanishingly small by increasing the number of rerandomizations. The MRPP (Mielke, 1979), on the other hand, is exact for small sample sizes (e.g., ≤25), approximates for larger samples, does not contain a Type I error, but presently is restrained to nonrestrictive randomizations.

Finally, it should be noted that the above techniques are also applicable to partitioned response values. Each subset of the partitioned data can be analyzed individually. (The partitioning typically is based on insight or meteorological knowledge.) If one then desires to combine subset results, a number of methods are available for combining independent test results (e.g., Kendall and Stuart, 1968). However, unless the particular subsets have been chosen a priori, the so-called multiplicity problem (e.g., Tukey, 1977a) arises, and then the probability results can only be viewed as suggestive.

(ii) Double difference or double ratio

This treatment effect model can be written as follows:

$$DD = (y_S^* - x_S^*) - (y_N^* - x_N^*) \tag{16.2}$$

or

$$y_S^* = x_N^* + T - A$$

where

y_S^* is a summary measure for the target response variable (y) when the target is treated (S),

y_N^* is the corresponding measure for the target response variable when the target is not treated (N),

x_S^* is the corresponding measure for the control variable (x) when the target is treated (S),

x_N^* is a corresponding measure for the control variable when the target is not treated (N),

T is an additive treatment effect, and

A is an adjustment ($y_N^* - x_N^*$) due to the natural precipitation difference between the two areas.

The treatment effect model is still an additive one but with an adjustment (A) term. The ratio form of this model would be a multiplicative effect model but would become an additive effect model in the log form. The applicable design is a two-area target–control design with only the target area available for treatment. The target and control areas may be fixed or variable.

A related two-area design, the crossover design (i.e., Moran, 1959), would have a slightly different treatment effect model and could be given in the ratio form as follows:

$$DR = \frac{y_S^*}{x_N^*} \cdot \frac{x_S^*}{y_N^*} \tag{16.3}$$

where

y_S^* is a summary measure for the area y response variable when area Y is treated (S),

y_N^* is the corresponding measure for the area y response variable when area Y is not treated (N),

x_S^* is the summary measure for the area x response variable when area X is treated (S), and

x_N^* is the corresponding measure for the area x response variable when area X is not treated (N).

It should be noted that Eq. (16.3) is simply the geometric mean of two ratios. However, the problem of bias is now enhanced and compounded (e.g., Flueck and Holland, 1976). Again, in the log form this double-ratio treatment effect model would become an additive effect model.

The corresponding statistical display, estimation, and testing techniques for the double-difference effect model are identical to those of the previous single difference. The double-difference model has been used in a number of precipitation augmentation experiments, including Whitetop (e.g., Flueck, 1971).

However, for the double-ratio effect model the options currently are limited to an asymptotic t-test and rerandomization. Gabriel and Feder (1969) applied a rerandomization test to the double ratio for the Israel I experiment, and Mielke et al. (1981b) applied it to the rerandomization of the Climax I and II experiments. Kempthorne (1980) recently reviewed the rerandomization technique for the crossover design and presented some ideas for estimating additive effects.

Statisticians typically prefer not to use ratio estimators if alternatives are available. This generally is due to their difficult and sensitive distributions (e.g., Lee et al., 1979). Furthermore, their variability is typically larger than the corresponding difference statistic. Lastly, a ratio of means (a typical summary measure utilized in past precipitation experiments) has the added problem of being nonresistant (i.e., one or two extreme values can greatly influence a mean and hence the ratio). This later problem should be mitigated by the use of medians, and Mielke and Medina (1983) have presented a median technique. For the Climax I and II data, this ratio of weighted medians has less variability than the corresponding double ratio of Eq. (16.3). Further investigation appears to be merited.

(iii) Statistical linear regression equation

The third type of treatment effect model is the statistical linear regression equation. This model can be written as follows:

$$y = y^* + e, \qquad (16.4)$$

and the predictive linear equation as

$$y^* = b_0 + b_1 x_1 + \cdots + b_k x_k, \qquad (16.5)$$

where

y the response variable for the target area,
y^* the linear predicted response variable for the target area,
x_i a predictor or auxiliary variable ($i = 1, \cdots, k$),
b_i the coefficient for the ith predictor variable, and
e the error or residual variable.

Thus, every target-area precipitation value can be separated into a predicted or modeled component (y^*) and an unpredicted or residual component (e).

One of the predictive variables becomes the additive treatment effect term (e.g., $b_1 x_1$), and hence x_1 is an index variable with value 0 for nontreated units and value 1 for treated units. Now, b_1 is the measure of the additive treatment effect. The other predictive terms (e.g., $b_2 x_2$, $b_3 x_3$, etc.) are utilized in an attempt at accounting for the natural variability of precipitation, and these variables can include control area precipitation, relative humidity, cloud-top temperature, wind speed, numerical model predictions, etc. In an exploratory experiment the statistical linear model may be constructed interactively (e.g., Henderson and Velleman, 1981), and meteorological theory is very important in guiding the selection of the predictive variables (e.g., Flueck et al., 1986).

The estimation of the coefficient values ($b_i s$) for Eq. (16.5) is based on fitting the selected linear function to the sample data using an objective technique. The most popular fitting (coefficient estimation) method has been the method of least-squares (e.g., Draper and Smith, 1981). Thus, the traditional ANOCOV or regression analysis (typically with transformations of both the response and some predictor variables) provides the statistical output, and a t-test usually is performed on the treatment coefficient (e.g., b_1) to provide a measure of significance. Spar (1957), Mielke et al. (1977), Flueck et al. (1981), Woodley et al. (1982b), and others have used this method. The display of residuals is highly recommended.

Recent statistical research has developed a competing approach for this linear estimation problem, the median or least absolute deviation method (e.g., Huber, 1973; Mosteller and Tukey, 1977). The most important characteristic of this method is its resistance to outlying response observations in the estimation of the linear model coefficients. A number of algorithms are now available for implementing this technique (e.g., Bloomfield and Steiger, 1980), and statistical computing packages are beginning to include this method.

A modification of the preceding linear model technique is to fit the statistical linear model to the data without a term for treatment and then examine the residuals for evidence of treatment [the analysis methods under (i) above would be appropriate]. This so-called sweep-out technique (i.e., sweep out the "background effects" and then investigate for a treatment signal; e.g., Flueck et al., 1981) has the advantage that it does not specify the form of the treatment effect, and thus it is a more general statistical linear model approach to assessing the effects of treatment. Flueck et al. (1981), Woodley et al. (1982b, 1983), and apparently, Hall (1957) have used this approach, in conjunction with the method of least-squares, to assess treatment effects in precipitation augmentation experiments. Wong et al. (1983) have applied the sweep-out method, in conjunction with median regression, to a hail project.

In reviewing the three general types of treatment effect models for the univariate response case, the statistical linear equation or model seems best. First and foremost, the other two types of treatment models can be transformed into the statistical linear model. Second, it provides for the removal of effects of other factors that might affect the response variable and thus allows a more precise assessment of the treatment effect. Third, this statistical linear model approach provides additional information on the potential treatment effect itself and on the accompanying predictor variables.

Lastly, the question of the fitting method depends greatly on the empirical distribution of the responses. If there are no "outliers," both the method of least-squares and the method of least absolute deviation appear to give very similar results. If there are outliers, the latter method is preferable as an overall technique.

Multivariate response variables. With the increasing emphasis on the chain-of-events approach to assessing treatment effects and the use of more than one response variable for a given physical process, the need for multivariate statistical analysis of response variables is becoming more important in precipitation experimentation. However, the preferred form of the multivariate treatment effect model currently appears to be uncertain. Some meteorological and statistical insight and theory are needed for this nontrivial problem.

On the statistical side, there are a number of multivariate methods that have been utilized in other research disciplines (e.g., Morison, 1976). Methods such as the multivariate (Hotelling) T^2, multivariate analysis of covariance, repeated measures, principal components, MRPP, etc. are all available for use on a multivariate response precipitation experiment. One can even consider analyzing each response variable separately and combining the tests or estimation results.

Which of the above techniques or methods are most appropriate for precipitation experiments is still unknown.

Only the MRPP method appears to have been given much attention (e.g., Mielke et al., 1984), and it has raised a number of meteorological and statistical questions. Clearly, there is a need for further research on multivariate techniques in precipitation augmentation experiments.

16.5. Summary and concluding remarks

This article has indicated that proper field comparative experimentation in precipitation augmentation is not an easy task. The stages, components, and all of the associated detailed activities must be given full attention and careful consideration if a proper experiment is to be performed. However, when feasible, we should carry out field experiments, for this is the best way to advance our knowledge of precipitation processes.

A formal structure for evaluating the performance of an experiment has been presented, and some principles and prescriptions for accomplishing quality comparative experimentation have been offered. Clearly, the ultimate goal of all experiments is to achieve a highly believable (successful) result. In this quest, the interplay between proposed theory or conceptual model and offered supporting statistical evidence is paramount in achieving a believable result and eventually a scientific law.

The hierarchy of preliminary, exploratory, and confirmatory experiments has been presented, and the contents and importance of each class have been discussed. This trichotomy should be carefully maintained in precipitation augmentation research. One is not a substitute for the others.

A detailed review of some of the components of an experiment (i.e., conceptual models, treatment design, and treatment effect models and analyses) has shown that some improvements in design, implementation, analysis, and interpretation have occurred. However, opportunities for further improvements remain, and the offered prescriptions have been presented with this in mind. It is believed that the increasing emphasis on the use of the statistical linear model, and its ramification in estimating treatment effects, is a desirable development.

Finally, the author has attempted to indicate that the role of statistics in general, and statisticians in particular, is pervasive in high-quality precipitation augmentation research. As such, the statistician should become a general partner in the research venture; gone, then, are the days of nursemaid or prosecuting attorney. It is in this shared responsibility role that experiments and science will be best served.

REFERENCES

AMS (American Meteorological Society Council), 1984: Planned and inadvertent weather modification. *Bull. Amer. Meteor. Soc.,* **65**, 1322–1323.

Armitage, P., 1975: *Sequential Medical Trials.* 2nd ed., Wiley and Sons, 194 pp.

Bloomfield, P., and W. L. Steiger, 1980: Least absolute deviations curve-fitting. *SIAM J. Sci. Statist. Comput.,* **1**, 290–301.

Braham, R. R., Jr., 1966: *Final Report of Project Whitetop, Part I—Design of the Experiment. Part II—Summary of Operations.* Dept. of Geophys. Sci., University of Chicago, 156 pp. [NTIS PB-176-622.]

——, 1981: Designing cloud seeding experiments for physical understanding. *Bull. Amer. Meteor. Soc.,* **62**, 55–62.

——, and P. Squires, 1974: Cloud physics—1974. *Bull. Amer. Meteor. Soc.,* **55**, 543–586.

——, L. J. Battan and H. R. Byers, 1957: Artificial nucleation of cumulus clouds. *Cloud and Weather Modification, Meteor. Monogr.,* No. 11, Amer. Meteor. Soc., 47–85.

Brier, G. W., 1974: Design and evaluation of weather modification experiments. *Climate and Weather Modification,* Chapter 5, W. N. Hess, Ed., Wiley and Sons, 206–225.

Bureau of Reclamation, 1979: *The Design of HIPLEX-1.* Div. of Atmos. Resour. Res., Bureau of Reclamation, U.S. Dept. of the Interior, Denver, 271 pp.

——, 1982: *The Design of SCPP-1,* Project Skywater, Div. of Atmos. Resour. Res., Rev. (December 1982), Bureau of Reclamation, U.S. Dept. of the Interior, Denver, 61 pp. plus Appendices.

Byers, H. R., 1974: History of weather modification. *Climate and Weather Modification,* Chapter 1, W. N. Hess, Ed., Wiley and Sons, 3–44.

Chappell, C. F., 1970: *Modification of Cold Orographic Clouds.* Dept. of Atmos. Sci. Rep. 173, Colorado State University, 196 pp.

Cotton, W. R., 1972: Numerical simulation of precipitation development in supercooled cumuli—Part II. *Mon. Wea. Rev.,* **100**, 764–784.

Cox, D. R., 1957: The use of a concomitant variable in selecting an experimental design. *Biometrika,* **44**, 150–158.

DeMott, P. J., W. G. Finnegan and L. O. Grant, 1983: An application of chemical kinetic theory and methodology to characterize the ice nucleating properties of aerosols used for weather modification. *J. Climate Appl. Meteor.,* **22**, 1190–1203.

Draper, N. R., and H. Smith, 1981: *Applied Regression Analysis.* 2nd ed., Wiley and Sons, 709 pp.

Fisher, R. A., 1926: The arrangement of field experiments. *J. Ministry Agric. Great Britain,* **33**, 503–513. [Paper 17 in Fisher, R. A., 1950: *Contributions to Mathematical Statistics.* Wiley and Sons.]

——, 1935 (1st ed.) to 1966 (8th ed.): *Design of Experiments.* Oliver and Boyd, 362 pp.

Flueck, J. A., 1971: *Statistical Analyses of the Ground Level Precipitation Data, Part V; Final Report of Project Whitetop.* Cloud Physics Laboratory, Dept. of Geophys. Sci., University of Chicago, 294 pp. [NTIS N72-13559.]

——, 1976: Evaluation of operational weather modification projects. *J. Wea. Mod.,* **8**, 42–56.

——, 1978: The role of statistics in weather modification experiments. *Atmos.-Ocean,* **16**, 377–395.

——, 1982: Comparative experimentation: Some principles and prescriptions. *Teaching of Statistics and Statistical Consulting,* J. S. Rustagi and D. A. Wolfe, Eds., Academic Press, 443–463.

——, and B. S. Holland, 1976: Ratio estimators and some inherent problems in their utilization. *J. Appl. Meteor.,* **15**, 535–543.

——, and P. W. Mielke, Jr., 1978: Design and evaluation of hail suppression experiments. *Hail: A Review of Hail Science and Hail Suppression,* Chapter 19, *Meteor. Monogr.* No. 38, G. B. Foote and C. A. Knight, Eds., Amer. Meteor. Soc., 225–235.

——, and R. J. Boik, 1982: Treatment Design. *The Design of SCPP-1,* Appendix G, Project Skywater, Div. of Atmos. Resour. Res., Bureau of Reclamation, U.S. Dept. of the Interior, Denver, 15 pp.

——, W. L. Woodley, R. W. Burpee and D. O. Stram, 1981: Comments on E. Nickerson's FACE rainfall results: Seeding effect or natural variability? *J. Appl. Meteor.,* **20,** 98–107.

——, W. L. Woodley, A. Barnston and T. Brown, 1986: A further assessment of treatment effects in the Florida Area Cumulus Experiment through guided linear modeling. *J. Climate Appl. Meteor.,* **25,** 546–564.

Gabriel, K. R., 1979: Some statistical issues in weather experimentation. *Commun. Statist.—Theory Meth.,* **A8,** 975–1015.

——, and P. Feder, 1969: On the distribution of statistics suitable for evaluating rainfall stimulation experiments. *Technometrics,* **11,** 149–160.

Hall, F., 957: The weather bureau ACN project. *Cloud and Weather Modification, Meteor. Monogr.,* No. 11, Amer. Meteor. Soc., 24–46.

Haurwitz, B., G. Emmons, G. Wadsworth and H. C. Willett, 1950: On the results of recent experiments in the artificial production of precipitation. *Bull. Amer. Meteor. Soc.,* **31,** 346–347.

Henderson, H. V., and P. F. Velleman, 1981: Building multiple regression models interactively. *Biometrics,* **37,** 391–411.

Hill, A. B., 1951: The clinical trial. *Brit. Med. Bull.,* **7,** 278–282.

Huber, P. J., 1973: Robust regression: Asymptotics, conjectures and monte carlo. *Ann. Statist.,* **1,** 799–821.

Kempthorne, O., 1955: The randomization theory of experimental inference. *J. Amer. Statist. Assoc.,* **50,** 946–967.

——, 1980: Some statistical aspects of weather modification studies. *Statistical Analysis of Weather Modification Experiments,* E. J. Wegman and D. J. DePriest, Eds., Marcel Dekker, 89–107.

Kendall, M. G., and A. Stuart, 1968: *The Advanced Theory of Statistics.* Vol. 3, 2nd ed., Griffin & Co., 557 pp.

Kulkarni, S. R., 1968: On the optimal symptotic tests for the effects of cloud seeding on rainfall. (1) The case of fixed effects. *Austral. J. Statist.,* **10,** 105–115.

——, 1969: On the optimal symptotic tests for the effects of cloud seeding on rainfall. (2) The case of variable effects. *Austral. J. Statist.,* **11,** 39–51.

Langmuir, I., 1948: The growth of particles in smoke and clouds and the production of snow from supercooled clouds. *Proc. Philos. Soc. London,* **92,** 167–185.

——, 1950: Control of precipitation from cumulus clouds by various seeding techniques. *Science,* **112,** 35–46.

Lee, R., B. S. Holland and J. A. Flueck, 1979: Distribution of a ratio of correlated gamma random variables. *SIAM J. Appl. Math.,* **36,** 304–320.

Malkus, J. S., and R. H. Simpson, 1964: Modification experiments on tropical cumulus clouds. *Science,* **145,** 541–548.

Mielke, P. W., Jr., 1979: Some parametric, nonparametric and permutation inference procedures resulting from weather modification experiments. *Commun. Statist.—Theory Meth.,* **A8,** 1083–1096.

——, 1985: Geometric concerns pertaining to applications of statistical tests in the atmospheric sciences. *J. Atmos. Sci.,* **42,** 1209–1212.

——, and J. G. Median, 1983: A new covariate ratio procedure for estimating treatment differences with application to Climax I and II experiments. *J. Climate Appl. Meteor.,* **22,** 1290–1295.

——, L. O. Grant and C. F. Chappell, 1970: Elevation and spatial variation effects of wintertime orographic cloud seeding. *J. Appl. Meteor.,* **9,** 476–488.

——, —— and ——, 1971a: Corrigendum. *J. Appl. Meteor.,* **10,** p. 1982.

——, —— and ——, 1971b: An independent replication of the Climax Wintertime Orographic Cloud Seeding Experiment. *J. Appl. Meteor.,* **10,** 1198–1212.

——, —— and ——, 1976: Corrigendum. *J. Appl. Meteor.,* **15,** p. 801.

——, J. S. Williams and S. C. Wu, 1977: Covariance analysis technique based on bivariate log-normal distribution with weather modification applications. *J. Appl. Meteor.,* **16,** 183–187.

——, K. J. Berry and G. W. Brier, 1981a: Application of multi-response permutation procedures for examining seasonal changes in monthly mean sea-level pressure patterns. *Mon. Wea. Rev.,* **109,** 120–126.

——, G. W. Brier, L. O. Grant, G. J. Mulvey and P. N. Rosenzweig, 1981b: A statistical reanalysis of the replicated Climax I and II wintertime orographic cloud seeding experiment. *J. Appl. Meteor.,* **20,** 643–659.

——, K. J. Berry, A. S. Dennis, P. L. Smith, J. R. Miller, Jr. and B. A. Silverman, 1984: HIPLEX-1: Statistical evaluation. *J. Climate Appl. Meteor.,* **23,** 513–522.

Moran, P. A. P., 1959: The power of a cross-over test for the artificial stimulation of rainfall. *Austral. J. Statist.,* **1,** 47–52.

——, 1970: The methodology of rain-making experiments. *Rev. Int. Statist. Instit.,* **38,** 105–119.

Morrison, D., 1976: *Multivariate Statistical Methods.* 2nd ed., McGraw-Hill, 415 pp.

Mosteller, F., and J. W. Tukey, 1977: *Data Analysis and Regression.* Addison Wesley, 588 pp.

NAS–NRC (National Academy of Sciences–National Research Council), 1959: *Skyline Conf. on the Design and Conduct of Experiments in Weather Modification.* Publ. 742, Washington, DC, 24 pp.

——, 1966: *Weather and Climate Modification: Problems and Prospects, II, Research and Development,* Publ. No. 1350, Washington, DC, 198 pp.

Neyman, J., 1967: Experimentation with weather control. *J. Roy. Statist. Soc.,* **A130,** 285–326.

——, and E. L. Scott, 1960: Correction for bias introduced by a transformation of variables. *Ann. Math. Statist.,* **31,** 643–655.

NRC (National Research Council, Committee on Atmospheric Sciences), 1973: *Weather and Climate Modification, Problems and Progress.* Natl. Acad. Sci., Washington, DC, 258 pp.

Pocock, S. J., 1977: Group sequential methods in the design and analysis of clinical trials. *Biometrika,* **64,** 191–199.

Reynolds, D. W., and A. S. Dennis, 1986: A review of the Sierra Cooperative Pilot Project. *Bull. Amer. Meteor. Soc.,* **67,** 513–523.

Schaefer, V. J., 1946: The production of ice crystals in a cloud of supercooled water droplets. *Science,* **104,** 457–459.

Simpson, J., J. R. Stinson and J. W. Kidd, 1966: Stormfury cumulus experiments: Preliminary results 1965. *J. Appl. Meteor.,* **5,** 521–525.

——, G. W. Brier and R. H. Simpson, 1967: Stormfury cumulus seeding experiments 1965: Statistical analysis and main results. *J. Atmos. Sci.,* **24,** 508–521.

Smith, P. L., A. S. Dennis, B. A. Silverman, A. B. Super, E. W. Holroyd III, W. A. Cooper, P. W. Mielke, Jr., K. J. Berry, H. D. Orville and J. R. Miller, Jr., 1984: HIPLEX-1: Experimental design and response variables. *J. Climate Appl. Meteor.,* **23,** 497–512.

Snedecor, G. W., and W. G. Cochran, 1967: *Statistical Methods.* 6th ed., Iowa State University Press, 593 pp.

Spar, J., 1957: Project SCUD. *Cloud and Weather Modification, Meteor. Monogr.,* No. 11, Amer. Meteor. Soc., 5–23.

Stigler, S. M., 1977: Do robust estimators work with *real* world data? *Ann. Statist.,* **5,** 1055–1098.

Todd, C. J., and W. E. Howell, 1985: *World Atlas and Catalog of Reported Results of Precipitation Management by Cloud Seeding.* Todd and Howell Publishers.

Tukey, J. W., 1977a: Some thoughts on clinical trials, especially problems of multiplicity. *Science,* **198,** 679–684.

——, 1977b: *Exploratory Data Analysis.* Wiley and Sons, 688 pp.

——, D. R. Brillinger and L. V. Jones, 1978: *The Management of Weather Resources, II, The Role of Statistics in Weather Resources Management.* Report of the Statistical Task Force to the Weather Modification Advisory Board, U.S. Govt. Printing Office, 118 pp.

Twomey, S., and I. Robertson, 1973: Numerical simulation of cloud seeding experiments in selected Australian areas. *J. Appl. Meteor.,* **12,** 473–478.

Vonnegut, B., 1947: The nucleation of ice formation by silver iodide. *J. Appl. Phys.,* **18,** 593–595.

Williams, M. C., and R. D. Elliott, 1985: Weather modification. *Facets of Hydrology, II,* Chapter 4, J. C. Rodda, Ed., Wiley and Sons.

Winer, B. J., 1971: *Statistical Principles in Experimental Design.* 2nd ed., McGraw-Hill, 907 pp.

Wong, R. K. W., H. Chidambaram and P. W. Mielke, Jr., 1983: Application of MRPP and median regression for covariate analyses of possible weather modification effects on hail responses. *Atmos.-Ocean,* **21,** 1–13.

Woodley, W. L., J. A. Flueck, R. Biondini, R. I. Sax, J. Simpson and A. Gagin, 1982a. Clarification of confirmation in the FACE-2 experiment. *Bull. Amer. Meteor. Soc.,* **63,** 263–276.

——, J. Jordan, J. Simpson, R. Biondini, J. A. Flueck and A. Barnston, 1982b: Rainfall results of the Florida Area Cumulus Experiment, 1970–1976. *J. Appl. Meteor.,* **21,** 139–164.

——, A. Barnston, J. A. Flueck and R. Biondini, 1983: The Florida Area Cumulus Experiment's second phase (FACE-2), Part II: replicated and confirmatory analyses. *J. Climate Appl. Meteor.,* **22,** 1529–1540.

Zelen, M., 1974: The randomization and stratification of patients to clinical trials. *J. Chron. Diseases,* **27,** 365–375.